CULTURAL VALUES AND HUMAN ECOLOGY IN SOUTHEAST ASIA

edited by Karl L. Hutterer, A. Terry Rambo, and George Lovelace

MICHIGAN PAPERS ON SOUTH AND SOUTHEAST ASIA
CENTER FOR SOUTH AND SOUTHEAST ASIAN STUDIES
THE UNIVERSITY OF MICHIGAN   NUMBER 27

Library of Congress Catalog Card Number: 84-45711

ISBN 0-89148-039-0 (cloth)
ISBN 0-89148-040-4 (paper)

Printed in the United States of America

CONTENTS

101880

PREFACE

In June 1983, a one-week conference on cultural values and tropical ecology was held at the East-West Center in Honolulu. Jointly sponsored by the East-West Environment and Policy Institute (EAPI), the Center for South and Southeast Asian Studies of the University of Michigan, and the Center for Asian and Pacific Studies of the University of Hawaii, the conference was attended by seventeen scholars from eight Asian and Pacific countries including the United States (a list of participants is appended). The meeting was the first in a planned series of jointly sponsored annual conferences dealing with issues in the human ecology of Southeast Asia.

The impetus for the conference grew out of discussions over the course of 1982 between the editors and other Research Associates and Fellows participating in EAPI's Program on Human Interactions with Tropical Ecosystems (HITE). Initiated in 1980 to encourage interdisciplinary human ecology research on problems of developing and managing natural resources in Southeast Asia, the HITE Program has worked closely with a number of younger Asian social scientists in developing such research. Most of these scholars were associated with member projects of the Southeast Asian Universities Agroecosystem Network (SUAN), a regional association of resource management research projects in Indonesia, the Philippines, and Thailand. As the extent of involvement of social scientists grew, the lack of adequate conceptual frameworks for studying human-environment interactions became increasingly evident. In particular, the poorly understood role of ideational factors (symbols, beliefs, values) in human ecology was seen as an obstacle to research.

Given the goals of the conference, it was decided to bring together participants with a wide range of cultural and academic backgrounds and professional interests, including the natural sciences, the social sciences, the humanities, and applied fields. The resulting papers reflect this diversity. All have undergone extensive revisions by the authors and editors. In particular, the editors have tried to bring out common themes and concerns linking the various contributions, and have imposed a moderate amount of homogeneity in format to stress the unifying interests underlying the diversity of approaches. At the

Preface

same time, however, the editors have tried to preserve the unique flavor and essence of the individual papers. The work of the editors notwithstanding, the final responsibility for the content of the papers rests with the respective authors.

A number of individuals and institutions were helpful in planning and executing the conferences and in bringing this volume to press. The Ford Foundation, through Dr. Tom Kessinger, provided a special grant that made the participation of Ho Ton Trinh possible. George Ellis, Director of the Honolulu Academy of Arts, agreed to organize a show to parallel the conference; an attractive and informative exhibit was put together by the Curator of European Art, James Jensen, and the Curator of Asian Art, Howard Link, under the title "Landscapes East and West." Fannie Lee Kai, Peter Brosius, and Wilma Fujii were most helpful during the conference, and Peter also assisted in editing some of the papers. Avery Dubay and Elizabeth Figel typed multiple drafts of the manuscripts. Sheryl Bryson did copyediting, while production editing was done by Jim Hynes and Janet Opdyke. Lorinda Grimshaw managed the typesetting program on the computer. We thank all of them for their assistance.

Conference Participants

Dr. Geoffrey Benjamin
Department of Sociology
Singapore National University
Kent Ridge Campus
Singapore

Ms. June Prill Brett
Division of Social Sciences
University of the Philippines College Baguio
Baguio City
Philippines

Mr. Peter Brosius
Department of Anthropology
1054 L.S.A. Building
The University of Michigan
Ann Arbor, Michigan   48109

Preface

Dr. Chavivun Prachuabmoh
Department of Sociology and Anthropology
Thammasat University
Bangkok, Thailand

Dr. Alice G. Dewey
Department of Anthropology
University of Hawaii at Manoa
Honolulu, Hawaii   96822

Ms. Wilma Fujii
Research Intern
East-West Environment and Policy Institute
1777 East-West Road
Honolulu, Hawaii   96848

Dr. Gerald C. Hickey
Fellow, East-West Center
1777 East-West Road
Honolulu, Hawaii   96848

Professor Ho Ton Trinh
State Committee of Social Sciences
Hanoi
Vietnam

Dr. Karl L. Hutterer
Curator, Division of the Orient
Museum of Anthropology
The University of Michigan
Ann Arbor, Michigan   48109

Dr. Neil Jamieson
East-West Environment and Policy Institute
1777 East-West Road
Honolulu, Hawaii   96848

Dr. Roger Long
Department of Drama and Theatre
University of Hawaii at Manoa
Kennedy Theatre 115
Honolulu, Hawaii   96822

Preface

Dr. George W. Lovelace
East-West Environment and Policy Institute
1777 East-West Road
Honolulu, Hawaii   96848

Mr. Pei Sheng-ji
Director, Yunnan Institute of Tropical Botany
Academia Sinica
P.O. Box 302
Xishuangbanna, Yunnan Province
China

Dr. Peter Pirie
East-West Population Institute
1777 East-West Road
Honolulu, Hawaii   96848

Dr. A. Terry Rambo
East-West Environment and Policy Institute
1777 East-West Road
Honolulu, Hawaii   96848

Dr. Otto Soemarwoto
Director, Institute of Ecology
Padjadjaran University
Jl. Sekeloa
Bandung, Indonesia

Dr. Aram Yengoyan
Department of Anthropology
1054 L.S.A. Building
The University of Michigan
Ann Arbor, Michigan   48109

CHAPTER 1

INTRODUCTION

Karl L. Hutterer

A. Terry Rambo

Concepts in Human Ecology

The social sciences and natural sciences have followed very different and largely separate intellectual paths, reflecting the fundamental conceptual dichotomy in Western thought between culture and nature. The discipline of ecology shares the conceptual tradition of its natural science background. Only recently have ecologists come to acknowledge that most ecosystems cannot be understood without taking into account human interventions that may have been of significant magnitude for thousands of years (Rambo 1979). Conversely, beginning around the turn of this century and with increasing emphasis since the 1960s, social scientists have recognized that much of human behavior cannot be understood in isolation from the environment in which it is acted out (Orlove 1980). Thus, there has been a growing concern by both natural and social scientists with the relationships and interactions between natural ecosystems and human social systems, a concern reflected in the ever-growing number of publications dealing with "human ecology."

The development of human ecology has been retarded, however, by the lack of suitable conceptual frameworks allowing natural and social scientists to link domains of inquiry that traditionally have been thought of as radically different (Rambo 1983). Lacking such frameworks, natural scientists generally have shied away from dealing substantively with the human component of ecosystems. When they do try to deal with human interactions with the environment, it is often by viewing *Homo sapiens* as simply another animal species, albeit a very prolific, highly successful, and usually dominant one. In this context, humans often are viewed negatively as having a tendency to run amok among other species of their ecological communities, degrading and destroying what many ecologists appear to assume would otherwise be homeostatic systems (Boughey 1971).

Social scientists who have become involved in ecological research have often abandoned their intellectual heritage and uncritically applied concepts and analytical procedures developed in biological ecology, also treating humans as just another animal population (Vayda and Rappaport 1968; cf. Anderson 1974). In practice, this approach has been successful in explaining certain specific aspects of human behavior. Birdsell (1953, 1979), for instance, has been able to show that population density and territory size of aboriginal groups in the arid interior of Australia are strongly correlated with rainfall patterns, and he has explained the relationship on the basis of Liebig's "law of the minimum." Similar observations and interpretations have been made on the !Kung San of southern Africa (Lee 1972). Other biological-ecology approaches, such as optimal foraging theory, also have been used with some success to explain the form of certain specific aspects of the organization of small-scale societies, primarily hunter-gatherers (Winterhalder and Smith 1981).

A number of serious problems remain, however, in treating human ecology simply as an extension of biological ecology. For one, virtually all of these approaches so far have been employed successfully only in analyzing highly specific organizational elements of human populations that are by and large homologous to aspects of the social organization of animal populations. These approaches generally have not been designed to deal with ideational elements of human organization (those aspects of human behavior often defined as "culture" in the narrow sense). On the contrary, they have tended to discount the influence of ideational constructs on certain areas of social organization that are deemed to be directly environmentally determined. This raises the question of whether the ideational realm of human organization is nothing more than a nonessential elaboration of basic social facts that are fundamentally directed toward the biological maintenance of a population (White 1959), or whether such analyses may be ignoring some essential relationships. Another observation should be made: virtually all successful human-ecology studies using biological-ecology concepts and analytical tools have dealt with small-scale societies characterized by relatively simple organizations. Most attempts to explain aspects of the organization of complex societies on the basis of environmental conditions and ecological relationships have been unsuccessful, often involving a trivialization of ecological concepts. A long list of examples could be cited, including most ecological studies trying to explain the rise and fall of prehistoric or historic civilizations (cf. Coe 1961; Meggers 1954; Raikes 1964; Culbert 1973; Sanders and Price 1968) or more specific investigations dealing with food taboos, human sacrifice, and cannibalism (cf. Diener and Robkin 1978; Harner 1977; Harris 1966, 1973, 1974, 1977; Ortiz de Montellano 1978; Sahlins 1979). This occurs partly because of the complexity of the organizations involved. Frequently, the analytical matrices are too large and complex to be easily manipulated, and often are greatly simplified on the basis of tenuous assumptions. More serious is the fact that complex societies generally modify their environments far more extensively than do small-scale societies (Bennett 1976). This modification clearly involves an interactive process between the natural environment and human organization in which ideational constructs play an important role in directing human action. Thus, unlike the case of plant and animal populations, human ecological relationships cannot be understood adequately without taking this ideational element into consideration. In order to do this it is necessary to reexamine conceptual frameworks used in the social sciences for analysis of ideational factors, particularly cultural values.

The Concept of Values in Human Ecology

To say that the human species is unique is a tautology. The definition of a species itself implies that a given population of organisms possesses distinct characteristics that set it apart from all other populations of organisms. The important question, therefore, is in precisely what ways is the human species different. Ecologically, the most striking difference is that *Homo sapiens*, unlike any other animal species, is a component of virtually every major terrestrial ecosystem. Further, this enormous ecological breadth is not based on any specific anatomical or physiological adaptation. The explanation for this unique global distribution can only be found in the immensely flexible human behavioral capabilities. Thus, any comparison between humans and other animal species must question the nature and organization of the behavior exhibited by the species being compared.

Many aspects of human behavior are closely comparable to aspects of animal behavior, for instance, actions that involve essential physiological functions or are directed toward the satisfaction of basic biological needs. As indicated previously, it is in these areas that traditional ecological analysis has been most successful. Even among small-scale societies having little internal differentiation, however, there are many aspects of human behavior that cannot easily be reduced to this level. Even those that seemingly can be comprehended from a purely biological view, such as acquisition of food and shelter, aspects of eating, drinking and sleeping, and biological reproduction, bear the imprint of a behavioral organization that transcends concerns with material needs and conditions. In other words, most human actions—including those directed primarily toward biological survival and physical well-being—are imbued with cultural meaning and are organized according to rules consistent with this system of meaning. This applies to simple societies as well as complex ones. All known hunter-gatherers, for instance, maintain extensive and complex ideational systems (described by anthropologists as mythology, cosmology, or religion) into which is integrated practical and efficacious knowledge (cf. Dunn 1975; Eliade 1959; Endicott 1979; Yengoyan 1976). This knowledge includes the organization and behavior of the plants and animals from which humans derive a living or from which they must protect themselves, the layout of the geographical space within which they live and move, the organization of social space and the place and role of specific individuals in it, and the rules of proper behavior by which they relate these aspects of the perceived environment.

Technology often is viewed as the aspect of behavior through which humans articulate most directly with the physical environment. But even here it has been said that "a technology is not comprehended by its physical properties alone" (Sahlins 1972,

79). An extensive set of cultural rules and social relationships must be considered in order to understand the formal and functional configuration of a particular technological system, even one as simple as a set of bows and arrows (cf. Watanabe 1975). It has been shown that with the Malaysian aborigines, for instance, it is their unique knowledge of the rain forest, rather than any specific tools *per se*, that allows them to occupy a habitat considered hostile by other Malaysian ethnic groups (Dunn 1975; Hutterer 1977; Rambo 1980). This knowledge is formulated and transmitted through cultural means; it is codified in a system of arbitrary symbols (Endicott 1979; Benjamin, this volume).

Although anthropologists have long realized that symbols play an important role in culture, it was, ironically, the self-declared materialist Leslie White (1949) who first pointed out their basic and overriding significance in human organization. Most human perceptions are processed and interpreted through symbols; symbols are used to constitute systems of knowledge concerning the external world; symbols form a basic element through which congruence is achieved between disparate realms of human experience and action; and symbols encode the messages transmitted in linguistic communication. Despite the conceits of some ethologists to the contrary, no animal species has ever conclusively demonstrated the the ability to generate and manipulate symbols. It is in this sense that *Homo sapiens* alone among all known species inhabits a culturally constituted world.

Symbols not only involve the representation of one reality through another essentially unrelated one, but they also are constituted on an arbitrary basis and are manipulated deliberately. The question needs to be posed as to the way in which symbols, systems of symbols, and symbolic structure are involved in human action. Action is not based directly on symbols, although whatever lies behind it as a motivating and directing force will be defined and expressed by symbols or an action itself may be intended to be symbolic. The immediate basis for action is provided by a different sort of mental construct—values.

In common language, the term "value" has many different connotations relating to emotional, ideological, economic, social, or political aspects of human life and experience. In the social sciences, the term made its entrance around 1920 (Williams 1968) and has since been much in evidence, although there is little consensus on its precise meaning and proper usage. In anthropology the study of values was pioneered by Clyde and Florence Kluckhohn (Kluckhohn 1951; Kluckhohn and Strodtbeck 1961), and for this discussion, we will, with some modifications, follow their work. In their terms, values may be defined as mental constructs that enable humans to execute comparative assessments of the state of the world with regard to certain qualities; i.e., things, actions, ideas, and so forth are assessed to be more or less good, true, ethical, pleasant, or beautiful.

Values contain both cognitive and emotive elements and form the base upon which the motivation for action is founded, but the values themselves should not be mistaken for motives (Devereux 1978). Motivation arises from a choice between courses of action relating to alternative, and often conflicting, values. Further, values should not be mistaken for explicit behavioral norms. Frequently, such manifest norms—usually highly specific in content and explicitly stated—are contradicted by values that are much broader in content, fuzzy in their definition, and not fully explicit in their conceptualization. Values themselves (as Benjamin states in this volume) are virtually always ideographic and nondiscursive, difficult to explicate to others, and difficult for researchers to elicit. Values are socially shared. As Benjamin further explains, this sharing comes from a process of subtle imposition, defined by him as a "political" process, that constitutes an essential aspect of the broader social process. This sharing is neither perfect nor complete; value concepts are subject to both individual and class variability within larger social groups.

How are values constituted? Modifying somewhat Rappaport's (1979) model of religion and ritual, it can be said that values are part of a larger, hierarchically ordered system of meaning that is expressed, professed, and confirmed in the liturgical order of ritual. At the apex of this system stand "ultimate sacred propositions" (concerning, for instance, the existence of God or spirits). Closely associated with these is a subordinate group of "cosmological axioms," which explain the nature of the universe, assign humans a place within it, and define the source and purpose of existence. By their very nature these axioms entail both explicit and implicit principles for behavior and action that are general in nature. Values, as they are defined here, would be part of these generalized principles (although Rappaport excludes them on the basis of a somewhat different definition). From these axioms flow sets of more specific rules governing social and other relations. Rappaport goes on to say that on a fourth and lower level liturgical orders also deal with symbolically constituted information about the external world. On a fifth level outside the liturgical order are secular meanings or understandings of the external world. In this scheme of things, as one moves down the levels of symbolic organization, the specificity of the context of axioms and rules increases at the same rate their immutability and permanence decreases.

The whole system of meanings is constituted symbolically. Values are part and parcel of this symbolic system, a necessary part of it and unthinkable outside of it. In this sense, values are a typical, unique, and essential element in the organization of human behavior and must be of concern not only to social scientists but also to natural scientists studying ecosystems that include humans as members of biological communities.

We believe that this view of values and their place in the larger intellectual order of societies is sufficiently broad to accommodate most of the contributors to this volume, although some differences will be noted among various authors. It should be obvious why, in organizing the conference that resulted in this volume, we focused on values rather than other concepts dealing with the ideational realm. Values represent an important and essential aspect of the intellectual organization of a society, integrated into (and ordained by) the over-arching cosmological system, and constituting the basis for meaningful action. Values are also intermediate between ultimate transcendental truths and the banal rationalizations of day-to-day action, in terms of both concreteness (or abstraction) of content and mutability (or permanence). Because of this balance, values lend themselves to the kinds of analyses of ecological relationships attempted here, those that demand a reasonable amount of specificity as well as historical stability.

Studies of values and value systems, however, face a number of serious problems, not the least of which has to do with the fact that values are ideographic and subliminal. This means that the empirical study of values is difficult, since it cannot rely on the overt meaning of statements elicited by interviews or questionnaires. Values must be inferred through the interpretation of discursive statements, as well as actions, or through analysis of myth, ritual, and artistic expressions. Because of this, the analysis, definition, and description of values entails a certain, often uncomfortably large, measure of subjectivity. This is particularly apparent in ecological anaysis, since values, unlike many other variables in ecological relationships, cannot be directly measured and quantified. Thus, a "humanistic" element is injected into the ecological analysis, making the analysis itself suspect to most natural scientists. Finally, it is not really possible to speak of values without extensive reference to other aspects of the ideational system. Thus, a number of papers in this volume that at first glance seem to have little to do with values do consider the broader framework within which values are embedded.

The Chapters in Context

Although the chapters that follow present a wide range of approaches and opinions, they are linked by a number of common threads. The chapters are grouped into three sections: the first contains statements of a general nature; the second includes case studies and more specific thematic discussions; and the third offers two commentaries on the foregoing proceedings.

In the first section the contribution by Jamieson and Lovelace, originally drafted as a background paper for conference participants, reviews opinions concerning the relationship between material and ideational aspects of human ecology. The chapter expounds particularly on the opposition between "materialist" and "ideational" approaches to the anthropological study of beliefs and values. The authors make reference to some of the more extreme positions, and in doing so draw the dichotomy sharply and starkly. The second chapter in this section, by Hutterer, discusses some general issues concerning the role of "cultural" variables in ecological relationships. He draws a brief sketch of tropical environments, indicating aspects of their ecological organization that should, in one way or another, be reflected in the cultural interaction with these environments.

In the final chapter of the first section, Sponsel reviews issues in the cultural ecology of Amazonia. An anthropologist specializing in Amazonia, Sponsel was invited to participate in the conference and contribute to the resulting volume in order to present information on another tropical region and thus provide a comparative perspective. There are many superficial similarities in the natural and human ecology of Southeast Asia and Amazonia, yet there are few detailed and worthwhile comparative studies. Sponsel's review of human ecological studies in Amazonia is extraordinary in its breadth and detail. Having no first-hand familiarity with Southeast Asia, however, he refrains from making comparisons himself. This is probably wise, given the uneven quality and quantity of empirical data available and the complexity of the issues involved. Yet, whatever the ecological similarities and differences may be, there is striking evidence of major differences in the ethnographic and anthropological approaches scholars have taken toward the two regions. Cultural materialism has played an important role in Amazonian studies, while this approach has been of relatively little significance in Southeast Asia. Within cultural-materialist studies, much attention has been focused on the apparent limitations in certain material resources available in Amazonian rain forests for the sustenance of human populations. This has led to a concern with specific environmental factors that may limit cultural development. In particular, a debate has ensued among Amazonian specialists over the factuality and importance of limited amounts of animal protein available on a sustained basis for human consumption. This question has hardly surfaced in Southeast Asia, and when the question of environmental limitations has arisen, the finger has generally been pointed at carbohydrates rather than protein (e.g., Hutterer 1982; Peterson 1978). It may also be noted that among Amazonian specialists a relatively clear division between materialists and "mentalists" is apparent, while such a dividing line is much more difficult to draw in Southeast Asia.

It would be interesting to ask what engenders these differences. Do they reflect academic responses to structural differences in the ecology of the two regions? Are they the result of different colonial histories that directly and indirectly affected both the methodology and findings of anthropological research among indigenous populations? Or are they simply the inevitable result of accidents in the history of anthropology? These questions must remain open at this time, but it is surely worth pondering to what extent the disparity in anthropological views of Amazonia and Southeast Asia is due to objective differences between the two regions, and to what extent it may represent an artifact of divergent anthropological approaches. Similarly, it is worth reflecting to what extent differences in anthropological attitudes themselves may be predicated on differences in empirical situations.

In his discussion of cultural values, Sponsel considers the value systems anthropologists have brought to Amazonian research, as well as clashes between the value systems of indigenous Amazonian populations and immigrant Western groups. The consideration of anthropologists' values is particularly appropriate in the context of this volume. It is well to keep in mind that our own work is directed and motivated by generally inarticulated and largely subconscious values that influence both our approaches and our findings.

Chapters in the second section are arranged roughly in geographical order, moving from the islands to mainland Southeast Asia, including southern China. Both the diversity of topics considered and the multiple thematic linkages among the papers suggested such an arrangement. The commentary here follows this order where possible but departs from it when linking themes make this desirable.

The section opens with a chapter by June Prill Brett that addresses both what may be considered traditional concerns in human ecology and some of the central issues at stake in this volume. Prill Brett deals with the ritual regulation of agricultural and irrigation activities among the Tukukan Bontok in northern Luzon, Philippines. She focuses on direct ecological relationships between ritual rules and activities on the one hand and agricultural practices on the other, interpreting ritual as providing supernatural sanctions for behavior that is seen as adaptive in the context of the traditional agricultural system. With this sort of interpretation, she moves in a respectable tradition of ecological thought (e.g., Moore 1957; Rappaport 1968). Her analysis goes further, however, as she shows that the regulation of agricultural activities through ritual is essential because, despite a considerable degree of social differentiation in Bontok society, crucial aspects of the agricultural system are managed according to egalitarian principles. Since regulation through centralized political force would not be possible, ritual supplies both the requisite authority and the mechanisms for effective regulation. The egalitarian aspects of Bontok society are explained as a social response to limitations in crucial resources, particularly water, that require cooperative management and control. Egalitarianism and cooperation are not only enforced through ritual sanction but are reflected in the ritual order itself and are thus propagated as values. It is interesting to note here that in water management, cooperation and competition appear as two aspects of the same process, an observation that touches on a point made by Dewey in her paper.

The following chapter by Yengoyan is concerned with a special problem pertaining to the role of cultural values in human ecology, namely, the mechanisms by which cultural coherence is achieved and maintained in the process of cultural reproduction. This issue is of considerable importance, for it relates both to the persistence of cultural values and to their basic roles in directing human action. Yengoyan points to particular forms of oral tradition—formal oratory and mnemonic structures—as preserving the memory of certain events, investing them with cosmological (and therefore social) meaning, and invoking them in the interpretation of contemporary ecological conditions and in considering choices for future action. For his empirical data, Yengoyan draws on his fieldwork among the Mandaya, a population of swidden farmers on the island of Mindanao, Philippines. In Mandaya society, a tradition of formal oratory has been specifically linked with a class of political and military leaders who play a particularly important role in propagating and perpetuating cultural values. This point relates closely to a contention promoted by Benjamin in his contribution—that values are always "politically" propagated and imposed.

Dewey's contribution moves on several levels, being as complex as the *batik* patterns she describes. In considering market relationships in eastern Java, Dewey interprets them as ecological interactions in a broad sense. Although the use of an ecological idiom in this context may seem at first startling, given the fact that she deals with organizational problems in the exchange of resources and the flow and transformation of energy, her use of an ecological framework is no less appropriate than it would be in any other realm of human organization. The ecological approach allows Dewey to make some pertinent observations about the importance of cooperation among small-scale vendors in eastern Java and the importance of this pattern for the survival of the individual trader as well as for mobilizing capital for the local market economy. From there she moves to a deeper analytical level and points to a basic structural characteristic in Javanese culture linking perception, symbolism, and behavior: the use of binary categories, the meaning of which can be reversed. The use of this device makes it possible to attain a very high degree of flexibility and complexity while preserving essential structure. Dewey elaborates on this structural relationship with reference to designs on Javanese *batik* cloth, the structure of gamelan music, and the conceptualization of characters in the *wayang kulit* dramas. In the case of the latter, she also touches on certain ritually encoded values that can be seen as being organized according to the same structural principles. Some of the most interesting points of her paper are both epistemological and ecological. She notes that most Western epistemologies are based upon rigidly defined conceptual categories that may deny the user important insights into the nature of social and natural relationships. She refers to the concepts of cooperation and competition (both central concepts in ecology) and proposes that, contrary to common perception, they may represent not polar opposites but complementary aspects of complex relationships.

The chapter by Long originated as a discussion contribution. It will be readily apparent that Long is not a natural or social scientist, but a performer and teacher of theater. Thus his approach differs from that of most of the other contributors, as he moves primarily on a symbolic and phenomenological level. Since the ritualized theater of Southeast Asia is an important storehouse and purveyor of cultural information, including cultural values pertaining to ecological relationships, we felt that his contribution was significant and valuable. In addressing some of the issues around which this volume is organized, Long reached conclusions that approximate those of other contributors, albeit from a different perspective. Long takes his departure from the Javanese theater, particularly the shadow puppets of the *wayang kulit*, and touches on the question of the transmission and propagation of values. In Java this occurs in the context of courtly drama, in marked contrast to the Mandaya who live in a much less complex social world and rely on oratory and mnemonics to accomplish these same ends. Long shows how the central message concerning basic values is maintained at the same time that details of the performance are continuously varied to suit particular needs and interests. The papers of both Long and Yengoyan contribute observation and elaboration to a statement made elsewhere in this volume by Benjamin concerning the ideographic nature of values. This particular characteristic is at least partly linked to the fact that values are transmitted and propagated through concrete images of speech and dramatic forms rather than in abstract conceptual form. The *wayang* scenes Long describes suggest that the Javanese are keenly aware of the interactive relationship between humans and nature, and, relating to Dewey's discussion, that their ecological values incorporate this complementary relationship.

In contrast to Long, Soemarwoto is not only a biological ecologist, but also has been much involved in ecological planning and *ex post facto* assessment of development projects in rural Indonesia. He brings a practical concern and experience to these discussions. Perhaps reflecting this applied commitment, Soemarwoto views values on a more specific level than do most of the other contributors. His involvement in applied ecology has made him well aware that specific values may persist beyond the point where they are ecologically "adaptive" and that they may either hinder desirable changes or promote undesirable ones. At the same time, however, he also shows that certain values themselves tend to change with broader social and economic transformations.

Benjamin's chapter is the widest ranging of the second group and proceeds with the most explicitly defined framework and the most specific definition of cultural values. It presents a broad chronological perspective, which is extremely important in an evolutionary conception of ecology, although Benjamin prefers to perceive his approach as history. He makes interesting and important points pertaining both to theoretical issues and to Malaysian cultural history and ethnography. He views values as broad organizational themes of cultures, "politically" imposed and propagated and related to several basic organizational aspects of the relationship of people to each other and to the land in which and from which they live. He argues persuasively for the long-term persistence of such values, notwithstanding continuously shifting details of social life. He suggests that the ethnographic mosaic of the Malay Peninsula can be explained as resulting from the operation of three such basic organizational themes rather than a series of discrete migratory waves, as is the conventional ethnological explanation. Within his extended diachronic view, values for Benjamin are arbitrary cultural elements that regulate individual interaction with the social and natural environment throughout the course of history rather than simply the cognitive aspect of continuously evolving, human, ecological relationships.

In certain respects, Prachuabmoh's study is related to that of Dewey, in that Prachuabmoh also uses an ecological idiom to discuss trade interactions. Prachuabmoh has conducted research on participation in trade activities by three ethnically distinct but geographically contiguous populations in southern Thailand. In her chapter in this volume, she considers one facet of her research, namely, involvement in market-trade activities by a Thai Muslim community. She discusses an apparent discrepancy between officially and explicitly stated views regarding trade as an occupation and the role of women in it, on the one hand, and actual behavior on the other. The latter seems to be governed by values that vary with the official view and are difficult to explicate. With this, Prachuabmoh touches on the important problem concerning the difference between specific manifest norms and broader latent values. Like Soemarwoto, Prachuabmoh deals with a situation in the process of change whereby it appears that shifts in values are closely associated with transformations in several areas of the cultural and social organization, including economy, patterns of spatial mobility, and social relations. She also demonstrates that the boundaries of human ecosystems are not necessarily coterminous with those of natural communities and thus must be defined in their own terms.

For many readers, the chapter by Trinh may be the most intriguing and even puzzling. It addresses cultural values and human ecology on several levels. Trinh presents the concerns of a developing nation undergoing social, economic, and environmental changes and of a nationally perceived need to guide these changes in a desirable direction. This concern is phrased in terms of a Marxist ideology with a particular view of integrating theory and practice. Without abandoning this framework, Trinh seeks to integrate into it perspectives derived from ecological science. In the process he makes several interesting points: he describes human interactions with the physical environment as dialectical in nature, a point also made by some non-Marxists (Rambo 1983); he states that social transformations entail ecological (and thus environmental) transformation; and he stresses the importance of ideology in promoting and guiding ecological change.

The chapter by Trinh may be difficult to understand for two reasons: he uses an intellectual framework that is essentially foreign to most of us, involving a terminology that imbues familiar words with unexpected meanings; and he himself is struggling to assimilate and integrate an intellectual framework (human ecology) that is, at this point, foreign to most social scientists in Vietnam. Among other things, Trinh's writing also reveals that the linkage between materialist and ideational perspectives is still problematical in Vietnamese Marxist thinking about social and environmental change. Thus, many of the things he says become fully understandable only to those who are familiar with the intellectual background on which these statements are predicated. Although it would have been desirable to append a lengthy commentary to Trinh's chapter, it was impossible in the present context. Nevertheless, the paper constitutes a poignant document and an interesting contribution, especially for students of contemporary Vietnam.

In confronting us with what is to many Western readers an unfamiliar conceptual framework, Trinh's chapter carries a very special, albeit implicit, message: scientific knowledge and insight are affected by the social context within which they are created. As Bourdieu (1977) points out, there are three forms of knowledge: phenomenological, objectivist, and explicit. Objectivist knowledge, although based on a reflective separation of subject and object, is not absolute but relative in the sense that it is predicated upon certain tacit assumptions that are given by the social context. In order to arrive at explicit knowledge, a further step of epistemological reflection and abstraction is necessary, which inquires "into the conditions of possibility, and thereby, into the limits of the objective and objectifying standpoint" (Bourdieu 1977, 3). Even explicit knowledge is not final, but it does entail an explicit recognition of the degree to which it is based on ideologically or culturally determined notions. It is important to realize that all scientific frameworks, including Marxist approaches to environmental relations and schools of human ecology presently popular in American and European universities, are bounded by certain conditions that individual scientists generally accept as given. Our various cultural and social backgrounds constitute relativistic frameworks within which we pursue our scientific work. The true value and validity of these frameworks cannot be assessed by judging one against the other but by stepping outside either of them.

While Trinh's contribution is very general, Pei, a scholar from the People's Republic of China, presents a specific case study. He, like Soemarwoto, is a natural scientist who has become deeply interested in human interactions with the environment. Among many other research activities, he has conducted ethnobotanical research among national minority groups, particularly the Dai, a T'ai-speaking people, in southwestern China. Pei makes several interesting observations concerning the use of the plant world by Dai people, whom he shows to be influenced both by their adherence to Buddhism and their animistic beliefs. Particularly noteworthy is his discovery that large areas of natural forests are being maintained and conserved under the influence of the concept of the "holy hill." In measuring the area covered by these preserves he has provided empirical substantiation for earlier, anecdotally based suggestions that such groves might be significant in conservation of genetic resources in Southeast Asia. The fact that the distribution of cultivated and domesticated plant species in southern Yunnan seems to be affected by the spread of Buddhism speaks once again to the importance of culture and ideology in shaping the environment.

In the final contribution to the second section, Lovelace deals with South China. Like Benjamin, he uses an extended time perspective in viewing interactions between humans and nature, although his approach and objective differ. Based on historical research in Hong Kong, Lovelace argues convincingly that the premodern history of land use in this area was strongly influenced, even governed, by a system of thought and ritual called *feng shui*. Like Anderson (1973), Lovelace argues that there is a direct relationship between principles and practices embodied by *feng shui* and settlement-subsistence behavior. Unlike Anderson, Lovelace concludes that the application of *feng shui* in the coastal areas around Hong Kong by agriculturalists moving in from interior China did *not* result in stable adaptations to the new environment. Rather, *feng shui* promoted a transformation of the environment that was coincident with changing patterns of demography and land use and proved to be adaptive in that context. Lovelace suggests that once the transformation of the landscape into a habitat suitable for wet-rice cultivation was completed, *feng shui* itself was transformed, shifting its emphasis from placement of settlements to siting of individual houses and graves.

The final section of the volume includes two commentaries. Hickey, an anthropologist with extensive experience in the field of rural development in Southeast Asia, responds to the papers from the perspective of a development specialist. He takes the conventional wisdom in development theories to task for its disregard for traditional, small-scale, socioeconomic systems that are often held to be stagnant, a hindrance to progressive development, and doomed to extinction. Hickey reviews the contributions made in this volume and shows the traditional tribal and peasant systems to be flexible, with the potential to furnish the social and economic basis for development processes that would involve a broad population base if planners were able and willing to understand these systems on their own terms.

Jamieson reviews some of the basic concepts underlying the preceding papers such as adaptation, evolution, change, and development. Writing from an evolutionary perspective, he sees an interaction between the state of ecosystems and human attempts to deal with them conceptually. He calls for a rethinking of familar concepts to take into account the magnitude and rate of ongoing change, because he feels that most contemporary ecosystems are undergoing such rapid change (reflected in the observations made by many of the contributing authors) that traditional concepts and paradigms cannot cope.

The chapters assembled here are exceedingly diverse. They include abstract theoretical discussions and specific case studies, ranging across the landscape of Southeast Asia from the islands to southern China. They deal with hunting-gathering populations as well as peasants operating within contemporary nation-states, and they are the work of natural scientists, social scientists, and humanists of Western and Asian origin. Diversity in the backgrounds of the authors contributes most to the varied approaches to the theme of this volume, because differences in cultural background and academic tradition will lead to different research interests and to differences in the empirical approaches chosen to pursue given problems. Once the basis of this diversity is understood, it is possible to see the common concerns and themes addressed here.

Practical Implications

The expectation of the organizers of this volume and the contributors to it was not that they would find final answers to the large and complex set of questions concerning the relationship between culture and the environment. While such a result is certainly desirable, it is clearly unrealistic to expect it to happen as the result of a single conference. As we said earlier, solutions are not likely to emerge on the basis of established conceptual frameworks. New concepts, even a paradigm shift, are needed, but these things cannot be achieved through determination alone. Neither the conference organizers nor the participants operated under the illusion that any of us might be the Copernicus or Einstein that the field so clearly awaits. If our aims were modest, why, then, did we engage in this exercise? There are three reasons.

First, the issues that concerned us here are of intrinsic interest. After all, we are dealing with the relationship between material and nonmaterial phenomena, two classes of being that in most Western thought have been held to be mutually exclusive yet patently are not. The question of how to explain this relationship constitutes a central problem in the interdisciplinary field of human ecology and takes us to the edge of contemporary knowledge and thought. Regardless of whether we believe that we can break through the intellectual boundaries that confine us, we want to peer across them, contemplate what lies beyond, and try to reach it.

Second, if we have not made a major breakthrough, we do think that some contribution has been made. Many of the papers offer considerable insight into the functioning, operation, and development of specific, Southeast Asian, social systems. There are many levels of understanding. Even if we may not grasp, in the final sense, how mind and matter interrelate, there are less abstract levels of analysis and explanation through which we can get a useful, if partial, view of ecological relationships involving ideational aspects. For the moment we have little choice but to proceed in small steps. It is often under the weight of a mass of evidence accumulated in this way that old paradigms collapse and new ones spring to life.

Finally, there is an eminently practical aspect to the concerns expressed in the papers of this volume. A worldwide system of social, political, and economic relationships has developed, dominated by industrialized societies but involving also thousands of small-scale, traditional, nonindustrialized populations (Wolf 1982). Under the impact of their relationships with industrialized nations, these populations are experiencing enormously rapid transformations in their social organization and natural environments. The rate, magnitude, and depth of these transformations is of concern not only to the affected societies but also to industrialized nations, many of whom are trying to influence this process in one way or another through various approaches to "development aid." On the basis of what has been said here, and of what is contained in the papers that follow, it must be clear that the true effect of monetary outlays, technology transfer, food and medical assistance, and expert advice in the area of urban planning cannot be predicted or gauged unless it is fully understood that development processes are not simply a matter of economics and technology. Many important questions must be considered, concerning the role of cultural traditions and values in environmental, economic, and technological change. These questions can be usefully and insightfully analyzed within the existing frameworks.

References

Anderson, E. N.
  1973   Feng shui: ideology and ecology. In *Mountains and Water: Essays on the Cultural Ecology of South Coastal China*, edited by E. N. and H. L. Anderson. Taipei: Orient Cultural Service. Pp. 127–46.
Anderson, J. N.
  1974   Ecological anthropology and anthropological ecology. In *Handbook of Social and Cultural Anthropology*, edited by J. J. Honigmann. Chicago: Rand McNally. Pp. 477–97.
Bennett, J. W.
  1976   *The Ecological Transition: Cultural Anthropology and Human Adaptation*. New York: Pergamon Press.
Birdsell, J. B.
  1953   Some environmental and cultural factors influencing the structuring of Australian aboriginal populations. *American Naturalist* 87:171–207.
  1979   Ecological influences in Australian aboriginal social organization. In *Primate Ecology and Human Origins: Ecological Influences and Social Organization*, edited by I. S. Bernstein and E. O. Smith. New York: Garland STPM Press. Pp. 117–51.
Boughey, A. S.
  1971   *Man and the Environment*. 2d ed. New York: Macmillan.
Bourdieu, P.
  1977   *Outline of a Theory of Practice*. Cambridge: Cambridge University Press.

Coe, M. D.
    1961   Social typology and the tropical forest civilizations. *Comparative Studies in Society and History* 4:65–85.
Culbert, T. P. (editor)
    1973   *The Classic Maya Collapse.* Albuquerque: University of New Mexico Press.
Devereux, G.
    1978   Two types of modal personality models. In *Ethnopsychoanalysis: Psychoanalysis and Anthropology, a Complementary Frame of Reference,* by G. Devereux. Berkeley: University of California Press. Pp. 113–35.
Dunn, F. L.
    1975   *Rainforest Collectors and Traders; A Study of Resource Utilization in Modern and Ancient Malaya.* Kuala Lumpur: Monograph 5 of the Malaysian Branch of the Royal Asiatic Society.
Diener, P., and E. Robkin
    1978   Ecology, evolution, and the search for cultural origins: the question of Islamic pig prohibition. *Current Anthropology* 19:493–540.
Eliade, M.
    1959   *Cosmos and History: The Myth of the Eternal Return.* New York: Harper and Row.
Endicott, K. A. L.
    1979   *Batek Negrito Religion.* London: Clarendon Press.
Hackenberg, R. A., and H. F. Magalit
    1983   *Demographic Responses to Development: Sources of Declining Fertility in the Philippines.* Boulder, CO: Westview Press.
Harner, M.
    1977   The ecological basis for Aztec sacrifice. *American Ethnologist* 4:117–35.
Harris, M.
    1966   The cultural ecology of India's sacred cattle. *Current Anthropology* 7:51–59.
    1973   The riddle of the pig. *Natural History* 82(2):20–25.
    1974   *Cows, Pigs, Wars, and Witches: The Riddles of Culture.* New York: Harper and Row.
    1977   *Cannibals and Kings.* New York: Random House.

Hutterer, K. L.
1977   Review of "Rain-Forest Collectors and Traders," by F.L.
        Dunn. *Journal of Asian Studies* 36:792–93.
1982   Interactions between tropical ecosystems and human
        foragers: some general considerations. Working Paper.
        Honolulu: East-West Environment and Policy Institute,
        East-West Center.
Kluckhohn, C.
1951   Values and value-orientations in the theory of action: an
        exploration in definition and classification. In *Toward a
        General Theory of Action*, edited by T. Parsons and E.
        Shils. Cambridge: Harvard University Press. Pp.
        388–433.
Kluckhohn, F. R., and F. I. Strodtbeck
1961   *Variations in Value Orientations*. Evanston, IL: Row
        Peterson.
Lee, R. B.
1972   !Kung spatial organization: an ecological and historical
        perspective. *Human Ecology* 1:125–48.
Meggers, B. J.
1954   Environmental limitations on the development of culture.
        American Anthropologist 56:801–23.
Moore, O. K.
1957   Divination—a new perspective. *American Anthropologist*
        59:69–74.
Orlove, B. S.
1980   Ecological anthropology. *Annual Review of Anthropology*
        9:234–73.
Ortiz de Montellano, B. R.
1978   Aztec cannibalism: an ecological necessity? *Science*
        200:611–17.
Peterson, J. T.
1978   *The Ecology of Social Boundaries: Agta Foragers of the
        Philippines*. Urbana, IL: University of Illinois Press.
Raikes, R. L.
1964   The end of the ancient cities of the Indus. *American
        Anthropologist* 66:284–99.
Rambo, A. T.
1979   Primitive man's impact on genetic resources of the
        Malaysian tropical rain forest. Malaysian Applied Biology

9(1):59–65.

1980   Of stones and stars: Malaysian Orang Asli environmental knowledge in relation to their adaptation to the tropical rain forest ecosystem. *Federation Museums Journal*, n.s., 25:77–88.

1983   *Conceptual Approaches to Human Ecology.* Honolulu: East-West Environment and Policy Institute, East-West Center. Research Report No. 14.

Rappaport, R. A.
1968   *Pigs for the Ancestors.* New Haven: Yale University Press.

1979   On cognized models. In *Ecology, Meaning, and Religion*, by R. A. Rappaport. Richmond, CA: North Atlantic Books. Pp. 97–144.

Sahlins, M. D.
1972   *Stone Age Economics.* Chicago: Aldine-Atherton.

1979   Reply to Marvin Harris. *New York Review of Books*, June 28:52–53.

Sanders, W. T., and B. J. Price
1968   *Mesoamerica: The Evolution of a Civilization.* New York: Random House.

Vayda, A. P., and R. A. Rappaport
1968   Ecology: cultural and non-cultural. In *Introduction to Cultural Anthropology*, edited by J. A. Clifton. Boston: Houghton Mifflin. Pp. 476–97.

Watanabe, H.
1975   *Bow and Arrow Census in a West Papuan Lowland Community.* Occasional Paper No. 5. Santa Lucia, Queensland: Anthropology Museum, University of Queensland.

White, L. A.
1949   *The Science of Culture: A Study of Man and Civilization.* New York: Grove Press.

1959   *The Evolution of Culture.* New York: McGraw-Hill.

Williams, R. M., Jr.
1968   The concept of values. In *International Encyclopedia of the Social Sciences* 10:283–87.

Winterhalder, B., and E. A. Smith (editors)
1981   *Hunter-Gatherer Foraging Strategies: Ethnographic and Archaeological Analyses.* Chicago: University of Chicago

Press.

Wolf, E. R.
1982   *Europe and the People Without History.* Berkeley, CA: University of California Press.

Yengoyan, A. A.
1976   Structure, event, and ecology in aboriginal Australia: a comparative viewpoint. In *Tribes and Boundaries in Australia*, edited by N. Peterson. Canberra: Australian Institute of Aboriginal Studies. Pp. 121–32.

I.   BACKGROUND

CHAPTER 2

CULTURAL VALUES AND HUMAN ECOLOGY: SOME INITIAL CONSIDERATIONS

Neil L. Jamieson
George W. Lovelace

The June 1983 conference on "Cultural Values and Tropical Ecology" was intended to address a very basic question: What is the nature of the relationship between the cultural values espoused by a people and the way that these people interact with their environment? This is a variation of the older and even more basic question: What is the relationship between ideas and the material conditions of existence?

This question has an illustrious and ancient history. Different societies have posed and answered it in many ways. Even within the traditions of what we customarily refer to as the "great civilizations," very different and sometimes diametrically opposed answers have been given.

Within the classical sources of Western thought, for example, one need only take note of the glaring disparity between the statements of philosophers such as Democritus and Lucretius, on the one hand, and Plato on the other. The controversy over the primacy of either mind or matter was displayed in well-developed form in Greek and Roman thought more than 2,000 years ago (Allen 1966, 5–6, 11–23, 243–255).

The major philosophical traditions of the East also exhibit considerable variation in the ways in which this question has been conceived and answered. Neo-Confucianist scholars, for example, posited the existence of one vast, all-encompassing cosmic order that was perceived to be both real and deterministic. This was a synthesizing view in which what we think of as mind and matter were but two different aspects of a single reality, a reality that had, however, an existence completely independent of the mind or ideas of any individual (Chan 1967b, 139–140; Fung 1966, 294–306). Although the Western, Cartesian emphasis upon two separate and different realities—one materialistic and the other idealistic—is generally absent in Chinese philosophy, the latter has on occasion displayed certain similarities, as, for example, in the idealist pronouncements of Lu Jiu-yuan and in the later reactions of empiricists like Tai Chen (Chan 1967a, 62–65).

The concept of ecology was developed in the late nineteenth and early twentieth centuries in the West in the context of biology— within an essentially Newtonian world view. It was and remains basically a science, and is materialist in its orientation. By and large, this strong materialist bias has continued to dominate the fields of cultural ecology, human ecology, and ecological anthropology. Regardless of the disciplinary origins of the investigators, research in these fields has placed the emphasis upon human beings as biological organisms, as part of the flow of energy and matter through larger biological systems.

Yet in recent years different studies have shown how belief systems and value systems are related intimately in human-environmental interactions. Since its development during the early 1960s, the field of ethnoscience has provided ample documentation of the ways different cultures employ categories for classifying various aspects of the natural world. Humans have devised many different taxonomic schemes for sorting plants, animals, soils, foods, colors, and diseases into categories (e.g., Conklin 1955; Frake 1961, 1962; Metzger and Williams 1963, 1966; Sturtevant 1964; Tylor 1969). The diverse ways peoples have categorized their experience of the world scarcely could be irrelevant to their behavior within and toward their environment.

There is considerable evidence to support the proposition that the human mind categorizes all experience, and, as structuralists have demonstrated again and again over the past thirty-five years, it tends to do this in terms of binary oppositions (e.g., male-female, natural-supernatural; sacred-profane, high-low, hot-cold). From these basic binary oppositions, members of particular cultures generate sets of shared assumptions about the nature of reality and the meaning of experience. Research indicates that people have a tremendous intellectual and emotional investment in the categories they use to organize their experience of the world. Psychological and anthropological research provides powerful evidence to support the assumption that human beings are in fact designed to be that way, to be preoccupied with categories and concerned with the maintenance of their boundaries. Further, it has often been demonstrated that anything defying or upsetting or contradicting these categories arouses strong emotion and that this emotion can be culturally channeled and expressed in many ways, including powerful taboos (e.g., Leach 1965; Turner 1964, 1967, 1969).

There are many examples of such taboos around the world. Certain categories of people either all of the time or some of the time are forbidden to enter certain places, engage in certain activities, or eat certain foods. In many cases, such taboos seem to have ecological implications. They exert some kind of direct or indirect influence upon the flow of energy and matter. They tend to conserve some resources and to put additional pressure on others.

To pick a well-known example, the Bible presents us with some interesting examples of food taboos. The book of Leviticus is full of them. Some animals were considered to be "unclean" or "polluting" and therefore unsuited for human consumption; others were deemed to be "clean" animals and fit for eating. Prominent among the prohibited animals was the pig, whose flesh was expressly judged to be polluting.

This was a deeply held religious belief, and even today orthodox Jews and Muslims do not eat the flesh of the pig. How does one explain this? Why should this particular source of food have been prohibited? In the early Neolithic villages of the Middle East, people raised and ate pigs as well as sheep and goats. Why should they have come to believe that pigs were polluting but not sheep or goats?

Mary Douglas, a British anthropologist, suggests that this was due to the system of categories the Semites employed for classifying animals (Douglas 1966). Let us review briefly what the Bible says about this matter. Chapter 11 of the book of Leviticus records the words of Jehovah on the dietary laws, or food taboos, to be observed by the people of Israel:

> These are the beasts which ye shall eat among all the beasts that are on the earth.
>
> Whatsoever parteth the hoof, and is clovenfooted, and cheweth the cud, among the beasts, that shall ye eat.

This would include sheep, goats, and cattle, all of which have cloven hooves and chew the cud. But the pig is excluded, and this exclusion is emphasized (Leviticus 11:7):

> And the swine, though he divide the hoof, and be clovenfooted, yet he cheweth not the cud; he is unclean to you.

Unclean animals, one discovers, are the anomolies of the classificatory system. They are polluting, even repulsive to eat, because they do not fit neatly into the cultural categories for classifying animals. One can find examples worldwide of people taking their systems of classification so seriously that the food they eat and even the words they speak are restricted by taboos related to an anomolous status in the categories of the culture.

Mary Douglas's explanation of why pigs were taboo for the ancient Semitic peoples thus seems highly plausible. Yet, Marvin Harris, another anthropologist, has a different, completely independent explanation of this phenomenon (Harris 1973, 1974) that is apparently equally compelling. Harris argues that the Semitic peoples stopped eating pigs because of ecological conditions in the ancient Middle East. Pigs were taboo, he claims, because ecologically they had become a great luxury, perhaps even a dangerous one. A summary of his argument follows.[1]

Pigs have no sweat glands and cannot tolerate high temperatures. They therefore require substantial amounts of water, not just to drink, but to wallow in. Because they often feed upon or are fed with food that can be consumed directly by human beings, pigs also compete with humans for food. For the ancient Semites at the time of Moses, it simply was not sensible to raise pigs under existing environmental conditions. It made much more sense for people to use the scarce arable land and water resources for agriculture and then to eat the grain themselves.

They could get their animal protein from cattle, goats, and sheep—from animals that have cloven hooves and chew the cud. These, Harris points out, are the "clean" animals. And, not coincidentally, they are admirably equipped to forage on the semiarid hillsides, eating plant life that is not suited to the digestive systems of either pigs or human beings. By and large, one finds, the "clean" animals are precisely those that exploit an ecological niche not directly exploitable by humans. They are, therefore, animals that convert energy from a form unavailable to people—because it is stored in plants with a high cellulose content—to the flesh of the herd, where it is stored in easily transportable, readily obtainable, and highly digestible form.

These two explanations of the Semitic pig taboo are specific examples of two very general and quite different ways of looking at and attempting to understand the nature of the world and human behavior within it. Mary Douglas presents what we call an "ideational" explanation; Marvin Harris, a "materialist" one.

From a broad materialist perspective, ideational phenomena such as beliefs, values, symbols, and categories are shaped by the material conditions of existence. Ideational phenomena are secondary, subordinate, and derivative. They are thought to justify, give meaning to, or motivate behaviors that have already been demonstrated by experience to be necessary for continued existence in a particular environmental setting. In other words, our categories, taboos, values, and symbols are often said to have been developed to encourage or condition us to do things that are adaptive. They discourage us from doing things that might appear tempting in the short run or from limited knowledge, but which the accumulated experience of the group has shown would be mistakes in the long run for the group as a whole. From the materialist perspective, ideational systems are often viewed as a shorthand form of accumulated wisdom, as a set of rules for doing what is ecologically and socially adaptive. They have their origins in, and are ultimately to be explained by, the material conditions of existence.[2]

From the ideational perspective, however, beliefs, values, symbols, and categories are systems of meaning. They often are depicted as relatively autonomous systems that develop according to an internal logic of their own. To the structuralists, this internal organization of concepts reflects the structure of the human brain itself. Ideational systems are believed to satisfy people's psychological needs, to provide meaning to life, and to help people organize and interpret their experience as sentient creatures. Many divergent views of human beings and human society are subsumed under the heading of ideational approaches, and many different dimensions of what we have labelled, with intentional ambiguity, psychological needs are emphasized by various schools of thought.

Increasingly over the past two hundred years, however, such diverse thinkers as Kant, Hegel, Marx, Weber, Freud, Jung, Durkheim, Lévi-Strauss, and Bateson have been making us more and more aware of how crucial, and how problematic, the role of ideas is in human life. Only recently have we become self-conscious enough to realize how little we know about our own minds and the relationship of the human mind and its products to the surrounding physical world.

In recent decades, research in many different fields has consistently confirmed the universal human need for ideational systems. The accumulating evidence suggests strongly that the evolution of the human species has involved cultural as well as biological factors and processes (Durham 1982, 290). This has significant and perhaps not yet fully appreciated implications for the way we view ourselves—for our ideas as to the nature of our species.

As members of human society, indeed to become human, we develop both cognitive and existential needs that are prerequisites for social life and are therefore prerequisites for human life. We need structures of both meaning and value to make our existence intelligible and satisfying. We need interlocking frameworks of questions and answers about the what and the why of the world and our existence in it. We have minimal requirements for coherent systems of ideas, categories, symbols, and values to maintain ourselves as functioning organisms and populations (as individuals and as groups), and these minimal intellectual and emotional requirements are as imperative to sustaining our viability as are the minimal needs we have for calories, protein, vitamins, water, or sleep. We are not only biological, but also cognitive and emotional, animals.

Our approach to the issue of cultural values and tropical ecology is based upon the fundamental premise that no system of which human beings are a part can be understood adequately without (1) taking into account the unique extent to which humans are multidimensional creatures, and (2) without acknowledging the basic validity, the partial truth, of both materialist and ideational perspectives.

Materialist Approaches

The materialist thinkers draw upon a philosophical tradition that extends far into antiquity. Democritus, Epicurus, and Lucretius are probably the clearest and best-known examples. But since the Renaissance, there have been many others who have revived, popularized, and developed this point of view: Thomas Hobbes, Herbert Spencer, Karl Marx, and Friedrich Engels, to mention a few prominent examples. Marvin Harris is perhaps the most vocal contemporary exponent of this point of view in Western social science.

Materialist frameworks focus broadly upon the "conditions of life" as the mainspring of human organization, with "conditions of life" referring to the natural environment or the social environment, or both. Much attention has been focused upon the economic and technological interface between the natural and the social realms. With this focus, ideational systems are generally depicted as relatively passive, malleable, *ex post facto* rationalizations of the relationships that result from the more important and deterministic sphere of "practical" concerns that relate directly to physical survival.

Even in those basically materialistic studies that seem to assign a more active role to beliefs and values, this role is seldom seen to involve any significant influence upon the nature of the supposedly more dominant material relationships. The most active roles one finds assigned to ideational factors in such writings are essentially regulatory in nature. That is, ideational factors are sometimes said to stabilize, maintain, or reinforce existing relationships between human populations and their environments (e.g., Rappaport 1967, 1968, 1971a, 1971b; Flannery and Marcus 1976; Harris 1966, 1973, 1974; Lindenbaum 1972).

Within the broad category of materialist approaches there is a wide variety of specific theories and methodologies. The most influential of all materialist theories, dialectical materialism as formulated by Marx and Engels more than a century ago, is still very much alive and well in contemporary intellectual life. Not only is it the dominant and sometimes compulsory approach within the socialist world, but it has numerous adherents and practitioners in many nonsocialist countries. Some recent theorists have made modifications in the original formulation (usually emphasizing certain aspects of the later works of Marx). This modified Marxist framework is often referred to as "Neo-Marxism."[3]

There are also modern scholars who bear the label of "structural Marxists."[4] Their work is predicated upon the assumption that both Marx and Lévi-Strauss discovered through independent investigations some "inner but concealed pattern" that is "very much different from those relations as seen on the surface, in their real existence" (Godelier 1977, 46). Within anthropology in the United States, "cultural materialism" has become very influential. This is the approach developed, named, and popularized by Marvin Harris, whose materialist explanation of the Semitic taboo on eating pigs was presented earlier. Harris's ability to generate such materialist explanations for the origin of well-known components of ideational systems has been a major factor in the dissemination of this theoretical viewpoint (e.g., Harris 1966, 1973, 1977). One of the more recent variants of materialist explanation is known as "operant conditioning" (e.g., Jochim 1981). This is basically an extension and development of "social learning theory" and an examination of "stimulus" and "response." The underlying assumption behind this work is that behavior is shaped over time in particular environments by positive and negative reinforcements. Certain behaviors and ideas are more likely to succeed at certain tasks than others under any given set of environmental conditions. As a result of differential rates of success, people eventually come to prefer (believe in, value) those behavioral and cognitive patterns that are the most appropriate for the conditions in which they live.

Ideational Approaches

In sharp contrast to these materialist frameworks are several approaches that tend to treat such cultural phenomena as beliefs, values, and symbols as if they constitute an autonomous system that is neither excessively constrained by, nor subordinate to, the physical or social conditions of existence. In these approaches, which we here refer to as "ideational," emphasis is placed upon the conceptual capacities and needs of the human mind in both its individual and collective forms.

The philosophical emphasis of these approaches is not new. Its persistent vitality is illustrated amply by the writings of Bishop George Berkeley, David Hume, and Immanuel Kant in the eighteenth century; Georg Wilhelm Friedrich Hegel and August Comte in the nineteenth century; and Max Weber, Ruth Benedict, Clifford Geertz, and Claude Lévi-Strauss in the twentieth century.

As in the case of the materialist approaches, we find considerable variation in philosophical, theoretical, and methodological stances and much controversy between different factions within the larger ideational camp. One ideational approach that is now common, especially in France but also in Great Britain and the United States, is known as "structuralism" (Lévi-Strauss 1963, 1966, 1969). This approach is based upon the premise that all human beings and all human groups organize their perceptions and experience of the world into categories in terms of binary oppositions: natural-supernatural, male-female, sacred-profane, etc. The ideational explanation of the Semitic pig taboo presented by Mary Douglas is essentially a structuralist explanation. At a more theoretical level, structuralism is generally directed toward uncovering the unconscious but structured regularities that underlie human thought. Much of this approach draws its primary inspiration from the field of structural linguistics, especially the groundbreaking work done by Roman Jakobson at the New School for Social Research during the early 1940s. The works of such disparate thinkers as Claude Lévi-Strauss, Jean Piaget, and Victor Turner illustrate the breadth of problems and diversity of emphases that usefully can be informed by the basic insights provided by Jakobson's work.

A group of American anthropologists has developed a field of study known as "ethnoscience." Like the structuralists, ethnoscientists concentrate upon cognitive structures and derive much of their inspiration from models that originated in the discipline of linguistics. By painstaking fieldwork directed toward documenting and analyzing the various ways in which people in different cultures classify parts of their world (colors, diseases, plants, animals, etc.), the ethnoscientists have aspired to discover cultural rules or principles of classification that will permit them to construct (or reconstruct) the "grammar" of a culture the way linguists have formulated the grammar of a language. Thus, the work of these ethnoscientists has often been called "new ethnography." Although the high hopes and enthusiasm that characterized this approach during the 1960s have abated considerably in recent years, such a demonstration of the extent to which different cultures employ quite different criteria to classify the phenomena of the world has potentially great significance for issues in cultural ecology.

Many other people, of course, have looked at ideational phenomena in quite different ways. Freud, Jung, and many others since have had a great deal to say about these phenomena from a psychoanalytic point of view. This particular approach has tended almost exclusively to deal with symbols from the standpoint of the individual psyche. Universal symbols have been asserted to exist and to retain their power because they are believed to represent elements that are universal in the human experience.

More recently, several people practicing what we might call "symbolic anthropology" have sought to analyze symbols as they are actually used in particular social and cultural contexts. Clifford Geertz (1964, 1966, 1971, 1972), Victor Turner (1967, 1969, 1974, 1975), David Schneider (1968, 1969), and others have studied particular sets of symbols as they were found to be associated with the interests, purposes, ends, and means of human action in concrete situations. In this body of work, symbols are viewed as embedded in the social processes and cultural contexts in which people use them and respond to them. They are, in fact, seen as forming an important and integral part of that context at one level of analysis and as taking part in the creation of that context at another level. Within this approach, the structure, properties, and precise meaning of a symbol (or, more often, a set of symbols) and its relationship to other symbols (or sets of symbols) are considered to be dynamic and flexible, functioning within certain appropriate contexts of action. Although symbolic anthropology offers us no single, well-developed model to use in analysis, it is an emerging tradition of great interest and one that may be of considerable significance for our purposes. It relates behavior, affect, and cognition and demonstrates the dynamic relationship betweeen affect, meaning, and context and action. Symbolic anthropology thus has the potential for revealing systematic linkages that might be overlooked by many of the more narrowly focused ideational approaches.

In a similar vein, some ideational approaches within the discipline of geography recently have been articulated in a still loosely structured subdiscipline known as "humanistic geography." Rather than concentrating upon the spatial organization of culture, society, and human activity within the environment, writers such as Tuan Yi-fu (1977), Anne Buttimer (1974; see also Buttimer and Seamon 1980), and Edward Relph (1976) emphasize the cultural organization of both space and time. These writers and others often use the concept of "place" as a way to organize the environment into subjectively meaningful units. The influence of such twentieth-century philosophical movements as existentialism and phenomenology is exhibited clearly in some of these studies.

Although much of the work within this subdiscipline is as yet tentative and self-consciously exploratory, humanistic geographers are in many ways at the forefront of efforts to explore actively the notion that the ways in which human beings perceive, conceive, and subjectively experience their environments significantly shape the nature of their influence upon and their interactions with the flows of energy and matter within their physical world.

The brief comments offered here serve merely to illustrate the great diversity of approaches included under the rubric of ideational viewpoints. There seems to be little substantive interaction between many different theorists who see ideas as playing a primary, or at least important, role in shaping the world. Some seem to ignore altogether the relationship between ideational and materialist factors. Some seem to see the physical or material world as extensively and rather easily shaped by human activity, as a passive variable that actually comes to acquire the attributes that people project onto it. Still other advocates of ideational or idealist approaches seem to suggest that many rules or taboos, values, symbols, or categories that directly or indirectly affect patterns of human behavior may nevertheless have no direct or significant ecological or even social consequences whatsoever. The ideational and materialist realms are treated by some as almost nearly independent universes. Even among those who would not go so far, there is a tendency to assume that the most important function of a symbol, value, or concept is to protect or maintain the integrity of the ideational system itself. It is held to be as important to preserve some balance in one's ideational system as in one's ecosystem.

Situational Approaches

In examining the literature, we also have encountered many writers who, although concerned with the materialist-ideational issue in some way, avoid taking any clear-cut position. These writers tend to be more concerned with specific and practical issues pertaining to current relationships between human populations and the physical environment. We have lumped these works into a third and somewhat arbitrary category that we label "situational approaches." Although these works are less well developed conceptually than most of those discussed previously, they have broad popular appeal. Many of them proceed from the tacit assumption that in some unspecified way ideational factors influence human perceptions of, and thus structure their interactions with, the environment. They then proceed to express, in an expansive fashion, a widely shared concern that the beliefs, attitudes, and values prevailing in the modern world have become dysfunctional within the larger web of ecological relationships. In particular, it is often argued that the current, Western, world view is harmful to the preservation of sound ecological relationships. Many authors explicitly state that their own beliefs, attitudes, and values have been changed by their perceptions that undesirable changes have been taking place in the physical environment. Most seek to induce a similar change in the minds of their readers.

Although forshadowed in the works of G. Marsh (1864) and some of the early conservationists (e.g., Muir, Leopold), the greatest outpouring of "situational" literature began with Rachel Carson's *Silent Spring* in 1962. Public attention was captured by this eloquent warning against the destructive effects of modern technology upon the environment. To Carson, the primary cause of this pathological behavior was a widely held but illusory belief that people could improve nature with short-term, single cause and single effect manipulations of small parts of the ecosystem. "The control of nature," she wrote, "is a phrase conceived in arrogance, born of the Neanderthal age of biology and philosophy, when it was supposed that nature exists for the convenience of man" (Carson 1962, 261).

More than twenty years have passed since the publication of *Silent Spring*, and Carson's message has been repeated and amplified in many dozens of books and hundreds of articles. A rising chorus of voices has been calling for reform in at least some portion of our current beliefs, attitudes, and values in order to avoid future social and environmental disaster. The environmental crisis we now face, wrote Barry Commoner (1971, 298), "is not the product of man's biological capabilities, which could not change in time to save us, but of his *social* actions—which are subject to much more rapid change." Gregory Bateson (1972, 487) has stated his belief that "this massive aggregation of threats to man and his ecological systems rises out of errors in our habits of thought at deep and partly unconscious levels."

In a related manner, the British economist E. F. Schumacher has explicitly advocated an entirely new system of thought in his compelling and controversial book *Small Is Beautiful* (1973). The title itself expresses a value that he believes should constitute an important component of a new ideology more appropriate for life in the modern world. We must, argues Schumacher (1973, 84), through a conscious effort of intellect and spirit, rid ourselves of the "bad, vicious, life-destroying type of metaphysics that now warps our thinking and threatens our existence."

It is clear that for all the attention that many of these "situationalist" writers, including Commoner and Schumacher, devote to biological and economic factors in their published works, they have strong ties to the ideational school of thought. Basically they are arguing that bad ideas are the source of our current ecological problems and that better ideas must be the basis of any solution. Yet this position does not derive from any belief that the ideational system is independent of environmental influence; on the contrary, it comes from a belief that it is not. Our ideas, says Schumacher (1973, 87), "must be true to reality, although they transcend the world of facts. . . . If they are not true to reality, the adherence to such a set of ideas must inevitably lead to disaster."

In what is probably a related phenomenon that closely parallels the emergence of this "situational" literature, one finds another body of literature that explicates with wonder and admiration the way in which so-called primitive human groups have allegedly lived in harmony with nature, enjoying its bounty and preserving the environment with sympathetic understanding. This body of literature may be said to have begun with the publication of Colin Turnbull's *The Forest People*, an ethnographic study of the BaMbuti pygmies. To this hunting and gathering band living deep in the tropical rainforests of Central Africa, "the forest was Mother and Father, Lover and Friend," and they lived and loved and shared "in a world that is still kind and good . . . and without evil" (Turnbull 1962, 3).

Much of this literature suggests that the cosmological beliefs and symbols of simpler societies are homeostatic in function, maintaining the human population in a relatively static equilibrium with its environment. Thus, implicitly, much literature produced over the past twenty years has decried the self-destructive tendencies of modern humans and compared them unfavorably with the supposedly innate, innocent wisdom of the "Noble Savage." Such comparisons, of course, hearken back to the writings of Rousseau, Wordsworth, and even Marx and Engels, all of whom wrote in reaction to the earlier phases of industrialization in Western Europe.

Observations on the Literature

Our consideration of a vast amount of literature has assured us that the mind-versus-matter controversy—ideational versus materialist approaches—in its dualistic form is still very much alive and well. It continues to dominate the ways most of us experience the world, and categorize and think about our experiences. Even though it is now widely acknowledged that both ideational and materialist phenomena are somehow and sometimes importantly involved in human-environmental interactions, both affecting them and being affected by them, there is little agreement as to the nature of this involvement and even less agreement as to how we should go about studying it.

There are at the present time, it seems to us, several interrelated impediments to intellectual progress in this very important and as yet poorly understood domain. One impediment is the sharp distinction that most of us continue to draw between the ideational and the materialist realms. We treat ideas as one kind of phenomena and the material conditions of life as something else. The acuteness of this dichotomy makes it difficult for us to consider how the two realms interact dynamically as parts of the larger reality that is both the basis and the context of human life.

It was this deep-rooted distinction between the realm of ideas and the realm of things that led to the historical separation and polarization of the sciences and the humanities. Each of these academic orientations, of course, has become divided into even narrower specializations that we have called "disciplines," and each discipline has its own further subdivisions, schisms, and areas of specialization. This active process of bifurcation and polarization in academic life has led to the formulation of bodies of distinct theoretical, epistemological, ontological, and methodological habits and assumptions, each of which is deemed to be appropriate for inquiry in one realm but not in the other. The dualistic philosophy proposed by René Descartes more than three hundred years ago has become an implicit and central element in our world view.

This means that the values and beliefs of the observers have constituted an important factor in determining the nature of the descriptions from which we must work in reconstituting our past and apprehending our present state of existence. In almost all of the sources it is taken for granted that one aspect of existence must take precedence over the other. One must be "cause," and the other must be "effect." One must determine and the other must be determined. One is seen as "hard" and inflexible, the other as "soft" and easily molded. Most of us seem to agree on this method of structuring the problem. We simply disagree about which side is which. If this is so, the fact that this disagreement exists may be the most significant piece of information we have for assessing past work and reformulating future endeavors.

Many of us may have private doubts about the validity of this dichotomous, deterministic view of the world, but when we are intellectually "on-stage," so to speak, we show no sign of it. In our professional capacities as writers and lecturers and researchers, we feel constrained to perpetuate the conventional views that we have been socialized to accept as part and parcel of our "professionalism." The French thinkers who have used the term "professional deformation" seem to have had some such phenomenon in mind. It is time to reassess this dimension of professionalism.

It is also time to question those cultural biases and professional habits that tend to lead us to concentrate on and magnify the differences between our particular approach and all others. Most intellectual exchange now seems to take place between people who think very much alike and who devote most of their energy to debating what are, from a larger perspective, essentially minor points of disagreement among themselves. Some of this time and energy might usefully be spent trying to expand the shared areas of agreement or attempting to initiate and maintain a dialogue with those who view the world in a different way and therefore have learned very different things about it. This tendency to communicate primarily with those whose views are closest to our own and to concentrate upon differences rather than congruencies is a primary mechanism by which the process of continual subdivision and polarization takes place. It is a product of the subculture of academia; it is a learned and shared way of performing academic work. And therefore, it is rewarded and reinforced by existing academic structures and policies. It is a powerful socialization process that can be recognized and resisted only by rigorous self-discipline.

This centrifugal process becomes even more exaggerated, and our differences more rigid, because we normally choose (or are assigned) to work on classes of problems where the expectation of success is very high. In doing so, we relinquish the opportunity to learn from and be broadened by our failures. This is because, as an integral part of our educational training and professional development, we learn to avoid situations in which we can fail, in which our favorite intellectual tools might truly let us down. The situation is even further exacerbated by the fact that we tend to report and share only our most successful efforts, relegating our false starts and disappointments (and the valuable lessons to be learned from them) to the wastebasket.

It should be obvious that many of us adopt a relatively neutral position with regard to the sort of metaphysical issues that relationships between the material and the ideational realms ultimately raise. We tend to focus upon more manageable and better-defined problems in an eclectic manner that relies upon some combination of "positivism" and "empiricism." We entertain the possibility that there may not be a single approach or a single explanation that is correct for all situations and prefer to be guided instead by the particular circumstances of each case. "Objectively" comparing and testing the evidence against the various theories, we eventually opt for the particular explanation or model that seems most satisfying in terms of parsimony and elegance. In doing so there is often an assumption that correct answers and sound theories will emerge through time as an expanded body of reliable and validated data is accumulated.

This eclectic approach is appealing in many ways. Its concerns for both objectivity and scientific validation help to ensure that our efforts and interpretations will be grounded in reality and not just extensions of our theories. Further, it offers the degree of flexibility that seems to be warranted by the complex and diverse nature of human phenomena.

Despite the attractiveness of this approach and the importance of flexibility, objectivity, and scientific validation to any framework we might devise, certain problems persist. There is, for example, no necessary connection between a scientific preference for parsimony and elegance in the evaluation of alternative explanations on the one hand and reality on the other. The very notion that it is possible to examine any phenomenon in purely objective terms is also problematic. Observations and their associated meanings are influenced by the world view and experiences of the observer. Although many of us may be able to agree as to the existence of certain observable or demonstrable realities, there is no guarantee that our descriptions, much less our interpretations, of these realities will be the same (Skolimowski 1973, 99–109). In a similar manner, our world view conditions our perceptions of which phenomena constitute problems, the degrees of significance we attach to them, and the methods we select to deal with them. In other words, "objectivity" is not only culturebound, it is value laden. It comes in different forms, each of which assumes its meaning in reference to a particular context.

The concern for objectivity and scientific validation further tends to orient our analyses toward those phenomena that seem more amenable to direct measurement, quantification, and replication. By and large, we have come to associate these elements more with the physical world than with the world of ideas. Thus, there is often an unwitting materialist orientation to the eclectic approach despite its espoused neutrality.

This common approach seems to be a viable, long-term research strategy only if one assumes that a comprehensive view of complex phenomena can be attained through purely inductive means. But, one must ask, will the piecemeal accumulation of "reliable" data lead to an understanding of the larger whole through some process of accretion? If we assume that the larger research universe in which we are operating is a mechanical world in which the whole is always and exactly equal to the sum of its parts, then the answer would appear to be "yes." In a Newtonian world, we can perhaps proceed inductively with considerable success. But if life in any form is part of the system under consideration, this approach rapidly becomes much more problematic.

We cannot significantly increase our understanding of a tropical rainforest *only* by observing individual trees more carefully. We cannot significantly increase our understanding of human society and culture *only* by doing more detailed and rigorous studies of particular individuals or institutions. When we go one step further and try to look at one part or dimension of a larger system that contains people and trees and many other species in an ongoing interactive process, any inductive, empirical, positivistic approach to the study of arbitrarily defined segments of that system seems to become a hopeless enterprise.

Detailed studies of the constituent parts of so complex a system can come together to produce a better understanding of the entire entity—of the whole as a whole—only if the separate studies are informed by some central, unifying idea or set of related ideas about the nature of the whole itself. What is required is a centripetal paradigm, a paradigm that pulls things toward a center rather than pushing them out from the center as is now too frequently the case.

Our examination of the literature strongly suggests that the tendency has been and remains centrifugal rather than centripetal. Convergence and synthesis of divergent approaches are seldom encountered. We have identified many different approaches, categorized them, and to some extent explored the assumptions, operating procedures, strengths, and weaknesses of each type. But this in itself has not advanced us much beyond the starting point. Much work remains to be done in bridging the gap between the materialist and ideational poles of the ancient and persistent dichotomy between mind and matter.

Although the objects of study, and of our discussion, are the ideational systems of other cultural groups and the relationship between these systems and the physical environments in which they are found, it should be obvious that these phenomena are found in our own cultures and subcultures as well. Less obvious, perhaps, but potentially of great significance, is the probability that subcultural and cross-cultural groups are also often linked by their common orientations, training, and situational factors into similar implicit, shared, and largely unrecognized world views. Students of human ecology, for example, probably share a large number of assumptions and cognitive categories and values that reflect shared ideational, social, and material factors. More so than we might realize, and more than we might wish, our own beliefs and values, our own cultural predilections, influence the sorts of alternatives we devise and the recommendations we offer.

For challenging both our own assumptions and those of others (be they members of an academic discipline or a remote tribal group), situations of ecological change would seem to offer the most fruitful area for research. It is under conditions of change that the systemic interplay taking place between the realm of ideas and the material conditions of existence is most likely to be revealed in ways that penetrate the veil of prior assumptions. There has been, we would suggest, too exclusive a focus for far too long on what have been (or have been perceived to be or suggested to be) relatively "homeostatic" situations involving less complex societies and exotic (and thus poorly understood) physical environments.

The models that have emerged from this dominant focus have led many people to believe that the best of all possible worlds is one in which the material and ideational aspects of human existence systematically reinforce and maintain each other. Our conditions keep our ideas in check and our ideas prevent us from overexploiting and destroying our environments. This is seen as good, and all change is by this definition considered to be pathological. A surprising number of writers now seem to feel that such homeostasis is by and large attainable, and these writers advocate, explicitly or implicitly, what amounts to a return to the static golden age associated with stereotypes of the "Noble Savage."

This would be a partial and selective return, to be sure, and it may or may not be possible. It may or may not be desirable. Even more germane, however, is the fact that even if such romantic regression were achievable and desirable, a continued research emphasis upon homeostatic or near-homeostatic configurations would not allow us any understanding of how to achieve this state. For the modern world, paradoxically, any return to even an approximation of homeostasis would involve massive change.

By focusing upon situations of change, we are provided with a context in which we can discern the results of interrelationships, one in which we can begin moving toward increased understanding of why and how transformations can and do take place in the overall system. Because change and interaction are processes rather than events, an extended temporal perspective, an evolutionary perspective, is called for. Synchronic or short-term studies necessarily involving arbitrarily selected variables remain incapable of challenging (let alone disconfirming) the *a priori* assumptions of the investigators. In contrast, an evolutionary perspective would seem to facilitate a more creative and, it is hoped, more effective overview, one in which we might envision meta-patterns that are not yet perceptible. It might also allow us to consider a number of other important aspects of continuity, change, and stasis that have thus far been elusive targets for analysis: how external inputs or internally generated change either can be ramified into major alterations of each realm or be absorbed by the resiliency of one or both realms; how and when thresholds of internal modification are reached and crossed, thereby increasing the likelihood of modifications in the other realm; and so on.

In planning for or operationalizing any such model for planned change, or "development," one tends to think in terms of multidisciplinary or interdisciplinary considerations. On an initial or interim basis, this is commendable. But we may need in the longer run to move toward a more pan-disciplinary approach, toward a framework that is overarching in nature, toward a framework that, in recombining and fusing characteristics and considerations of a number of different and now diverging orientations, begins to generate a new and distinctive set of strategies, tactics, and concepts better suited for examining the dynamic interplay between the ideational realm and the material conditions of existence. To study successfully so multidimensional a system, nothing less than a truly multidimensional model can be adequate in the long run.

If we hope to broaden our perspectives along these lines, however, we must be fully aware of the relationships that exist between the various epistemological values that are embedded in existing approaches. If, for example, we wish to increase the scope of our work, we may have to sacrifice a certain degree of precision, and vice versa. The total pool of epistemological values in contemporary intellectual life is considerable. While few of these values are actually incompatible, not all can be maximized at the same time, and a heretofore rare degree of epistemological flexibility will be required of all who might participate in the construction of any new pan-disciplinary framework (Wilson 1973). To increase significantly an understanding of the sort of complex systems under discussion, there must be a convergence of science and humanism. Scientists must consider the less tangible and less easily measured aspects of human-ecological interactions without sacrificing a scientific spirit of inquiry. Similarly, humanists must display a greater willingness to explore the potential offered by the scientific method. We must all free ourselves of the unnecessary narrowness and rigidity arising from the polarization of science and humanities that we carry around as part of our historical heritage.

These comments and general ideas are offered to encourage individuals who are interested in and concerned about such issues to approach them with greater self-awareness, more conceptual and methodological flexibility, and with more attention to dynamic interplay instead of cause and effect. We advocate greater diversity and experimentation at this time, not the replacement of someone else's model with an equally well-defined one of our own.

Futhermore, we would emphasize that the envisioned goal is not—and *cannot* be—to develop some illusory capacity to identify the roles and effects of ideational phenomena in order that they may be negated and ultimately eliminated from the considerations of human ecology or environmental policy. We should study the origins, roles, and effects of ideational systems to increase our awareness and understanding of how they interact with other dimensions of human existence, including but not limited to the material conditions of existence. Only then will it become possible to move toward a fuller appreciation of the human ecological experience in all its richness. Only then can we hope to devise a framework for analysis that is, in the words of Flannery and Marcus (1976, 383), "neither mindless ecology nor a glorification of the mind, divorced from the land."

We might begin to rephrase some of these considerations by suggesting that we consider the larger research universe to be a dynamic mega-system consisting of several interrelated subsystems: one environmental, one ideational, and one social.[5] Each of these subsystems is potentially acting upon and reacting to the others through various human perceptions, conceptions, and actions, both individual and collective. It may well be the case that there are situations in which one subsystem will exert greater influence over, and thereby cause readjustments directly or indirectly in, the others. But there can be no *a priori* assumption that it will always be the same subsystem that does this. Nor can we assume that the connections between these various subsystems are such that a change in one must always bring about changes in the other or that a change in one can only occur as a result of change in another. We must allow for certain, possibly quite variable, amounts of autonomy, flexibility, and resilience in each subsystem. We must allow for the possibility that in any of these subsystems there may be internally generated change as well as varying degrees of external inputs that may produce pressure leading to change. In different contexts, an increase in the influence of any subsystem upon any of the others may be allowed, encouraged, or resisted.

Complex systems require complex models, especially when dynamic processes remain poorly understood. One way to move toward greater understanding of the complex mega-systems in which we live is to describe and analyze more fully processes of interaction within and between its subsystems in well-defined contexts. We need to document and ponder particular instances of change in ecological, ideational, and social systems with the closest possible attention directed toward the specific mechanisms of feedback, decision making, and readjustment by which change is either induced within any subsystem and then either spreads throughout the larger system, inducing further change, or is dampened or extinguished by some homeostatic process.

By tracing linkages between the material, ideational, and social aspects of human existence, we envision the possibility of building upon and moving toward the integration of many if not all of the approaches that have been discussed. This mergence will not be a rejection of past works or alternative approaches. But neither will it be even tacitly an acceptance of the *status quo* arrangements in which various approaches continue to be pursued independently. By granting, at least provisionally, some validity to both the materialist and the ideational positions, we allow the possibility of constructing a broader, more inclusive, and more dynamic paradigm.

To construct such a model we must begin to learn and think in new and different ways. This will doubtless entail much frustration, many setbacks, and some painful periods of apparently fruitless flailing about. But the potential rewards are greater than the discomfort to be endured in the quest. This is an opportunity to acquire not just more knowledge, but perhaps greater wisdom in setting and pursuing our human goals in this ever more complex and rapidly changing world.

Modern technology continues to accelerate change in both the ideational and material realms at the same time it intensifies interaction between the two. More than ever before, the two realms are one. They are one in practice. We must make them one in theory and come to understand them more effectively as one. If we do not we will pay a terrible price for our continuing failure to understand them as arbitrarily isolated categories of our experience.

Notes

1. Harris has been challenged on the grounds that his ecological analysis is superficial and misleading (Diener and Robkin 1978) as well as on the grounds that it is superfluous to invoke religious taboos against obviously impractical activities (Alland 1975). Harris (1979, 194–95) has maintained his position however, and the issue remains essentially unresolved.

2. The classic statement of this principle, of course, comes from Karl Marx in *A Contribution of the Critique of Political Economy* (1970). "The mode of production in material life determines the general character of the social, political, and spiritual processes of life. It is not the consciousness of men that determines their existence, but on the contrary, their social existence determines their consciousness." In a similar vein, commenting on Mary Douglas's writing on the Lele, an African tribal group, Yehudi Cohen (1974, 221) remarks that: "Douglas suggests that symbolism and ritual determine people's behavior in respect to the habitat. This point of view has many proponents. . . . My own view is the opposite, that these beliefs serve to justify or give meaning to activities made necessary by an adaptive strategy."

3. Neo-Marxism is such a widespread, diffuse, and often contradictory phenomenon that it defies accurate description in any brief form. See O'Laughlin (1975), Markus (1978), and Giddens (1981) for examples and discussion of a variety of emphases and issues in recent Marxist thought.

4. Structural Marxism is a relatively recent and highly controversial development that is more influential in Europe than elsewhere. Two leading practitioners and spokesmen of this approach are Maurice Godelier (1972, 1975, 1977) and Jonathan Friedman (1974a, 1974b, 1975).

5. The interactive social system-ecosystem model used to organize research within the East-West Environment and Policy Institute's program on Human Interactions with Tropical Ecosystems (Rambo 1979, 1982) offers a preliminary formulation of such an approach.

References

Alland, A.
1975   Adaptation. *Annual Review of Anthropology* 4:59–73.
Allen, R. E.
1966   *Greek Philosophy: Thales to Aristotle.* New York: The Free Press.
Bateson, G.
1972   *Steps to an Ecology of Mind.* New York: Ballantine.
Buttimer, A.
1974   *Values in Geography.* Commission on College Geography Resource Paper No. 24. Washington, D.C.: Association of American Geographers.
Buttimer, A., and D. Seamon (editors)
1980   *The Human Experience of Space and Place.* New York: St. Martin's Press.
Carson, R.
1962   *Silent Spring.* New York: Houghton Mifflin Co.
Chan Wing-Tsit
1967a   Chinese theory and practice, with special reference to humanism. In *The Chinese Mind: Essentials of Chinese Philosophy*, edited by C. A. Moore. Honolulu: East-West Center Press. Pp. 11–30.
1967b   The story of Chinese philosophy. In *The Chinese Mind: Essentials of Chinese Philosophy*, edited by C. A. Moore. Honolulu: East-West Center Press. Pp. 31–76.
Cohen, Y.
1974   *Man in Adaptation: The Cultural Present.* Second edition. Chicago: Aldine.
Commoner, B.
1971   *The Closing Circle: Nature, Man, and Technology.* New York: Alfred A. Knopf.
Conklin, H. C.
1955   Hanunoo color categories. *Southwestern Journal of Anthropology* 11:339–44.
Diener, P., and E. Robkin
1978   Ecology, evolution, and the search for cultural origins: the question of Islamic pig prohibition. *Current Anthropology* 19:493–540.

Douglas, M.
1966   *Purity and Danger: An Analysis of Concepts of Pollution and Taboo.* New York: Praeger.

Durham, W. H.
1982   Interactions of genetic and cultural evolution: models and examples. *Human Ecology* 10:289–323.

Flannery, K. V., and Joyce Marcus
1976   Formative Oaxaca and the Zapotec cosmos. *American Scientist* 64:374–83.

Frake, C. O.
1961   The diagnosis of disease among the Subanun of Mindanao. *American Anthropologist* 63:113–32.
1962   Cultural ecology and ethnography. *American Anthropologist* 64:53–59.

Friedman, J.
1947a   Marxism, structuralism, and vulgar materialism. *Man* 10:444–69.
1947b   The place of fetishism and the problem of materialist interpretation. *Critique of Anthropology* 1:26–62.
1975   Tribes, states and transformations. In *Marxist Analyses and Social Anthropology*, edited by M. Bloch. London: Malaby Press. Pp. 161–202.

Fung Yu-lan
1966   *A Short History of Chinese Philosophy.* New York: The Free Press.

Geertz, C.
1964   Ideology as a cultural system. In *Ideology and Discontent*, edited by D. E. Apter. New York: The Free Press. Pp. 47–56.
1966   Religion as a cultural system. In *Anthropological Approaches to the Study of Religion*, edited by M. Banton. London: Tavistock. Pp. 1–46.
1972   Deep play: notes on the Balinese cockfight. *Daedalus* 101:1–37.

Geertz, C. (editor)
1971   *Myth, Symbol, and Culture.* New York: Norton.

Giddens, A.
1981   *A Contemporary Critique of Historical Materialism: Vol. 1. Power, Property and the State.* Berkeley: University of California Press.

Godelier, M.
    1972   *Rationality and Irrationality in Economics.* New York:
            Monthly Review Press.
    1975   Modes of production, kinship, and demographic
            structures. In *Marxist Analyses and Social Anthropology,*
            edited by M. Bloch. London: Malaby Press. Pp. 3–27.
    1977   *Perspectives in Marxist Anthropology.* Cambridge:
            Cambridge University Press.
Harris, M.
    1966   The cultural ecology of India's sacred cattle. *Current
            Anthropology* 7:51–59.
    1973   The riddle of the pig. *Natural History* 82(2):20–25.
    1974   *Cows, Pigs, Wars, and Witches: The Riddles of Culture.*
            New York: Harper and Row.
    1977   *Cannibals and Kings.* New York: Random House.
    1979   *Cultural Materialism: The Struggle for a Science of Culture.*
            New York: Random House.
Jochim, M.
    1981   *Strategies for Survival: Cultural Behavior in an Ecological
            Context.* New York: Academic Press.
Leach, E.
    1965   Anthropological aspects of language: animal categories
            and verbal abuse. In *New Directions in the Study of
            Language,* edited by E. Lenneberg. Cambridge: MIT
            Press. Pp. 23–63.
Lévi-Strauss, C.
    1963   *Structural Anthropology.* New York: Doubleday.
    1966   *The Savage Mind.* Chicago: University of Chicago Press.
    1969   *The Raw and the Cooked.* New York: Harper and Row.
Lindenbaum, S.
    1972   Sorcerers, ghosts and polluting women: an analysis of
            religious belief and population control. *Ethnology*
            11:241–53.
Markus, G.
    1978   *Marxism and Anthropology: The Concept of "Human
            Essence" in the Philosophy of Marx.* Assen: Van Gorcum.
Marsh, G.
    1864   *Man and Nature.* London: Sampson, Low and Son.
Marx, K.
    1970   *A Contribution of the Critique of Political Economy.* New

York: International Publishers. Originally published: 1859.

Metzger, D., and G. E. Williams
1963   A formal ethnographic analysis of Tenejapa Ladino weddings. *American Anthropologist* 65:1076–1101.
1966   Some procedures and results in the study of native categories: Tzeltal "firewood." *American Anthropologist* 68:389–407.

O'Laughlin, B. O.
1975   Marxist approaches in anthropology. *Annual Review of Anthropology* 4:341–70.

Rambo, A. T.
1979   *Development of a Conceptual Framework for Human Ecology.* Working Paper No. 4, Department of Anthropology and Sociology, University of Malaysia, 54 pp.
1982   Human ecology research and tropical agroecosystems in Southeast Asia. *Singapore Journal of Tropical Geography* 3:86–99.

Rappaport, R. A.
1967   Ritual regulation of environmental relations among a New Guinea people. *Ethhnology* 6:17–30.
1968   *Pigs for the Ancestors: Ritual in the Ecology of a New Guinea People.* New Haven: Yale University Press.
1971a   The sacred in human evolution. *Annual Review of Ecology and Systematics* 2:23–44.
1971b   Ritual, sanctity and cybernetics. *American Anthropologist* 73:59–76.

Relph, E.
1976   *Place and Placelessness.* London: Pion, Ltd.

Schneider, D. M.
1968   *American Kinship: A Cultural Account.* Englewood Cliffs: Prentice-Hall.
1969   Kinship, nationality, and religion in American culture: toward a definition of kinship. In *Forms of Symbolic Action,* edited by V. Turner. New Orleans: Tulane University. Pp. 116–25.

Schumacher, E. F.
1973   *Small Is Beautiful.* New York: Perennial.

Skolimowski, H.
1973   The twilight of physical descriptions and the ascent of

normative models. In *The World System*, edited by E. Lazlo. New York: George Braziller, Inc. Pp. 99–118.

Sturtevant, W. C.
1964   Studies in ethnoscience. *American Anthropologist* 66(3, pt. 2):99–131.

Tuan Yi-fu
1977   *Space and Place: The Perspectives of Experience.* Minneapolis: University of Minnesota Press.

Turnbull, C. M.
1962   *The Forest People: A Study of the Pygmies of the Congo.* New York: Simon and Schuster.

Turner, V.
1964   Betwixt and between: the liminal period in rites de passage. In *Proceedings of the 1964 Annual Spring Meeting of the American Ethnological Society*, edited by J. Helm. Seattle: University of Washington Press. Pp. 4–20.

1967   *The Forest of Symbols.* Ithaca: Cornell University Press.

1969   *The Ritual Process: Structure and Anti-Structure.* Chicago: Aldine.

1974   *Dramas, Fields, and Metaphors: Symbolic Action in Human Society.* Ithaca: Cornell University Press.

1975   Symbolic studies. *Annual Review of Anthropology* 4:145–61.

Tylor, S. (editor)
1969   *Cognitive Anthropology.* New York: Holt, Rinehart and Winston.

Wilson, A.
1973   Systems epistemology. In *The World System*, edited by E. Lazlo. New York: George Braziller, Inc. Pp. 121–40.

CHAPTER 3

PEOPLE AND NATURE IN THE TROPICS: REMARKS CONCERNING ECOLOGICAL RELATIONSHIPS

Karl L. Hutterer

In an essay entitled "Social Dramas and Ritual Metaphors," Victor Turner (1974, 32) called attention to a distinction made by the Polish sociologist Znaniecki (1936) between natural and cultural systems. According to Znaniecki, natural systems are objectively given and exist independently of the experience and action of humans. Cultural systems, on the other hand, depend for their existence and their meaning not only on the presence and activity of conscious and volitional human agents but also on the relations existing among them. Znaniecki referred to this as the "humanistic coefficient," which he saw precisely in the importance of the presence and activity of humans who bring their minds and wills to bear in creating a reality that is distinct and separate from the physical world. For this reason, Znaniecki insisted that sociological methodology must include accounts of personal and "subjective" experiences of investigator as well as subject, a demand which put him at odds with the mainstream of American sociology of the time.

I refer here to Turner's account of Znaniecki not only because of the apparent influence Znaniecki had on Turner, a leading thinker and writer concerning the nature of symbols and their role in human affairs, but also because of the explicit statement regarding the distinction, or even opposition, between natural and cultural systems. Znaniecki was not the only or first sociologist to perceive such a profound difference, nor was he the first to suggest that sociological investigation demanded a methodology of its own, fundamentally different from the methodology of the natural sciences. Indeed, the epistemological debate behind this concern can be traced several hundred years. In the present context, however, it is perhaps most appropriate to point to Èmile Durkheim (1965) who, several decades before Znaniecki, came to the conclusion that the social world could be properly investigated and interpreted only in its own terms.

The question concerning the relationship between nature and culture has, of course, become the very special battleground of anthropology, as the chapter by Jamieson and Lovelace in this volume attests. I am not aware of any anthropologist who has ever gone so far as to consider culture or social organization to be but a trivial side product, an epiphenomenon, of human biology or human material needs (although some of the attackers of sociobiology accuse the proponents of that approach of saying, or at least implying, just that [cf. Sahlins 1967b]). Yet there are certainly some anthropologists who see social and cultural forms, if not their content, strongly determined by environmental and biological conditions. In particular, scholars working in the tradition of cultural materialism, while admitting that human cultural behavior is qualitatively different from animal behavior, maintain that its only major purpose is the satisfaction of material needs and that it can therefore be adequately analyzed and understood by reference to these needs alone (e.g., Harris 1979; White 1949). On the other hand, quite a few students of human affairs, as diverse as Boas and his student Kroeber (1917), Geertz (1973), and Sahlins (1976a), were and are convinced that culture is fundamentally a unique phenomenon, essentially autonomous of tangible natural conditions that at best provide only a container or imperfect vehicle for it. Within such a framework, culture becomes wholly unapproachable and inexplicable to the methods of the natural sciences; it becomes an esoteric concern.

There are many reasons for rejecting Znaniecki's strict dichotomy between natural and cultural systems. The most important one has to do with the nature of human knowledge concerning external reality. Any thoughtful analysis of human perception makes it clear that a naive positivism cannot do justice to the complexity of the processes involved. A range of positivistic epistemologies has dominated much of Western science, particularly the natural sciences, over the past 150 years. It has been said that the value and intellectual power of positivism as an approach to knowledge and scientific reasoning is demonstrated by the remarkable achievements of Western science and its applications in engineering contexts. It can also be argued, however, that the true value of contemporary scientific theories and explanations cannot be assessed fully and adequately in a contemporary context alone and simply on the basis of the fact that these theories "work" by being predictive within certain limits. In the final analysis, the validity of scientific frameworks can only be judged adequately not from within but from without (cf. Bourdieu 1977). The only thing we know indubitably at this time is that there are many things we do not know and understand. It stands to reason that our lack of understanding is attributable not so much to the nature of these issues and problems, but to our lack of appropriate conceptual tools. In many respects, the latter may well be linked to the conceptual rigidity and narrowness of a positivistic stance.

In any event, an analysis of perception shows that we never grasp reality directly and *per se*, but always indirectly, by way of subjective perceptions of it. Conversely, it is also the case that the human mind operates within an objectively given external world, and most perceptions (perhaps all) have reference to this objective external reality. The analysis of sensory (or scientific) data must therefore include an awareness of, and attention to, the subjective dimension in these data while an analysis of subjective data cannot usefully proceed without some reference to the external world as it is understood.

Moving these considerations into an anthropological perspective, we can go one important step further and say not only that we reach the external world through subjective and imperfect perceptions, but that these perceptions are socially conditioned. In turn, the social universe itself is part of the larger objective reality and operates within its potentials and constraints. On the basis of their perceptions, humans act on this reality and change it, and again approach and interpret the changed world through their perceptions.

Before considering particulars concerning the nature of the relationship between human subjects and the external world in and of which they live, a very general issue needs to be discussed. If there is any validity to the thoughts discussed above, then it must be evident that a soundly conceived anthropological enterprise needs to include environmental and ecological analyses and considerations. The role of environmental factors in social and cultural organization is neither trivial nor secondary; it is essential to the extent that culture simply cannot be understood adequately without careful reference to it. On the other hand, it will also be clear that culture itself cannot be understood by reference to environmental factors and relationships alone. Nor are ecological analyses adequate or complete, unless they incorporate perspectives that have traditionally been classified under the "cultural" rubric in the narrow sense: cognition, symbolism, ideology, ritual and religion, and so on.

It is unlikely that many anthropologists would disagree completely with these propositions, although there are wide differences in emphasis and interest. However, surprisingly few anthropologists have to date concerned themselves explicitly with the nature of this relationship between environmental and cul-tural/ideational variables and the way it expresses itself in social processes. The contributors to this volume, and the conference on which it is based, share an interest in this topic and feel that some insight may be gained by reviewing and analyzing a series of specific situations involving the relationship between ideational forms and contents on the one hand, and human interactions with the social and natural environment on the other. They also share an interest in the tropics, particularly the Southeast Asian tropics, and they use this interest to give their discussions a specific focus.

The first step must be to sketch briefly a rough outline of tropical environments and ecology, particularly as they concern Southeast Asia, as a general baseline for the inquiries in this volume. Based on what I have said previously, it is not possible to consider specific environmental conditions simply as objectively given, since human environments themselves are, at least partially, culturally created. It is possible, however, to separate analytically and delineate certain broad organizational properties of environments that humans must take account of if they want to be successful (i.e., survive) as a biological species. Once this environmental background has been sketched, it will be useful to return once more to the issue of environment and culture and deal with it somewhat more specifically.

Tropical Ecosystems

Somebody once wrote (I have unfortunately lost the source of this quote) that "there is no such thing as tropical ecology; rather there is only ecology in the tropics." What is implied by this statement is that, inasmuch as ecology represents a scientific interpretation of environmental states and processes, and inasmuch as it employs a general theoretical framework, ecological concepts and principles apply equally to the total range of environmental conditions, in the tropics as well as elsewhere. This admonition may serve as an occasion to clarify what we mean when we use the term "tropical ecology" in the admittedly somewhat sloppy context of everyday academic discourse. When we speak of a distinctive "tropical ecology," we generally refer to two different things: (1) we imply that tropical *environments* can be distinguished from nontropical environments on the basis of certain fundamental biotic and nonbiotic conditions prevailing in them (such as levels of light radiation, temperature, humidity, soils, flora, and fauna); and (2) we mean to say that *ecological* relationships as well as structural properties (such as species diversity, nutrient cycling, successional characteristics, productivity, complexity of energy pathways, and systemic stability) of tropical ecosystems tend to have different values from equivalent ecological relationships and structures of nontropical environments.

It has been pointed out by numerous investigators that the tropical regions contain the most diverse and complex ecosystems in the world. This makes it most difficult to provide a concise account of tropical environments and ecology; adequate treatment of tropical environments and their ecology would be far beyond the limits of this volume (for more extensive treatments see, among others, Farnworth and Golley 1974; Golley et al. 1975; Holdridge et al. 1971; Hutterer 1982; Janzen 1975; Richards 1973; Schnell 1970; UNESCO 1978; Walter 1979; Whitmore 1975). The following thumbnail sketch is by necessity extremely superficial and will consequently be of only limited utility. It is meant to be no more than a stimulus to activate readers' memories of reading and experiences regarding tropical environments.

Geographically, the word "tropics" denotes a sector of the globe straddling the equator and stretching from the Tropic of Cancer (23°30' N) in the northern hemisphere to the Tropic of Capricorn (23°30' S) in the southern hemisphere. This definition tells us little, if anything, however, about what we are interested in here, namely, the life conditions that can be described as "tropical." It is generally assumed that tropical life conditions prevail within the geographical zone designated above, but it is also known that these latitudinal boundaries delineate the distribution of tropical environments only in the crudest of terms. What are the most salient characteristics of tropical environments? There is no simple answer because, among other things, we are not dealing with a single environmental type but rather a broad range.

The single most important factor characterizing all tropical environments and distinguishing them from nontropical ones is the high level of solar radiation (insolation) they receive throughout the year. In practical terms, this translates into high levels of light during the day and generally high temperatures with only relatively minor daily and seasonal fluctuations compared to the climates of nontropical areas. More significantly, since solar radiation constitutes virtually the only energy source for natural ecosystems, this factor constitutes the most important controlling element of tropical ecosystems. It influences the nature of tropical climates, and through the climates the formation of soils; through both it influences the composition and structure of biotic communities in tropical latitudes. The total amount of solar radiation received is highest, and its seasonal distribution most even, near the equator, while values for both of these variables decline with increasing latitude. Corresponding with the importance of solar radiation in regulating other kinds of environmental phenomena, the latitude affects all dependent variables.

Next to light and sunshine, the most important of the climatic variables is precipitation. With some specific exceptions, the total amount of yearly rainfall received is highest and its distribution throughout the year most uniform near the equator. As one moves away from the equator, annual precipitation decreases in total volume and becomes more seasonally distributed. As dry seasons become longer in the outer tropics, one eventually encounters climates with year-long droughts. This situation is reflected in a coarse subdivision of tropical nature into "humid," "seasonal," and "arid" tropical environments. Patterns of precipitation are strongly influenced by seasonal fluctuations in hemispherical heat exchange that in turn are affected by the distribution of major landmasses and bodies of water (Nieuwolt 1977; Riehl 1979).

Since Southeast Asia lies on the border between the Eurasian landmass and the Pacific Ocean, and about one-half of its area is made up of island arcs, the region's pattern of precipitation is affected not only by its geographic position within the tropical latitudes but also by the strongly seasonal pattern of monsoonal air exchanges between the Asian mainland and the Pacific. In addition, specific local climates are influenced by mountains. The rainfall pattern of the region is therefore quite complex (Gaussen et al. 1967). Very broadly, however, the southernmost border of the mainland portion of Southeast Asia, the Malay Peninsula, most of Sumatra, Borneo, westernmost Java, much of Celebes (except the southern part of the island), and the eastern portion of the Philippines can be considered to have humid climates. Dry seasons are of relatively long duration and considerable severity in much of eastern Indonesia and the northern areas of mainland Southeast Asia. Because of a rain-shadow effect between parallel mountain systems, central Burma is the most arid portion of the region. The nearby Assamese highlands in India, on the other hand, are exposed to the rain-bearing Indian monsoon, and, because of the orographic effect, receive one of the highest amounts of rainfall of any place in the world.

Under the influence of high temperature and humidity, both weathering of inorganic rock and breakdown of organic litter is relatively rapid in the equatorial tropics. Although this should make an ever-increasing amount of nutrients available for use by organisms, the same conditions favor rapid leaching of many of these nutrients and the oxidation of others. In reality many tropical soils tend to be rich in iron and aluminum oxides, high in pH, and poor in other mineral nutrients, and therefore relatively infertile (Sanchez 1976). In higher tropical latitudes, weathering and decomposition processes tend to be slower, but so are processes that remove nutrients through leaching. These soils are therefore generally richer in native nutrients, and the controlling factors in their productivity are usually the availability of water and a variety of problems relating to a high evapotranspiration ratio. The distribution of Southeast Asian soils conforms with this picture in very broad terms, although there are numerous exceptions. The alluvial soils of major and minor river basins as well as the soils derived from geologically recent volcanism on many of the islands (Sumatra, Java, Bali, Celebes, parts of the Philippines) are often extremely fertile.

In general, the equatorial flora is one of the richest and most luxuriant in the world. A characteristic vegetation type of the humid tropics is the rain forest. It is a vegetative type dominated by very tall trees, extremely large phytomass, and very high species diversity. Because the dense canopy filters out a large portion of the useful sunlight, undergrowth is limited and most green plant matter is found some distance above the ground. Typical rain forests are also characterized by an amazing number and diversity of epiphytes (clinging and climbing plants), which often maintain parasitic or mutualistic relationships with other plants. There are also many, often highly specialized, mutualistic interactions between animals (particularly insects) and plants. Again, plant communities change drastically with increasing distance from the equator. There tends to be a decrease in the height of forest trees, a decrease in species diversity, and generally a decrease in stand density. Toward the outer tropics, forests eventually break up into open woodlands, scrub forests, savannas and steppes, and eventually deserts. As height and density of the forest decline, there is a concomitant decrease in plant biomass.

Although the natural vegetation cover of Southeast Asia has been affected by a long history of often intensive human interference predating colonial influences by thousands of years, it can be reconstructed that the distribution of major vegetation types originally (at least within Holocene times) followed closely what one would predict on the basis of the regional climatic pattern: various forms of rain forest vegetation in areas characterized by perhumid climate; seasonal and monsoon forests in most of the rest of the region; and open woodlands to scrub forests in the arid portions. It is still not clear to what extent savannas and grasslands may have existed in Southeast Asia under completely natural conditions (not caused by human interference), although it has been suggested that native grazing animals may have played an important role in maintaining a forest-grassland mosaic originally created by agricultural populations (Wharton 1968).

At first glance, animal life in the tropics holds some surprises (Bourlière 1973; Coe et al. 1977; Owen 1976). Animal density in rain forests is far from what one might expect given the luxuriance of the vegetation. Indeed, expressed as a ratio of plant biomass to animal biomass, rain forests contain fewer animals per kilogram of plant matter than any other terrestrial ecosystem. It is also interesting to note that, in terms of biomass, invertebrates constitute the largest proportion among rain forest animals. Less surprising is the fact that among vertebrates, flying, arboreal, and scansorial animals predominate. In relative terms the rain forest fauna mirrors the species diversity of the flora. From the viewpoint of human hunters, it is of some significance that the majority of vertebrates in the rain forest are solitary or are organized in relatively small family groups. In more seasonal tropical environments, the diversity of animal life decreases, while vertebrates constitute an increasingly important proportion of the total animal biomass of these ecosystems. In woodlands, savannas, and steppes, large herds of grazing mammals account for the bulk of animal biomass (Bourlière and Hadley 1970; Harris 1980).

The animal world of Southeast Asia is influenced by the region's position between the Eurasian continent and the Oceanic area. In general, animals of the Malesian faunal region show relationships to both the Paleartic (northern Eurasian) and Abessynian (African and South Asian) faunal regions. In the islands to the east, however, there is a gradual faunal transition to the Australian region: because of island conditions, the number of Southeast Asian land vertebrates declines rapidly with distance from the Asian mainland, while there is a simultaneous increase in the number of endemic species as well as a gradual increase in the number of marsupials and other typically Australian species represented. The ecological structure of animal communities broadly corresponds to this, although there are some differences when compared with comparable habitats in the African or South American tropics. To single out one specific point: the seasonal environments of Southeast Asia are essentially devoid of the major herds of grazing ungulates typical of the seasonal and dry tropics of Africa (cf. MacKinnon and MacKinnon 1974; Marshall 1979).

Much could and probably should be said about ecological relationships in various tropical environments, but the framework of this volume does not permit it. I will, therefore, discuss only two specific points that seem to be of particular interest in the present context.

The first has to do with the questions of diversity and complexity. As previously stated, tropical environments are generally characterized by great diversity both within and between biotic communities and by a great complexity of energy pathways. Both diversity and complexity are highest in humid tropical ecosystems and decline with increasing seasonality away from the equator. It seems evident that this diversity and complexity (both terms connote complicated ecological relationships and need to be explored in greater detail) must impose some peculiar demands on the conceptual approach to and management of these environments. If this is true, then it also stands to reason that once the native environments have been transformed and simplified by extensive human manipulation such as agriculture, conceptual approaches to the new environments will have to be very different. None of these issues has so far been explored adequately by anthropologists or ecologists. For an understanding of the cultures and societies of Southeast Asia, such considerations could have far-reaching implications. On the one hand, it seems clear that populations engaging in permanent field agriculture have essentially "locked themselves out of the forest" conceptually. This fact is quite easily demonstrated by the mythological conceptualization among such societies that the forest is dangerous and fearful, and by their practical reluctance to enter the jungle (e.g., Lombard 1974). On the other hand, wherever forests have survived they have remained important sources of wild animal protein, medicinal substances, industrial raw materials, and trade goods, even for agricultural populations. The demand for forest products is usually satisfied by specialists, be they slightly marginal individuals within agricultural communities or small communities or societies of forest dwellers. That such forest-oriented groups maintain cultures and ideologies that differ in certain respects from their agricultural counterparts is not only or even necessarily a reflection of distinct and independent cultural traditions but first and foremost a reflection of the need to approach different environments from different conceptual frameworks. Here is an area of obvious interest for further comparative study, not only because of its implications for the reconstruction of cultural history on the basis of contemporary ethnographic facts, but also for the potential it holds for illuminating the relationship between natural habitats and symbolic systems.

The second point has to do with the productivity and fragility of tropical environments. Again, humid tropical ecosystems, particularly rain forests, show some of the highest values for primary productivity of any terrestrial ecosystem. The figures for high primary productivity (plant growth) stand in striking contrast to the relatively poor animal life in these environments. The explanation for this apparent discrepancy between large plant biomass and small animal biomass lies in the fact that tropical forests often grow on very poor soils and have been able to develop only by evolving highly efficient systems of nutrient cycling that involve rapid turnover, quick decomposition, and rapid and efficient uptake of nutrients. Within such nutrient-limited ecosystems, predation by herbivores is extremely costly to plants because it would delay, or even interrupt, the nutrient cycling process. Typically, plants in such environments have developed many defenses against herbivores, thus depressing the level of consumers (animals). In terms of human interaction with such environments, it needs to be noted that these ecosystems are also relatively poor habitats for human populations. Equally important, once rain forests have been cleared and the efficient nutrient cycle broken, they recover slowly and with difficulty, and the underlying soils are subject to rapid degradation. Thus, many tropical environments, particularly rain forests, are highly fragile and consequently demand special managerial skills from humans depending on them over long periods of time. Traditional societies in Southeast Asia have, of course, successfully maintained themselves in such environments for thousands of years. Among the topics of major interest within the broader issues of human interactions with the environment are not just those concerning environmental and technical skills, but the conceptual and symbolic forms through which various societies have rationalized, directed, and reproduced their practices and integrated them into broader cosmological and transcendental world views.

The problems of the complexity and fragility of tropical environments and human interaction with them raises a series of fundamental questions. These questions could also be raised with regard to any other nontropical environmental context, of course, but the notion of complexity and fragility in tropical ecosystems seems to make them particularly poignant. Among other things, we have to ask: how do traditional societies store and retrieve the extensive body of information necessary to live and survive in those environments? How do they monitor diverse and complex environmental situations in light of that information and how do they make decisions about impending or future action? How is this knowledge safeguarded and passed between generations yet kept open and flexible enough to remain adaptable in the face of continuously occurring minor or major stochastic fluctuations in environmental systems?

These and other questions would be a good deal easier to answer if the human species were to inhabit, like most other animals, a relatively specialized ecological niche. Humans are, however, by far the most ecologically generalized of all animals and at the same time have evolved into the most dominant species in virtually all terrestrial environments. It is clear that humans have not achieved this by virtue of particular biological endowments. Rather, it is culture that has invested them with the potential for their peculiar ecological status, culture not just in the sense of the ability to make and use tools, but in the sense of being able to use and manipulate symbols and through them to interact with the environment (as well as among themselves) in ways that would otherwise be inconceivable. What, then, is the role of this ability to engage in symbolic action in the relationship of humans to their environment?

Culture in Ecology

Few anthropologists have conducted explicit and detailed studies of the role of symbolic and ideational matters in ecological relations. Not that there is a dearth of references to the issue. In most cases, however, the relationship is anecdotally seen only on an overt and mechanistic level. For instance, myth and ritual are seen as long-term-memory-storage devices for environmental knowledge (e.g., Goodale 1970); taboos and divination rites are interpreted as effecting conservation of crucial resources (e.g., Moore 1957); and ritual interactions are described as mechanisms to exchange important resources between individuals or across group boundaries (cf. Healey 1978). As long as the analysis moves on this level, however, the relationship between the ideational realm of societies and their ecology is perceived as essentially incidental, and there is the danger for human organization itself to be seen as consisting of two parts that are only loosely and superficially connected. Yet it must be postulated that the relationship between the external world (the environment) and subjective perception (the ideational aspects of human organization) must be deep and fundamental.

One of the few anthropologists who has tried to define some of the essential aspects of this linkage and to illuminate it systematically through a series of writings is Roy Rappaport (1968, 1971, 1976, 1978a, 1978b, and 1979d). Working within the framework of systems analysis, he has proposed that rituals and symbolic systems be seen as playing a crucial role in the cybernetic regulation of human ecosystems. By doing so, he has been able to suggest a level of integration between environmental and cultural-ideational concerns that few, if any, other human ecologists have approached. In viewing Rappaport's work, it is important to be mindful of his premises, taken from systems analysis. Indeed, it is well to note that he has been taken to task by several critics (Brown 1979; Friedman 1974, 1979; Whyte 1978; see also Rappaport 1978b, 1979c). However, one would be hard pressed to find another anthropologist who has, from an ecological viewpoint, equally profound things to say about object-subject interactions and about the relationship between environment and culture.

The title of this volume refers to cultural values rather than to culture, symbols, ideology, or other similar concepts. The reason for this lies in the fact that, in the most immediate sense, it is values that motivate human decisions and actions (see Chapter 1). While this may be so, it is nevertheless necessary to see cultural values in a larger social and cultural context: values do not exist as isolated cultural phenomena but are part and parcel of broader intellectual and emotional structures. Values rest on ideological systems that ascribe a place to individuals within the social and natural universe. These ideologies are of great emotive import because they explain the origin and destination of individual existence and thereby give meaning to individual life. This is the ultimate source for the rationale and power of the values that motivate actual behavior. The ideological systems as a whole derive their justification, and sanctification, from ultimate transcendental propositions that are themselves completely beyond empirical verification. While values constitute the immediate motivation for social and environmental action, this causal linkage is only a reflection of a much deeper relationship between ideology and the social and natural environment or, even more broadly, between culture and the environment.

In one of his essays, Rappaport (1979c, 59) reminds us that ecosystems may be described as systems of *matter and energy flow* among populations of organisms and between biological populations or organisms and nonbiotic things and processes. Culture, on the other hand, belongs to a category of phenomena that are not contingent on matter and energy flows (although these occur—and are essential—among culture-bearing organisms) but on symbols. Rappaport points out that early ecological anthropologists used "culture" as unit of analysis. This led to various problems, since the concept of culture is not commensurable with the concept of ecosystem. Rappaport proposes instead that human populations should be used as units of analysis, since this concept is commensurable with that of ecosystem. In this framework, cultures and constituents of cultures are seen as *properties* of human populations. Culture is, therefore, at least in part, analogous "to the distinctive means by which populations of other species maintain their environmental relations" (1979c, 62). However, while culture is thus conceived of as an essential means for humans to meet their physical needs, it is in no way said to be merely an instrument for survival. Cultures surely transcend their biological roles in many ways and maintain purposes of their own. This is indicated by the fact that cultural ideas and values are often inimical to biological demands. Thus, culture is not fully understandable without an ecological pespective, but neither is it completely explicable from an ecological perspective alone!

Here, of course, is the area of specific interest to the contributors of this volume: the areas where ecosystemic and cultural concerns overlap and where they clash. The statements here make the assumption that *systemic* relationships exist within environmental organizations. By systemic, I mean (following Rappaport) that environmental organizations are self-organizing and self-regulating. This conceptualization of environmental organizations as systemically organized (although not necessarily constituted as fully equilibrated stable systems! [Rappaport 1979a]) remains, of course, an assumption that is ultimately not verifiable but nevertheless furnishes a useful model to explicate many important properties of ecological relationships. Such a broad conceptualization of systemic organization does not necessarily stand in the way of evolutionary or other perspectives but is in many cases integrative and complementary. The important point here is that if one subscribes to the assumption of the systemic nature of ecosystems, and if cultures play a role in the operation of human ecosystems, then cultures too, at least in certain respects, must behave systemically.

How, then, do cultures participate in the operation of ecosystems? Rappaport (1979b) suggests that social scientists should prepare two models of a society's approach to its environment: a "cognized" model and an "operational" model:

> Nature is seen by humans through a screen of beliefs,

knowledge, and purposes, and it is in terms of their
images of nature, rather than of the actual structure
of nature, that they act. Yet, it is upon nature itself
that they do act, and it is nature itself that acts upon
them, nurturing or destroying them. Disparities
between images of nature and the actual structure of
ecosystems are inevitable. Humans are gifted
learners and may continually enlarge and correct
their knowledge of their environments, but their
images of nature are always simpler than nature and
in some degree or sense inexact, for ecological
systems are complex and subtle beyond full
comprehension. (1979b, 97)

According to Rappaport, cognized models see nature populated by
spirits, directed by supernatural forces, and divided into various
linguistic and conceptual categories unique to a given culture.
Operational models are interpretations of environmental relationships in terms of Western ecological science. Cognized models are not
simply wrong or ignorant renderings of reality. In fact, they are not
necessarily any more wrong or ignorant than scientific models are.
They are simply conceptual schemes to interpret external reality,
conceived within the broader framework of given cultures. As such,
their adequacy is to be judged on the basis of their contribution to the
"biological well-being of the actors and of the ecosystem in which
they participate" (1979b, 98). It is in terms of these cognized models
that the state of the world is judged and consequent action is taken.
Although these cognized models are in constant interaction with
external reality, they may nevertheless maintain certain ideas or
concepts ("reference values") that disagree with the range of
variability permitted to keep a system in equilibrium. Conceptual
factors may therefore be causal in promoting systems change.

However, knowledge and models are not enough to explain the
structure and organization of human action. Rappaport sees ritual
as playing an extremely important role in maintaining and
actualizing this knowledge. In his view, ritual acts to transmit
information, standardize knowledge, and particularly, to "transduce"
information between disparate aspects of human organization that
are not directly commensurable with each other. Perhaps most
significant, in submitting to the liturgical order and structure of

ritual, an individual accepts a body of concepts and rules that are encoded in the ritual and socially sanctioned. From this perspective, ritual can be viewed as having not just an incidental relevance in ecological relationships but as playing a central role in the organization of all human thought and action, including that relating to the environment.

Conclusions

Rappaport goes far in allowing us to perceive belief systems, mythology, conceptual structures, and rituals as playing a central role in human interactions with the environment by integrating all these aspects of human organization within a single system of structural relationships. In this way, it is possible not only to list the ecological functions of myth and ritual, for example, but to perceive the functions themselves as organized and structured rather than random, and thus accessible to systematic scientific inquiry. Rappaport's approach, while insisting on the adaptive significance of "purely cultural" phenomena, has the virtue of allowing us to examine their adaptive value rather than forcing us to assume either that all cultural values and ideas should be equally adaptive or that they have no intrinsic adaptive significance whatsoever.

In the final analysis, the important issue is not whether one accepts Rappaport's model of human organization as it now stands. But one point that pervades his writing is crucial: that environment and culture are *intrinsically* and deeply interrelated, and that social and environmental processes affect each other in fundamental ways. Neither can be adequately explained without the other, although neither can be explained by reference to the other alone. The relationships and interactions are so deeply embedded and complex that their nature will not easily be revealed in full. However, we must persist in the attempt. The problem is of more than academic interest, particularly with regard to the tropics. Most of the developing nations are found in tropical regions, and every year hundreds of millions of dollars are spent in development aid, deliberately fostering economic and ecological change while at the same time attempting to create conditions favorable to the preservation of valuable and nonrenewable biotic resources of importance to all of humankind. It is no secret that much of this money is spent in vain and often the opposite of what was intended is achieved. The fault lies not simply with mismanagement and political conflict but primarily with the perception that economic development is essentially only a technological, economic, and environmental problem. The price being paid for this folly is to be measured not only in terms of dollars, but in terms of human sweat and tears.

References

Bourdieu, P.
  1977   *Outline of a Theory of Practice.* Cambridge: Cambridge University Press.
Bourlière, F.
  1973   The comparative ecology of rain forest mammals in Africa and tropical America: some introductory remarks. In *Tropical Forest Ecosystems in Africa and South America: A Comparative Review*, edited by B. J. Meggers, E. S. Ayensu, and W. D. Duckworth. Washington, D.C.: Smithsonian Institution Press. Pp. 279–92.
Bourlière, F., and M. Hadley
  1970   The ecology of tropical savannas. *Annual Review of Ecology and Systematics* 1:125–52.
Brown, P.
  1979   Change and the boundaries of systems in Highland New Guinea: the Chimbu. In *Social and Ecological Systems*, edited by P. C. Burnham and R. F. Ellen. New York: Academic Press. Pp. 235–51.
Coe, M. J., D. H. Cumming, and J. Phillipson
  1977   Biomass and production of large African herbivores in relation to rainfall and primary production. *Oecologia* 22:341–54.
Durkheim, È.
  1965   *The Elementary Forms of the Religious Life.* Translated by J. W. Swain. New York: Free Press. (Originally published in 1915.)
Farnworth, E. G., and F. B. Golley (editors)
  1974   *Fragile Ecosystems.* New York: Springer Verlag.

Friedman, J.
  1974   Marxism, structuralism and vulgar materialism. *Man*,
          n.s., 9:444–69.
  1979   Hegelian ecology: between Rousseau and the World Spirit.
          In *Social and Ecological Systems*, edited by P. C.
          Burnham and R. F. Ellen. London: Academic Press. Pp.
          253–70.
Gaussen, H., P. Legris, and F. Blasco
  1967   *Bioclimats du Sud-Est Asiatique*. Institut Francais de
          Pondichery, Travaux de la Section Scientifique et
          Technique, vol. 3, pt. 4. Pondichery: Imp. de la Mission.
Geertz, C.
  1973   *The Interpretation of Cultures*. New York: Basic Books.
Golley, F. B., J. T. McGinnis, R. G. Clements, G. I. Child, and M.
J. Duever
  1975   *Mineral Cycling in a Moist Tropical Forest Ecosystem*.
          Athens: University of Georgia Press.
Goodale, J.
  1970   An example of ritual change among the Tiwi of Melville
          Island. In *Diprotodon to Detribalization: Studies of Change
          among Australian Aborigines*, edited by A. R. Pilling and
          R. A. Waterman. East Lansing: Michigan State
          University Press. Pp. 350–66.
Harris, D. R. (editor)
  1980   *Human Ecology in Savanna Environments*. London:
          Academic Press.
Harris, M.
  1979   *Cultural Materialism: Cultural Behavior in an Ecological
          View*. New York: Academic Press.
Healey, C. J.
  1978   The adaptive significance of systems of ceremonial
          exchange and trade in the New Guinea Highlands. In
          Trade and Exchange in Oceania and Australia, edited by J.
          Specht and J. P. White. Mankind 11:198-207.
Holdridge, L. R., W. C. Grenke, W. H. Hatheway, T. Liang, J. A.
Tosi, Jr.
  1971   *Forest Environments in Tropical Life Zones*. Oxford:
          Pergamon Press.
Hutterer, K. L.
  1982   *Interaction between Tropical Ecosystems and Human*

Foragers: Some General Considerations. East-West Environmental and Policy Institute Working Paper. Honolulu: East-West Center.

Janzen, D. H.
1975   *Ecology of Plants in the Tropics.* The Institute of Biology's Studies in Biology No. 58. London: Edward Arnold.

Kroeber, A. L.
1917   The superorganic. *American Anthropologist* 19:163–213.

Lombard, D.
1974   La vision de la forêt a Java (Indonesie). *Etudes Rurales* 53–56:473–85.

MacKinnon, J., and K. MacKinnon
1974   *Animals of Asia: The Ecology of the Oriental Region.* London: Peter Lowe.

Marshall, A. G. (editor)
1979   *The Abundance of Animals in Malesian Rain Forests.* Miscellaneous Series No. 22. Hull, England: University of Hull, Department of Geography.

Moore, O. K.
1957   Divination—a new perspective. *American Anthropologist* 59:69–74.

Nieuwolt, S.
1980   *Tropical Climatology.* London: John Wiley and Sons.

Owen, D. F.
1976   *Animal Ecology in Tropical Africa.* London: Longman.

Rappaport, R. A.
1968   *Pigs for the Ancestors.* New Haven: Yale University Press.

1971   Ritual, sanctity, and cybernetics. *American Anthropologist* 73:59–76.

1976   Adaptation and maladaptation in social systems. In *The Ethical Basis of Economic Freedom*, edited by I. Hill. Chapel Hill, NC: American Viewpoint. Pp. 39–82.

1978a   Maladaptation in social systems. In *The Evolution of Social Systems*, edited by J. Friedman and M. Rowlands. Pittsburgh, PA: University of Pittsburgh Press. Pp. 49–71.

1978b   Normative models of adaptive process: a response to Anne Whyte. In *The Evolution of Social Systems*, edited by J. Friedman and M. Rowlands. Pittsburgh, PA:

University of Pittsburgh Press. Pp. 79–87.

1979a Adaptive structure and its disorders. In *Ecology, Meaning, and Religion*, by R. A. Rappaport. Richmond, CA: North Atlantic Books. Pp. 145–72.

1979b On cognized models. In *Ecology, Meaning, and Religion*, by R. A. Rappaport. Richmond, CA: North Atlantic Press. Pp. 97–144.

1979c Ecology, adaptation, and the ills of functionalism. In *Ecology, Meaning, and Religion*, by R. A. Rappaport. Richmond, CA: North Atlantic Books. Pp. 43–95.

1979d *Ecology, Meaning, and Religion*. Richmond, CA: North Atlantic Books.

Richards, P. W.

1973 The tropical rain forest. *Scientific American* 229(6):58–67.

Riehl, H.

1979 *Climate and Weather in the Tropics*. London: Academic Press.

Sahlins, M. D.

1976a *Culture and Practical Reason*. Chicago: University of Chicago Press.

1976b *The Use and Abuse of Biology*. Ann Arbor: University of Michigan Press.

Sanchez, P. A.

1976 *Properties and Management of Soils in the Tropics*. New York: John Wiley and Sons.

Schnell, R.

1970 *Introduction a la phytogeography des pays tropicaux*. Paris: Gauthier-Villars. 2 vols.

Turner, V.

1974 Social dramas and ritual metaphors. In *Dramas, Fields, and Metaphors: Symbolic Action in Human Society*. Ithaca, NY: Cornell University Press. Pp. 23–59.

UNESCO

1978 *Tropical Forest Ecosystems*. Paris: UNESCO.

Walter, H.

1979 *Vegetation of the Earth and Ecological Systems of the Geo-biosphere*. New York: Springer-Verlag.

Wharton, C. H.

1968 Man, fire and wild cattle in Southeast Asia. *Annual Proceedings of the Tall Timbers Fire Ecology Conference*

8:107–67.

Whitmore, T. C.
1975   *Tropical Forests of the Far East.* Oxford: Clarendon Press.

White, L. A.
1949   *The Science of Culture: A Study of Man and Civilization.* New York: Grove Press.

Whyte, A.
1978   Systems are perceived: a discussion of "Maladaptation in social systems." In *The Evolution of Social Systems*, edited by J. Friedman and M. Rowlands. London: Duckworth. Pp. 73–78.

Znaniecki, F.
1936   *The Method of Sociology.* New York: Farrar and Rinehart.

CHAPTER 4

ECOLOGY, ANTHROPOLOGY, AND VALUES IN AMAZONIA

Leslie E. Sponsel

Amazonia is a stimulating region for ecological and anthropological research, given the richness and diversity of its natural and cultural life. It offers an ideal place to explore the role of values in tropical ecology. The indigenous cultures and natural ecosystems of Amazonia are increasingly threatened by Western civilization and development. A fundamental conflict is evident between the values of the so-called primitive and civilized worlds. Cultural ecology can contribute to the improvement of this situation by documenting the adaptive role of indigenous culture and knowledge and by evaluating the influence of Western civilization and development in this light. But first it must examine its own values.

This chapter explores these and related problems. To provide a background, it describes first the Amazonian ecosystem as a whole and gives some of the highlights in the study of its cultural ecology. Then it turns to an examination of the role of values in the cultural ecology of Amazonia in the contexts of indigenous adaptations and economic development.

Amazonia

Geographically, Amazonia refers to the tropical lowland region of South America that is dominated by rain forests (*hylaea*) and bounded by the Andean, Brazilian, and Guiana highlands. Amazonia extends over 7,050,000 km^2 (Crist 1982, 653). Politically, the region includes substantial portions of Venezuela, Colombia, Peru, Ecuador, Bolivia, and Brazil.

The Amazon is one of the longest navigable rivers in the world. Estimates of its total length range up to 6,771 km (Sioli 1975, 461). The width along most of its course is more than 8 km, increasing to 320 km at its mouth. The Amazon River and its more than one thousand affluents drain the Amazon Basin, a hydrographic region that covers 40 percent of the continent of South America. About 20 percent of all the water that runs off the earth's surface is transported by the Amazon River (Crist 1982, 653).

Amazonia is the product of 60 million years of geological and biological evolution (Meggers 1971, 8; Richards 1973, 59). It accounts for six out of the eight million km^2 of moist forest in tropical America (Myers 1979, 130). Indeed, Amazonia is the largest continuous area of tropical forest on earth, composing 54 percent of that biome (Grainger 1980, 11). It is also probably the richest biological area on earth (Myers 1979, 23, 131). Of the planet's five to ten million species of plants and animals, possibly one million are Amazonian (Myers 1979, 23). Yet, according to some estimates, at least one species per day is becoming extinct through deforestation and other disturbances by humankind (Myers 1979, 5; Myers and Ayensu 1983, 72). Nearly 30 percent of Amazonia's forest has been destroyed, much of it within the past twenty-five years. In this region, even if the present rate of deforestation does not increase, the forest will be eliminated in about fifty years (Myers 1979, 132–33; cf. Hecht 1980).

Archaeologically, South America is one of the least-known regions of the world, and this is especially so for Amazonia. The earliest pottery appears by 980 B.C. on the island of Marajo in the mouth of the Amazon River. However, there is no direct evidence of human occupation in Amazonia during the preceramic period (Meggers 1982, 485). Lynch (1978, 473) interprets the absence of evidence as indicating that Amazonia was not inhabited by the palaeoindians. In contrast, Meggers (1982, 488–89; 1979, 124) attributes this absence to a combination of factors such as indigenous reliance on perishable materials for tools because of the rarity of stone resources, the transitory nature of sites, unfavorable conditions for preservation (tropical climate and soils, fluvial geomorphology), unfavorable conditions for fieldwork (e.g., dense cover of vegetation), and the limited amount of archaeological research in the region. Meggers (1979, 124) thinks that it is quite reasonable to suppose that food collectors may have occupied parts of Amazonia long before the introduction of pottery, as was the case in adjacent biomes that paleoindians colonized by 12,000 B.P.[1]

Although *Homo sapiens* is relatively new to Amazonia, and the environment appears to be at least superficially rather homogeneous, the region is linguistically and culturally one of the most diverse in the world. Today more than 325 indigenous human cultures survive in Amazonia with varying degrees of acculturation, and many more existed at the time of European contact (Dostal 1972; Ribeiro 1967). Despite the diversity of languages and cultures, there have been several attempts to characterize Amazonia as a single culture area, in contrast to the Andean and circum-Caribbean areas (Weiss 1980). Common elements are evident in the material culture, technology, subsistence, and economy of the cultural system, while other aspects of culture are more variable (Steward 1948, 885–86; Zerries 1969, 378). At least on a general level the different cultures appear to be variations on a complex set of basic themes.

Each Amazonian culture exhibits some combination of the following traits. The principal source of food, especially carbohydrates, was slash-and-burn cultivation that centered on the raising of root crops, mainly manioc. The chief source of protein in the riverine zone was fishing, but aquatic game also was hunted. In the interior forest, hunting was more important than fishing. There the common prey were wild pig, deer, tapir, rodents, monkeys, and birds. Gathering of insects, turtles, and other small game as well as plants, especially palm fruits, supplemented the diet. Domesticated plants included manioc, maize, yams, sweet potatoes, beans, peanuts, avocado, papaya, pineapple, peach palms, bottle gourds, cotton, and tobacco. No animals were domesticated. Each village was self-sufficient as the unit of resource acquisition, processing, distribution, and consumption. In most areas trade was very limited and usually more important for social and political functions than for economic ones. Craft specialization was absent. Manufactured items were hammocks, waistbands, loincloths, baby slings of woven cotton, pottery, basketry, wooden stools (frequently in zoomorphic shapes), drinking and eating containers cut from gourds, ceramic griddles and basketry presses for processing manioc, blowguns, bows and arrows, and dugout canoes. Metallurgy was absent prior to European contact. Many plants were used for various purposes such as barbasco to "poison" sections of streams to harvest fish, curare for the tips of arrows and blowgun darts, tobacco, hallucinogens, and medicines. Beer brewed from manioc in hollow log drums was taken during festivals and rituals that often lasted several days. Most people wore little if any clothing but meticulously decorated their bodies with paint, feathers, and/or flowers. Village size varied from about 100 to 1,000 persons, and most villages were relocated every few years. The smaller villages were often a single, large, circular building with a conical roof, constructed from poles and palm thatch. A single building housed several nuclear or extended families, or lineages. Polygyny was fairly common. Most of these societies were egalitarian, and social relations were based on kinship. There were, however, exceptions: some societies showed incipient stratification and some practiced slavery. Usually, each village was politically autonomous with a prominent personality serving by consensus as headman. Warfare was endemic in many areas. Religion centered on the shaman communicating with animal-like spirits through chanting, smoking tobacco, or consuming hallucinogens, sometimes assisted by gourd rattles and other paraphernalia. Rituals focused on major subsistence events such as garden clearing, planting, and harvesting, and on the life events of birth, puberty, marriage, sickness, and death.[2]

Estimates of the size of the total pre-Columbian population of Amazonia range from 2,188,970 (Steward and Faron 1959, 53) to 5,100,000–6,800,000 (Denevan 1976, 232–34). Denevan (1976, 226) estimates aboriginal population densities in precontact time at 28.0/km^2 for large floodplains and 0.2/km^2 for lowland interior forests. Although such statistics may be based on careful research, they entail many speculations, assumptions, and tenuous data and can thus only suggest an order of magnitude.[3] More certain is that most indigenous societies experienced demographic collapse as a result of contact with European civilization (Ashburn 1947; Crosby 1972), a process that still continues (Bodley 1982). Microbial shock is a major component in this depopulation and the ensuing social disruption (Neel 1982).[4] Thus, although Amazonia is the largest area of tropical rain forest in the world, it has today the lowest population size and density for this biome. Djalma Batista estimated the current total population (including nonindigenous persons) at 4,841,000. He calculated a density of 0.7/km^2 based on the area of 6,288,000 km^2 (Tocantins 1974, 22–23). On the basis of such statistics, some refer to the Amazon as if it were a demographic vacuum, with the implication that the region is open for the taking, i.e., ripe for colonization and development.

The contemporary circumstances of indigenes in Amazonia may be appreciated from Ribeiro's (1967, 96) study of Brazil. In that country, from 1900 to 1957, eighty tribes came into contact with the national society and perished. During the same period, the country's indigenous population declined markedly, from about one million to 200,000. In other Amazonian countries, the situation is similar to varying degrees (Dostal 1972; Bodley 1982).

After sixty million years of geological and biological evolution, and after at least three thousand years of cultural evolution, Amazonia's tremendous richness and diversity in biological and human (genetic, cultural, and linguistic) terms is now being rapidly depleted in the course of a few centuries, often for the quick profit of a few transient individuals and organizations. This background helps one to appreciate the role of cultural values in the tropical ecology of Amazonia. The region provides a tragic example of the conflict between the cultural values of the so-called primitive and civilized worlds, the human and environmental consequences of that conflict, and the roles of anthropology and ecology (see Bodley 1976, 1982) in understanding the changes that are occurring.

The Amazonian Ecosystem

Although for more than a century the biological ecology of Amazonia has been explored by naturalists such as Humboldt, Bates, Wallace, Agassiz, Spruce, and Beebe (Goodman 1972), Amazonia has remained largely a mystery. Only recently have biologists started to unlock the secrets of the region (Farnworth and Golley 1974, 2). Their studies are usually very technical treatments of narrow topics, however, and it may be years before any comprehensive synthesis is available.[5] Nevertheless, a few tentative generalizations can be offered at this stage. Amazonia shares many features of the structure and function of other regions of tropical rain forest, even though it differs in species composition, latitudinal and altitudinal characteristics, and other aspects (National Research Council 1982, 38–39; UNESCO 1978). In particular, its geology, hydrology, and limnology are distinctive.

For centuries, Amazonia has been viewed as a tropical paradise, an idea that persists to this day, except among specialists. The notion reflects the luxuriant plant growth of the forest. Like other tropical rain forests, Amazonia has provided optimum conditions for plant growth. Insolation (2,900 kcal/cm^2 annually at canopy level), temperature (+24° C mean monthly), rainfall (+2000 mm annually), and humidity (+60 percent mean daily relative) are all very high (National Research Council 1982, 53–54). The tropical rain forest is usually so moist that Odum (1971, 401) referred to it as the only major terrestrial biome where fire is not a significant ecological factor. Because of its ideal climate, the tropical rain forest may be considered the norm for vegetation, and anything else seen as a deviation in response to stresses such as seasonal drought or frost (Walter 1971, 72).

The tropical rain forest is one of the most productive biomes on the planet, producing annually at least three times as much organic matter as a rich temperate forest (Lieth 1975, 205). Tropical rain forests account for 50 percent of the biomass of the earth (Longman and Jenik 1974, 120), although they cover only 10 percent of its land surface (National Research Council 1982, 39).

High species diversity is another index of the luxuriance of this biome. Typically, there are one hundred plant species per ha, but as many as four hundred species per ha can be found (Longman and Jenik 1974, 68). This high species diversity means that, except under very special edaphic or hydric conditions, tropical rain forests do not have dominant species, unlike their temperate counterparts (Billings 1970, 96). In any area there are few members of each species (low population density), and the distribution of members is highly dispersed (Richards 1973, 60–61).

Yet, as noted in the previous chapter, while the tropical climate is ideal for plant growth, it is also ideal for soil weathering and erosion. Especially active is chemical weathering, which leaches nutrients from the soil to great depths. High temperature and rainfall are also optimum for the destruction of humus. The great geological antiquity of Amazonia has allowed ample time for these processes to operate to their full extent. The geological stability of this region has not produced fresh parent material for soil enrichment, and the vast majority of soils derive from parent material of Tertiary age (Meggers 1971, 8–15). These conditions have led to "demineraliza-tion," the impoverishment of the very ancient soils of long-quiescent land masses, where climate and time have allowed extensive leaching of minerals (Schwabe 1968, 121, 127). By comparison, floodplain soils are relatively fertile, since they are derived from

Quaternary parent material and annually rejuvenated by inundation. Floodplains, however, represent at most only 10 percent of Amazonia (Sponsel 1981, 128).

In Amazonia, then, most soils are poor in nutrients.[6] The forest grows in spite of soils that serve for little more than physical support. The forest has evolved numerous mechanisms to maintain itself under such conditions, acting as a giant sponge to maximize nutrient capture and minimize nutrient loss (Herrera, et al. 1978; Jordan 1982). The forest is a tight, closed system of nutrient cycling, deriving nutrients mostly from litter and rainfall. The groundwater reflects this cycling in that, unlike the rainwater, it is nearly as pure chemically as distilled water (Sioli 1975, 475). Such considerations led Meggers (1971; 1973, 7) to label Amazonia a "counterfeit paradise."

Red soils (oxisols and ultisols) are the most common type in Amazonia, where they cover proportionally more area than they do in Africa and Asia (National Research Council 1982, 49). "Their main limitations are chemical: high soil acidity; aluminum toxicity; deficiency of phosphorus, potassium, calcium, magnesium, sulfur, zinc, and other micronutrients; and low effective cation-exchange capacity, which last indicates a high leaching potential" (National Research Council 1982, 44).

This impoverishment is also reflected in the patterning of Amazonian fauna.[7] Several factors are involved. As with plants, the diversity of animal species is very high in Amazonia compared with rich temperate forests. Thus, there are few members of each species in any given area, and the distribution of members is dispersed. Yet, when species diversity is compared within the same biome for different regions, Amazonia is lower for at least some taxa than similar regions of the Old World. For example, there are nine species of ungulates in Amazonia while there are twenty-seven species in African forests (Bourlière 1973, 281). Moreover, the species in Amazonia are smaller in body size than their African counterparts (Hershkovitz 1972, 372). These differences may be related to the very low nutrient value of litter in Amazonia compared to other tropical rain forests (Fittkau and Klinge 1973, 2).

The ratio of animal to plant biomass is very low in Amazonia. A study in one area found that animals compose only 0.02 percent of the total biomass (Fittkau and Klinge 1973, 8). This figure is even more striking since 78.6 percent of the animal biomass consists of soil fauna, mostly invertebrates. This situation probably reflects the limited availability of edible plant material. Most of the plant biomass is wood, which herbivores cannot consume. Leaves account for only 2 percent of the living plant biomass, most of it accessible only to arboreal herbivores like primates, sloths, and some birds. About half of the animal biomass feeds from the litter on the forest floor (Fittkau and Klinge 1973, 9, 11–12). Another complication is presented by the numerous antipredator defense mechanisms that plants have evolved, such as the presence of toxic substances in leaves (Janzen 1974, 277–78; 1975, 35–44).

Even if animals were fairly abundant, they would not necessarily be readily available for exploitation by hunters. A survey of the common prey species for indigenous hunters reveals that the majority are arboreal (43.9 percent), solitary (53.6 percent), small in body size (less than five kg) (39.4 percent), nocturnal (73.2 percent), and well camouflaged. Most reproduce with low frequency and have small litter sizes (Sponsel 1981, 156–97).

The peak of the wet season is a period of food scarcity for many indigenous communities. Wet-season inundation of extensive floodplain areas far into the forest results in widespread dispersion of fish and game, rendering them less available for indigenous exploitation. This flooding is a result of the almost negligible gradient of the Amazon River (only 2 m above sea level 1,000 km from its mouth) combined with annual fluctuations of 5–20 m in its water level (Sioli 1975, 467, 471–72). Agassiz (1896, 256) aptly described the scene: "Its watery labyrinth is rather a fresh-water ocean, cut up and divided by land, than a network of rivers."

In some regions, especially the Rio Negro of the northwest Amazon, so-called black water predominates. These waters are very low in productivity because of such factors as high acidity and dark, tea-like color, the latter limiting light penetration for photosynthesis. Floating vegetation mats ("floating meadows") are absent, in contrast to other water types ("white" and "clear" waters). Even aquatic insects are negligible (Sioli 1975, 473–75; Janzen 1974). As a consequence, both the productivity and the diversity of fish species are both very low (Goulding 1980, 16). Likewise, the flora and fauna of the adjacent terrain are impoverished in these oligotrophic ecosystems. Black-water areas are notorious as "hunger" or "starvation" rivers.

In spite of the luxuriance of the Amazonian forest as determined by such measures as high productivity and high species diversity, therefore, the soils and fauna are impoverished. Wet-season flooding, and in some areas black waters, also contribute to the counterfeit nature of this apparent paradise.

Recently, Jordan and Herrera (1981) among others have challenged such stereotypes, as, for example, Meggers's view of Amazonia as a counterfeit paradise. These authors argue that there is a continuum from eutrophic to oligotrophic ecosystems in the humid tropics. Regardless which side one follows, the foregoing discussion indicates clearly that Amazonia presents serious challenges to human adaptation.

Cultural Ecology[8]

While occasional discussions of the cultural ecology of Amazonia can be found in the writings of the early naturalists, not until the 1950s did the subject really begin to take shape. Sauer (1958) delimited many of the major perennial issues in the study of the cultural ecology of the neotropics. First, he pointed to soils and the riverine habitat as limiting factors in Amazonia. Second, he was concerned with human impact on the environment: the aboriginal alteration of plants, animals, and habitats, especially through fire and shifting cultivation. A third, but closely related, issue was whether savannas are of natural or human origin. Finally, he noted the resemblance between the subsistence economy and ecology of indigenous cultures in the rain forests of the neotropical and oriental biogeographic regions.

The issues of limiting factors and carrying capacity have become the dominant focus for anthropological studies of the ecology of indigenous cultures in Amazonia. Only a few studies have directly addressed indigenous environmental impact. Savanna formation is discussed by Scott (1977) and Smole (1976). The relationship between game depletion and the introduction of Western hunting weapons is considered by Hames (1979) and Saffirio and Scaglion (1982). No study has concentrated on the fourth issue, a comparison of the cultural ecology of indigenous societies of Amazonian and Asian rain forest ecosystems. Such research would likely produce stimulating insights into convergent adaptations.

During the 1950s and the immediately preceding decades, Steward was pioneering in the development of cultural ecology (Steward 1955; Steward and Murphy 1977) and in synthesizing existing knowledge about South American Indians (Steward 1946–59; Steward and Faron 1959). Steward was the major stimulus in the study of cultural ecology in Amazonia and elsewhere. He was largely responsible for the use of cultural ecology as a means to the end of cultural evolutionism (Anderson 1973; Hanc 1981). Thus, the theoretical framework for the study of the cultural ecology of Amazonia was constructed around (1) the ecological concepts of carrying capacity and limiting factors, and (2) the anthropological concepts of cultural complexity and cultural evolution.[9] The logic of the cultural evolutionists assumes, but does not demonstrate, a causal link between carrying capacity and cultural complexity.[10] That is, carrying capacity is taken to restrict the level of cultural development by limiting the size, distribution, and permanence of the indigenous population.

Since the 1950s, Meggers (1954, 1957, 1971, 1979) has been one of the most outstanding proponents of this evolutionary view for Amazonia. For example, Meggers (1954, 814) writes, ". . . the level to which a culture can develop is dependent upon the agricultural potentiality of the of the environment it occupies. As this potentiality is improved, culture will advance. If it can not be improved, the culture will become established at a level compatible with the food resources." Meggers (1954, 809) did not restrict her argument to the aboriginal situation but applied it to criticize contemporary economic development: "Even modern efforts to implant civilization in the South American tropical forests have met with defeat, or survive only with constant assistance from the outside." At question here is the very nature of the Amazonian ecosystem, the human niche in it, and its influence in determining the evolutionary level of complexity that indigenous societies could achieve. Lessons are also drawn from this for modern economic development.

Carneiro (1956, 230–32) pioneered a quantitative approach to test empirically cultural-ecological hypotheses through fieldwork in Amazonia. Using a formula to calculate carrying capacity (in reference only to agriculture), he refuted Meggers's hypothesis that soils are the limiting factor in Amazonia. Carneiro (1961, 77) also anticipated the "protein hypothesis": "Among tribes for whom hunting still constitutes an important part of subsistence, the depletion of game animals in the vicinity of the village may dictate moving long before other conditions would warrant it."[11]

From the mid-1970s to the present, the major focus of research on cultural ecology in Amazonia has been on protein capture. The turning point was a comprehensive analytical review of the ethnographic and ecological literature pertinent to the subject by Gross (1975). Gross explicitly articulated the hypothesis that animal protein is the chief limiting factor concerning the size, distribution, and permanence of indigenous populations in Amazonia. He argued that the staple plant foods, especially manioc, are very low in quality protein; that indigenous societies lacked domesticated animals as an alternative source of protein; and that hunting is difficult, especially in the interior of the forest. While this hypothesis met considerable criticism, it has yet to be invalidated on theoretical or empirical grounds. It seems quite plausible that animal protein is at least one major limiting factor challenging human adaptation in many ecosystems of Amazonia (Sponsel 1981, 304–21).[12]

This hypothesis about animal protein as a limiting factor has become the foundation for a number of secondary hypotheses that attempt to offer cultural-materialist explanations for such Amazonian institutions as warfare (Harris 1974, 1977, 1979a, 1979b, 1979c; J. B. Ross 1980), faunal taboos (McDonald 1977; E. B. Ross 1978, 1979, 1980; E. B. Ross and J. B. Ross 1980), and intravillage sexual politics (Siskind 1973a, 1973b). Of these secondary hypotheses, the one that has received the most attention, both in theoretical debate and field research, is the explanation of the origins and functions of Yanomama aggression.[13]

The Yanomama are celebrities thanks to Chagnon (1968a, 1968b, 1977, 1983) who has promoted their image as "the fierce people" in numerous academic and popular writings as well as other media. Most of the other anthropologists who have worked with the Yanomama believe that Chagnon greatly exaggerates the violent aspects of Yanomama culture and daily life. Some writers, however, carried the exaggerations even further. For example, Harris (1974, 87–88) characterizes the Yanomama as follows:

> By the time a typical Yanomamö male reaches maturity, he is covered with the wounds and scars of innumerable quarrels, duels, and military raids. Although they hold women in great contempt,

> Yanomamö men are always brawling over real or
> imagined acts of adultery and broken promises to
> provide wives. Yanomamö women are also covered
> with scars and bruises, mostly the result of violent
> encounters with seducers, rapists, and husbands. No
> Yanomamö women escapes the brutal tutelage of the
> typical   hot-tempered,   drug-taking   Yanomamö
> warrior-husband. All Yanomamö men physically
> abuse their wives. Kind husbands merely bruise and
> mutilate them; the fierce ones wound and kill.[14]

Although Chagnon has in recent years embraced sociobiology
(e.g., Chagnon and Irons 1979), his early explanations of Yanomama
aggression were essentially mentalistic in orientation. Chagnon
(1968a, 1) begins with the emic explanation of the Yanomama: "It is
in the nature of man to fight, according to one of their myths,
because the blood of 'Moon' spilled on this layer of the cosmos,
causing men to become fierce." This myth, then, is proposed to
contain the charter of Yanomama society, including their ideology of
fierceness. Chagnon (1968b, 112) summarizes his explanation (fig.
4.1):

> The hypothesis I put forward here is that a militant
> ideology and the warfare it entails function to
> preserve the sovereignty of independent villages in a
> milieu of chronic warfare. The origin of such a
> political milieu seems to be the result of the failure of
> Yanomamö political institutions to govern effectively
> the conflicts arising within villages, conflicts that give
> rise to internal fighting and village fission with the
> ultimate establishment of mutually hostile, indepen-
> dent villages.[15]

This explanation follows the position of the ancient Epicureans who
believed that the primitive state of humankind was one of violence,
and that sociopolitical relations were born of the necessity of alliance
for mutual defense and cooperation (Barnes 1923, 38). Chagnon's
explanation is also culturological in many respects. It contrasts
sharply with the cultural ecology of Steward (1955, 36), which ". . .
introduces the local environment as the extracultural factor" to

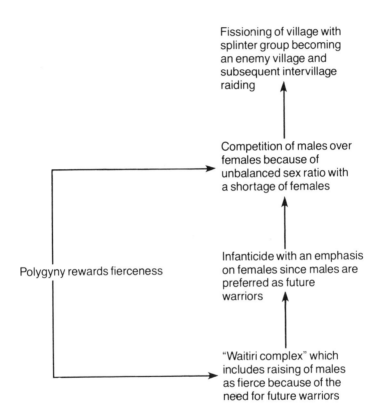

Fig. 4.1.   Chagnon's model of Yanomama aggression (after Sponsel 1981, 323).

counter "the fruitless assumption that culture comes from culture."

Here is exactly where Harris (1974, 1977, 1979a, 1979b, 1979c) applies his cultural materialism. He offers an ingenious theoretical explanation of the functional relationship between Yanomama aggression, population, production, and resources (fig. 4.2). Harris (1977, 49) summarizes his thesis:

> I believe that it is possible to show that the Yanomamo have recently adopted new technology or intensified a preexisting technology; that this has brought about a veritable population explosion, which in turn has caused environmental depletion; and that depletion has led to an increase in infanticide and warfare as part of a systemic attempt to disperse settlements and to prevent them from growing too big.[16]

Unfortunately, Harris's cultural materialism, and the animal protein hypothesis together with the secondary hypotheses it has engendered, are often equated with cultural ecology as if they were isomorphic (e.g., Chagnon and Hames 1979, 1980). This confusion hardly helps matters. Certainly there is overlap between these arenas. Not all cultural ecologists are materialists, however, and the converse holds as well. Not all cultural ecologists accept the materialist tenet of the primacy of the technological and economic components of the cultural system. Most cultural ecologists are simply intellectually attracted to exploring the legitimate question of how culture articulates human populations with their ecosystems (Sponsel 1982, 1983b).[17] Moreover, there is often very little ecology in the materialist approach. Finally, cultural materialists, in contrast to cultural ecologists, tend to offer armchair explanations instead of explanations based on empirical fieldwork.

In the case of anthropologists subscribing to the animal protein hypothesis and the secondary hypotheses, some are working within the framework of cultural ecology, others within cultural materialism, and still others in the conceptual area where these frameworks overlap. While carrying capacity and limiting factors are pivotal concepts in ecology, it should be clear from this review of the highlights of cultural ecology in Amazonia since 1950 that protein is only the latest candidate for the role of limiting factor, and

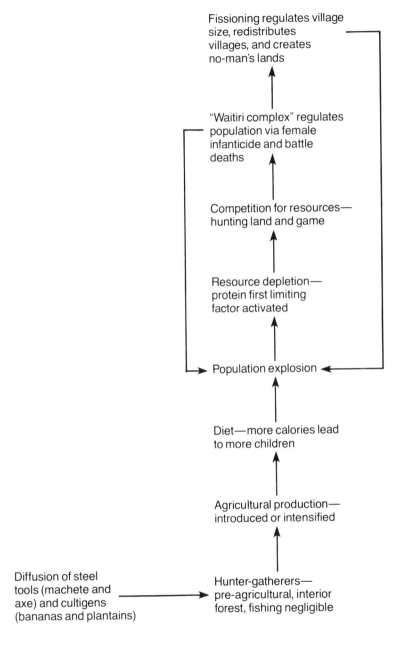

Fig.   4.2.   Harris's model of Yanomama aggression (after
                  Sponsel, 1981, 328).

it is not likely the last. Indeed, a combination of limiting factors may be involved, actually or potentially, depending on the level of the population and on different local ecosystems. This is quite different from the usual claim of a single factor for the entire region of Amazonia (cf. Vickers 1979; J. B. Ross 1980, 38). In any case, limitations of the intrinsic rate of population increase have been recognized since Malthus, and this notion has been an important principle in ecology from the beginning. The fact that a continuous exponential rate of population growth is not observed in Amazonian societies indicates the operation of intrinsic or extrinsic limiting factors, the demographic disruptions of European contact notwithstanding.

Although the controversies over the protein hypothesis and the secondary hypotheses connected with it have often generated more heat than light, one thing has been illuminated, namely, the biological and cultural ecology of indigenous hunting and fishing, an aspect previously neglected. A number of researchers on both sides of these controversies have contributed quantified empirical field data on distribution, capture, and consumption of protein, and to a lesser extent on other aspects of the behavior and ecology of indigenous predation in Amazonia. Certainly, the protein hypothesis and the secondary hypotheses have stimulated substantial progress in our understanding of the cultural ecology of Amazonia. These hypotheses have proved to be of great heuristic utility, regardless of their validity.[18]

The debate between Chagnon and Harris, and their respective supporters, exposes a deep and fundamental rift in anthropology between the materialist and mentalist positions (see Jamieson and Lovelace, this volume). Unfortunately, the extremists on both sides tend to deny each other's position *a priori*, rather than grappling with *Homo sapiens* as an integrated organic, cultural, and intellectual being, something that cultural ecology may well achieve by elucidating the relationship between particular cultures and their local ecosystems. While cultural ecology may not be able to reconcile the extremes of materialism and mentalism, it may render such extremes obsolete.

Values

Most appropriate here is a broad definition of values taken from Pepper (1958, 7): "Values refer to interests, pleasures, likes, preferences, duties, moral obligations, desires, wants, needs, aversions and attractions, and many other modalities of selective orientation." This definition allows a consideration of the role of values in Amazonian cultural ecology in at least three contexts: indigenous adaptation, anthropological research, and economic development.

One example of the role of values in indigenous adaptation has already been discussed with reference to the Yanomama. Chagnon (1968a, 3) writes, "The fact that the Yanomamö live in a state of chronic warfare is reflected in their mythology, values, settlement pattern, political behavior, and marriage practices." Clearly, the "waitiri complex," at least as described by Chagnon (1968b, 124–39), involves values; the fierceness of males is preferred by the community for its defense and survival. Thus, according to Chagnon (1968a, 1), the Yanomama are "the fierce people." He continues, "That is how they conceive themselves to be, and that is how they would like others to think of them."

Faunal prohibitions are a second example of values in the cultural ecology of Amazonia. These taboos dictate which animal species cannot be killed or eaten by members of a society. General taboos apply to all members of the society, whereas specific taboos apply only to particular individuals therein, depending on their age, sex, or social or physiological condition (Basso 1973, 16). Obviously such practices involve cultural values as a determinant force orienting behavior. As E. B. Ross (1978, 2) observed, taboos, like other aspects of myth, ritual, and symbol, have traditionally been the province of the cultural mentalists. They have seen these phenomena as fortuitous products of human creativity, albeit culturally rationalized (e.g., Kensinger and Kracke 1981). However, since the mid-1960s, various anthropologists working within the framework of cultural ecology or cultural materialism have challenged this mentalistic interpretation. They argue that these phenomena are cultural adaptations within the context of technology, economy, demography, and ecology (Harris 1974, 1977, 1979a; E. B. Ross 1980). For example, Eric Ross (1978) argues that prey species are not of equal significance for the human predator, since they have different behaviorial and ecological attributes. Through selective predation, an indigenous society takes these differences into account. Taboos direct predation away from the less accessible and more vulnerable species, and toward those which provide a better cost/benefit ratio combined with a sustained yield in the long term. In essence, faunal taboos are part of an indigenous strategy for wildlife management and conservation.[19]

Reichel-Dolmatoff (1971, 243) provides the most direct and explicit study relating values to indigenous cultural ecology in Amazonia. He summarizes Desana ethnoecology as follows:

> According to the Desana, the goal of life and of all human activities and attitudes is the biological and cultural continuity of their society. This goal can only be achieved by a system of strict reciprocity in all relationships that man establishes in the biosphere, be they in the framework of his own society or with the animals. The paradigmatic model is that of the two sexes, and in their interrelationship sexuality is compared to nutrition: the man fertilizes the woman who, in turn, produces the child, a cyclical phenomenon that implies a smaller and accelerated cycle of nutrition. Man is the producer of one category of foodstuffs, the proteins, while woman produces complementary foodstuffs, the carbohydrates. Its daily culinary transformation produces new energy in which man participates, thus guaranteeing the continuity of the wider procreative cycle. This reciprocity is based, in part, on the division of labor according to which men are mainly occupied in the masculine sphere of the forest while the women are in the fields. On the social plane this involves exogamic marriage among men identified as hunters and women identified as horticulturalists.

He concludes (p. 252) by noting that ". . . the world we label 'primitive' contains values we can ill-afford to deprecate."[20]

While Reichel-Dolmatoff devotes more attention to ethnoecology than to cultural ecology and very little attention to biological ecology, thus falling short of a true synthesis of these arenas, he does suggest one significant point. Values are a very important key to understanding the processes whereby *Homo sapiens* adapts simultaneously as an organic, cultural, and intellectual being. Indeed, values may provide the crucial link between the ideas and actions of individuals and systemic processes. This link has been missing in the field research in cultural ecology and in cultural materialism.[21] Thus Reichel-Dolmatoff's approach is a refreshing contrast to the usual mentalist and materialist studies.

While the role of values in indigenous adaptation has been neglected, except in the few studies just reviewed, the values held by anthropologists and their roles in the conduct of anthropological research, as well as the values held by Western societies engaged in the penetration of Amazonia, are even more neglected topics. One of the earliest discussions of how values enter anthropology is by Honigmann (1959, 113–17). He asserts that the anthropologist cannot help "valuing" as he/she works; that the ethics of the anthropologist are values; that anthropologists evaluate cultures; that cultural relativity itself may be viewed as a value judgment; and that some anthropologists are moralists. Hatch (1983, 92–93) comments on the Yanomama case:

> The moral principle of tolerance that is proposed by Boasian relativism carries the obligation that one cannot be indifferent toward other ways of life—it obligates us to approve what others do. So if missionaries or government officials were to interfere in Yanomamö affairs for the purpose of reducing violence, the relativist would be obligated to oppose these moves in word if not action. . . . The functionalists are led to a similar conclusion: Harris contends that Yanomamö warfare and infanticide are ecologically adaptive in that they keep the population within the carrying capacity of the land. Approval is warranted on practical grounds, even though the practical value of the institutions is asserted but not proved.[22]

Fieldwork is associated with values by anthropologists in numerous, often subtle, ways. The research design is evaluated by academic and granting committees. The first field trip is valued as a rite of passage marking the transition from novice to professional ethnographer. According to Fischer (1969, 13), the selection of the region and group and the final report are evaluated by several criteria:

1. Distance from the U.S. (both geographical and cultural)
2. The degree of cultural shock value of the culture studied from the point of view of Westerners
3. The primitiveness of the culture studied
4. The value of the culture in terms of theoretical problems of interest to anthropologists
5. The number of anthropologists who have done fieldwork in the culture or in adjacent regions (the fewer fieldworkers, the higher the status of the experience)
6. The physical difficulties involved in doing fieldwork in the culture
7. The time spent in the field
8. The comparative excellence of the fieldwork, which may be disguised to an extent by excellence and speed in reporting if no other experts are available to dispute the report

Amazonian indigenous cultures like the Yanomama fit many of these criteria perfectly. Indeed, this is one of the reasons why the Yanomama are such celebrities in anthropology. Take culture shock, for example. In the foreword to Chagnon's (1977, vii) ethnography, Spindler writes:

> Yanomamö culture, in its major focus, reverses the meanings of "good" and "desirable" as phrased in the ideal postulates of the Judaic-Christian tradition. A high capacity for rage, a quick flash point, and a willingness to use violence to obtain one's ends are considered desirable traits. Much of the behavior of the Yanomamö can be described as brutal, cruel, treacherous, in the value-ladened terms of our vocabulary.[23]

During much of its history, anthropology has been identified closely with cultural evolutionism, an interest in primitive cultures, and salvage ethnography (Braroe and Hicks 1967; Diamond 1974). Gruber (1970, 1293) describes salvage ethnography: "Throughout the century and with whatever theoretical framework, the refrain was the same: the savage is disappearing; preserve what you can; posterity will hold you accountable."

The value that anthropologists place on the traditional in "primitive" culture and on salvage ethnography is reflected in studies of the Yanomama. The first edition of Chagnon's ethnography viewed the Yanomama as significant because, as one of the largest unacculturated tribes remaining in all of South America, they were characterized by cultural purity and active warfare (Chagnon 1968a, 1). It was only in the second edition of the ethnography that Chagnon (1977, 138-64) addressed the question of "the beginning of Western acculturation." Yet the closing sentence of the book refers to the Yanomama as "our contemporary ancestors."[24] Other researchers working among the Yanomama have been similarly motivated.

This discussion has dwelled on the orientation of cultural evolutionism, the anthropological focus on primitiveness, and salvage ethnography for one reason: they have shaped the anthropological record of Amazonia. These orientations and their associated values in anthropology resulted in selectivity in the descriptive ethnography of the Yanomama. The accounts actually ignore some of the most critical facets of Yanomama culture, ecology, and daily life, distorting the record accordingly.

There is a curious paradox in the subjects of Yanomama warfare and the problem of limiting ecological factors. While in the halls of academe anthropologists like Chagnon and Harris are battling over the explanation of chronic internal warfare among the Yanomama, the very existence of this society is threatened by another war, the continuing global struggle between the so-called primitive and civilized worlds.[25] The Yanomama have apparently adapted fairly successfully to their tropical rain forest ecosystem for over two millenia (Spielman, et al. 1974), but their culture may be unable to cope with a new environmental hazard—Western civilization. Although warfare accounts for 24 percent of adult male deaths, epidemic diseases result in 54 percent of all adult deaths (Chagnon 1977, 20). Davis (1977a, 19) suggests that instead of the causes of internal warfare among the Yanomama, other topics are of greater immediate relevance, given the threat of extinction these people face. For example, he lists the cultural, political, and psychological effects of the encroachment of outsiders such as missionaries into

Yanomama territory; the repercussions on Yanomama culture and ecology of introduced Western technology (e.g. steel axes, machetes, shotguns); and disease and epidemics spread through contact.[26] As Gross (1982, 7) notes, indigenous cultures are adaptations to recent conditions and will continue to adapt as conditions change (of course, this does not apply when a society is faced with genocide or ethnocide). From this perspective, some traditional orientations in anthropology might well be revised. Thus, Schefold (1982, 13) writes:

> All cultures are continuously changing; the practical task of Urgent Anthropology [the modern descendant of salvage ethnography] must be to work against violent, forced upheavals and for conditions which allow an active and conscious development within the framework of one's own cultural identity.[27]

Here is the relevance of an analysis of values in anthropological research for understanding the cultural ecology of Amazonia. Values, which are largely implicit and unexamined, bias the design of research; they influence the collection, interpretation, and reporting of data; and, accordingly, they affect our understanding of cultures and their ecology. Traditional research on cultural adaptation is largely irrelevant to the present and future survival, adaptation, and welfare of the people studied!

A third context in which values enter the cultural ecology of Amazonia is that of economic development. For centuries, the world view of Western civilization has been put to action in the exploitation of Amazonia.[28] Only in recent decades, however, has this process become sufficiently extensive and intensive that the whole of Amazonia, including indigenes and their ecosystem, is endangered. This applies to Brazil more than any other country. In the 1960s the Brazilian government initiated programs that amounted to a technological and economic war to conquer the Amazon. Instead of the spontaneous colonization and exploitation that persisted on a relatively small scale for centuries, the government launched systematic efforts on a massive scale. There were several fronts to this campaign: the acceleration of the extraction of resources such as minerals and timber; agricultural development such as colonization schemes along new highways to resettle peasants from the poverty- and drought-stricken northeast; agribusiness developments including ranching; highway construction criss-crossing Amazonia to make the region accessible and integrate it; and, recently, the initiation of the construction of vast hydroelectric dams. Thus, the so-called civilized and primitive worlds confront one another in Amazonia. The process involves not only markedly different cultures, technologies, and economies, but, more fundamentally, divergent if not antithetical values regarding human relations within society and with the environment.[29]

Some typical statements from Brazil illustrate these points. In 1975, General Fernando Ramos Pereira, then governor of Roraima, the northeastern province that includes Yanomama territory, told the press, "I am of the opinion that an area as rich as this—with gold, diamonds, and uranium—cannot afford the luxury of conserving a half dozen Indian tribes who are holding back the development of Brazil" (Quoted in Davis 1977b, 103). More recently, General Democrito de Oliveira, who is in charge of the Brazilian agency for indigenous affairs, stated that a large reserve for the Yanomama would take up too much valuable land, and that they are "physically and possibly intellectually decadent" as a result of "incestuous" practices. He thus implied that the Yanomama would not be worth saving (Maurer 1979, 33). Finally, Nugent (1978, 102) notes:

> For many, especially in the south of Brazil, the Amazon is a reminder of what is culturally devalued in modern Brazil: the backwoods *mestico*, industrial stagnation, Indians, and the apparent domination of nature over man. Unfortunately, for some Brazilians, success in destroying the Amazon is a straightforward measure of progress.

What is progress and development? In contrast to the usual positive view, Appell (1975, 31) offers a critical definition: "Every act of development involves, of necessity, an act of destruction. This destruction—social, ecological, or both—is seldom accounted for in development projects, despite the fact that it may entail costs that far outweigh the benefits arising from the development." Appell (1975, 33) goes on to observe that, "A development act is any act by an individual who is not a member of a local society that devalues or displaces the perception by the members of that society of their relationship with their natural and social world."

Development is closely related to what J. W. Bennett (1976, 13) calls the ecological transition from equilibrium to disequilibrium societies. Most traditional indigenous cultures in Amazonia were equilibrium societies and many remain so, depending on the extent of their acculturation. These two types of societies are contrasted in table 4.1. While this dichotomy is rather simple and has the limitations of any generalization, it also has considerable heuristic utility.[30]

The key to the maintenance of an equilibrium society is most aptly stated by Kozlovsky (1974, 106): "Live as simply and as naturally and as close to the earth as possible, inhibiting only two aspects of your unlimited self: your capacity to reproduce and your desire for material things." Although Kozlovsky is an ecologist, he arrives at a conclusion that many anthropologists draw from cultural ecology. His statement is a description of the traditional indigenous cultures that still survive in some areas, including Amazonia. Indigenous societies in Amazonia have had millenia to experiment with the development of cultures that are ecologically viable in their ecosystem. The experience and success of Western civilization in the Amazonian ecosystem are both extremely limited so far. Thus, although a nation's right to develop and many of its explicit goals of development may not be questionable, it is less clear whether Western civilization has the technological and scientific expertise as well as the values required to develop Amazonia without destroying it.

An increasing number of writers from various fields are embracing this perspective. LaBastille and McIntyre (1979, 101) write:

> The bottom line is that no matter what development takes place, it should be done by keeping the Amazon in its natural state. This may mean abandoning super-technological types of ranching and farming and sticking with the age-old native systems. This will seem to some like cultural back-stepping. Yet the ultimate goal is not to make millions on quick tradeoffs like pepper, but to achieve environmental health and the self-sustainment of human life in Amazonia.

Table 4.1.   The Ecological Transition

Component	Equilibrium	Disequilibrium
Population	Small, controlled	Large, expanding, weakly controlled
Needs and wants	Minimized, limited mostly to satisfaction of physiological requirements	Maximized to include culturally defined and promoted needs and wants which are expanding
Resource network	Self-sufficient exploitation of local environment	Depending on importation from distant environments through an extensive exchange system
Technological capacity	Limited, based largely on human energy	Highly developed, industrial, based on multiple sources of energy including fossil fuels
Environmental contact on a daily basis	Large percentage of the individuals in society	Small percentage
Environmental impact	Relatively negligible for the most part	High, including resource depletion, species extinction, environmental transformation, degradation, pollution

Source: Modified after Bennett (1976, 13)

Ecologists Goodland and Irwin (1975, 65) note:

> Amerindians are the only societies with the necessary
> knowledge, expertise and tradition to prosper in the
> Amazon jungle. Amerindians not only profoundly
> appreciate what exists, but also understand ecological
> interrelations of the various components of the
> Amazonian ecosystem better than do modern
> ecologists. Indians perceive specific relationships
> which biologists are only now discovering to be
> accurate. And since the Amazon jungle is the most
> complex, richest and the least understood ecosystem
> in the world, the Amerindians' knowledge of it is of
> inestimable value.

Finally, given the almost antithetical nature of equilibrium and
disequilibrium societies, and the ecological rationality of the former,
it must be asked what leads indigenes to accept elements from
Western civilization so readily? Allen Johnson (1978, 57) offers part
of the answer, in stating that, "It is characteristic of a developed
culture's contact with small, isolated societies that the developed
culture is not met as a whole, but rather in highly selective ways
that emphasize manufactured goods and the aura of the great,
mysterious power that made them." The advantages of a steel axe
over one made of stone, a shotgun over a bow and arrow, or a boat
motor over a canoe paddle are clear in most circumstances. What is
not so clear is all the ramifications of these advantages.
Technological advances require equipment and supplies for operation
such as gasoline, oil, and spark plugs. To obtain these things, the
indigene sacrifices autonomy and works to accumulate trade items or
cash. In this way the ecological transition may be entered.
Individuals react to civilization rapidly and flexibly in favor of
short-term benefits for themselves, their family, or their community.
In contrast, the indigenous cultural system as a whole, which has
evolved over centuries or even millenia, is much slower and less
flexible in response. The system may actually become maladaptive
as it is faced with such changes.[31]
Cultural ecology has at least two contributions to make toward
the improvement of the situation in Amazonia. It can document the
ethnoecology and cultural ecology of indigenous societies not only for

the scientific record but also (1) to demonstrate from a relativistic perspective the adaptive nature of these cultures and their knowledge in the ecosystem, and (2) to evaluate and criticize Western civilization and the monolithic model of economic development that assumes universal applicability regardless of cultural and ecological variations.

Summary and Conclusions

The study of the cultural ecology of Amazonia developed around the concepts of carrying capacity, limiting factors, cultural complexity, and cultural evolutionism. Since the mid-1970s, the hypothesis that animal protein limits the size, distribution, and permanence of indigenous populations in particular has dominated much of the research. This in turn has provided the foundation for various hypotheses to explain phenomena such as warfare, faunal taboos, and intravillage sexual politics. Regardless of their validity, these hypotheses generated advances in theory, method, and data, especially for the previously neglected subject of indigenous hunting and fishing.

Explanations of Yanomama warfare reveal the fundamental split between cultural materialism and mentalism in anthropology. Cultural ecology is not isomorphic with either position. In contrast to these positions, cultural ecology has the potential to develop a holistic framework recognizing that *Homo sapiens* adapts simultaneously as an organic, cultural, and intellectual being.

Values may be considered in the cultural ecology of Amazonia in at least three contexts: indigenous adaptation, anthropological research, and economic development. As discussed earlier, the "waitiri complex," faunal taboos, and the Desana world view provide three illustrations of the role of cultural values in indigenous adaptations. Values might be the crucial link between indigenous ideas and actions and systemic processes that has been missing in cultural materialism, mentalism, and cultural ecology.

The values of anthropology include adherence to cultural evolutionism, the notion of the primitive, and salvage ethnography. These have shaped research and the ethnographic record, sometimes actually distorting cultural and ecological reality through biasing the researcher against considering changes resulting from contact, colonization, and development. Contemporary change deserves much more attention by cultural ecologists, since adaptation is an ongoing process and change threatens the survival and welfare of indigenes. Cultural ecology may help indigenes by asserting the value of their culture and knowledge through demonstrating their adaptive functions in the ecosystems of Amazonia. Cultural ecology may also contribute by applying concepts such as the ecological transition to critically evaluate Western approaches to economic development in Amazonia.

Acknowledgments

This paper is a synthesis of ideas and information developed through interaction with faculty and students at Cornell University, University of Massachusetts (Amherst), Instituto Venezolano de Investigaciones Cientificas (IVIC), and University of Hawaii. I am particularly indebted to Thomas Gregor (Vanderbilt University) and Nelly Arvelo-Jimenez (IVIC) for their encouragement over the years, and to the Yanomama and other indigenes with whom I had the pleasure to work in Amazonia. Of course only I am responsible for any deficiencies in this chapter.

Notes

1. While earlier evidence is very limited, a few sites beyond Amazonia are dated about 20,000 years ago (Meggers 1982, 485).

2. While this description provides some insight into the character of Amazonian culture, it is offered with caution in recognition of the problems and limitations of such generalizations and trait lists. It is synthesized from Lathrap (1970, 45–67), Steward and Faron (1959, 8–9), Willey (1971, 398–99), and Zerries (1969, 376–82).

3. Newman (1976) is among those who suggest that actual population size was about ten times higher than usually estimated, because European diseases often reached and decimated indigenous communities years in advance of direct contact when census records might be made.

4. Neel (1982) argues that epidemiological and psychosocial factors are more important than genetic ones in accounting for the great susceptibility of Amerindians to diseases

introduced from the Old World.

5.   The National Research Council (1980) discusses research priorities for tropical biology.

6.   Moran (1981, 219–21; 1982, 7) argues that large-scale mapping has obscured variation, screened areas of richer soils, and resulted in a gross underestimation of soil fertility in Amazonia.

7.   Since Buffon, the fauna of the New World in general and especially its tropics has been viewed as impoverished (Gerbi 1973).

8.   The literature on the cultural ecology of Amazonia is immense, with more than one thousand citations. This discussion will be restricted to selected highlights for each decade since 1950, with an emphasis on limiting factors, especially the availability of animal protein and allied matters. Criticisms of studies will be referred to only by citations in notes. Other important aspects not reviewed here include agroecosystems (Moran 1982; Norgaard 1981); and caboclos, colonists, and development (Moran 1981; Nugent 1981; Smith 1982). A volume edited by Hames and Vickers (1983) is also of interest. Fairly comprehensive reviews of the cultural ecology of Amazonia are provided by Roosevelt (1980) and Sponsel (1981).

9.   Brush (1975), Glassow (1978), and Hayden (1975) criticized the anthropological use of these ecological concepts. These evolutionary concepts have been reviewed and criticized by Carneiro (1973); Diener, Nonini, and Robkin (1980); Seagraves (1974); Wagner (1977); and Wylie (1971). Williams's (1898) early statements remain perceptive.

10.   For example, see Cowgill (1975a, 1975b).

11.   For criticisms of these matters, see Ferdon (1959), Hirschberg and Hirschberg (1957), and Torres-Trueba (1968).

12.   Beckerman (1979), Chagnon and Hames (1979), Hames (1980), Lizot (1977), and Yost and Kelly (1983) are among the critics of the animal protein hypothesis. While animal protein has not been perceived as an issue in other ecosystems of this biome, it is discussed for the Peruvian Andes (Bolton 1979), Polynesia (Beckerman 1977), and New Guinea (Dwyer 1974). Dickinson (1973) sees protein as a

problem throughout Latin America.

13. For convenience the term "Yanomama" is used here, following Migliazza's (1972, 26) suggestion, since it refers to all variants of this indigenous group.

14. Lindholm and Lindholm (1982) is another example of exaggeration to the extreme (Sponsel 1982, 1983a). Others have questioned the fierce image of the Yanomama (e.g., Davis 1977a; Smole 1976, 14–15).

15. Figure 4.1 diagrams what I believe to be a fair representation of Chagnon's model.

16. Figure 4.2 diagrams what I believe to be a fair representation of Harris's model. Critics of Harris include Chagnon (1983, 81–89); Chagnon and Hames (1979, 1980); Diener and Avery (1979); Diener, Moore, and Mutaw (1980); Hames (1979); Hara (1981); Price (1982); Sponsel (1981, 326–49; 1983a), and Yost and Kelley (1983).

17. Nietschmann (1972, 1973) and Reichel-Dolmatoff (1971, 1976) are clearly within the realm of cultural ecology but not cultural materialism.

18. A comparison of the literature before and after the landmark studies of Gross (1975), E. B. Ross (1978), and others clearly demonstrates this progress. It is noteworthy that Murdock (1968, 19), in his global survey of hunter-gatherers, said very little about Amazonia. Much more could be said today.

19. McDonald (1977) developed a similar but less elaborate explanation for faunal taboos of indigenes in South America. Among the critics of these explanations are Eichinger Ferro-Luzzi (1978) and Lizot (1979).

20. A summary of this book can be found in Reichel-Dolmatoff (1976).

21. Jochim (1981, 21–31), Moran (1979, 17–20, 97–101), and Orlove (1980, 245–61) discuss this problem.

22. I do not completely agree with Hatch. A functionalist or ecological explanation of a cultural phenomenon does not necessarily imply approval of the phenomenon. Nevertheless, the issues raised are important (see Bagish 1983).

23. Writers who have played on the cultural shock value of "the fierce people" beyond Chagnon, include Biocca (1970), Harris

(1974, 1977), and Lindholm and Lindholm (1982). This practice continues despite Chagnon's (1977, 162–64) attempt to balance the image of fierceness.

24. Actually Chagnon (1968a, 4) was first introduced to the Yanomama by a missionary who had worked with them for nearly fifteen years. Moreover, the Yanomama have experienced sporadic contact with Western civilization to varying degrees since at least 1758 (Smole 1976, 15).

25. See Bodley (1982), Davis (1977a), Davis and Wright (1981), Kellman (1982), Lizot (1976), and Ramos and Taylor (1979) for descriptions of the plight of the Yanomama.

26. It is noteworthy that in the latest edition of his ethnography, Chagnon (1983) does not bother to update the last chapter on acculturation even though there were many serious developments in the six years since the last edition and these are well documented.

27. Whitten and Whitten (1977) is an excellent example of such a new, urgent anthropology.

28. See Bodley (1982), Crosby (1972), Hemming (1978), and Parry (1979).

29. Discussions of various aspects of this process in contemporary Brazil may be found in Bourne (1978), Caufield (1983), Davis (1977a, 1977b), Goodland and Irwin (1975), Gross (1982), LaBastille and McIntyre (1979), Maurer (1979), Moran (1981), Ramos and Taylor (1979), Smith (1982), and Wagley (1974).

30. Similar schemes are primitive/consumer (Bodley 1976), paleotechnic/neotechnic (Clarke 1977), and ecosystem/biosphere (Dasmann 1976, 283). Also see Wilkinson (1973) for a discussion of the equilibrium and disequilibrium concepts in the context of economic development. There is some debate as to whether traditional indigenous cultures are equilibrium societies (Goldsmith 1973; Guthrie 1971; Heizer 1955; B. Johnson 1977).

31. The introduction of Western technology into the hunting and fishing activities of the Yanomama illustrates some of these points (Hames 1979; Saffirio and Scaglion 1982).

References

Agassiz, L.
1896   *A Journey in Brazil.* New York: Houghton Mifflin.
Anderson, J. N.
1973   Ecological anthropology and anthropological ecology. In
        Handbook of Social and Cultural Anthropology, edited by
        J. J. Honigmann. Chicago: Rand McNally. Pp. 179–239.
Appell, G. N.
1975   The pernicious effects of development. *Fields Within
        Fields* 14:31–45.
Ashburn, P. M.
1947   *The Ranks of Death: A Medical History of the Conquest of
        America.* Toronto: Coward-McCann.
Bagish, H. H.
1983   Confessions of a former cultural relativist. In
        Anthropology 83/84, edited by E. Angeloni. Guilford:
        Dushkin Publishing Group. Pp. 22–29.
Barnes, H. E.
1923   The natural state of man: an historical resume. *The
        Monist* 33:33–80.
Basso, E. B.
1973   *The Kalapalo Indians of Central Brazil.* New York: Holt,
        Rinehart and Winston.
Beckerman, S.
1977   Protein and population in tropical Polynesia. *Journal of
        the Polynesian Society* 86:73–79.
1979   The abundance of protein in Amazonia: a reply to Gross.
        American Anthropologist 81:533–60.
Bennett, J. W.
1976   *The Ecological Transition: Cultural Anthropology and
        Human Adaptation.* New York: Pergamon.
Billings, W. D.
1970   *Plants, Man, and the Ecosystem.* Belmont: Wadsworth.
Biocca, E.
1970   *Yanomama: The Narrative of a White Girl Kidnapped by
        Amazonian Indians.* New York: E. P. Dutton.
Bodley, J.
1976   *Anthropology and Contemporary Human Problems.* Menlo
        Park: Cummings.

1982   *Victims of Progress.* Menlo Park: Cummings.

Bolton, R.
1979   Guinea pigs, protein, and ritual. *Ethnology* 18:229-52.

Bourlière, F.
1973   The comparative ecology of rain forest mammals in Africa and tropical America: some introductory remarks. In *Tropical Forest Ecosystems in Africa and South America,* edited by B. J. Meggers et al. Washington, D.C.: Smithsonian Press. Pp. 279-92.

Bourne, R.
1978   *Assault on the Amazon.* London: Gollancz.

Braroe, N. W., and G. L. Hicks
1967   Observations on the mystique of anthropology. *The Sociological Quarterly* 8:173-85.

Brush, S. B.
1975   The concept of carrying capacity for systems of shifting cultivation. *American Anthropologist* 77:799-811.

Carneiro, R. L.
1956   Slash-and-burn agriculture: a closer look at its implications for settlement patterns. In *Men and Cultures: Selected Papers of the Fifth International Congress of Anthropological and Ethnological Sciences,* edited by A. F. C. Wallace. Philadelphia: University of Pennsylvania Press. Pp. 229-34.

1961   Slash and burn cultivation among the Kuikuru and its implications for cultural development in the Amazon Basin. In *The Evolution of Horticultural Systems in Native South America,* edited by J. Wilbert. Caracas: Fundacion La Salle de Ciencias Naturales. Pp. 47-67.

1973   The four faces of evolution. In *Handbook of Social and Cultural Anthropology,* edited by J. J. Honigmann. Chicago: Rand McNally. Pp. 89-110.

Caufield, C.
1983   Dam the Amazon, full steam ahead. *Natural History* 92(7):60-67.

Chagnon, N.A.
1968a   *Yanomamö: The Fierce People.* New York: Holt, Rinehart and Winston.

1968b   Yanomamö social organization and warfare. In *War: The Anthropology of Armed Conflict and Aggression*, edited by M. Fried. Garden City: Natural History Press. Pp. 109–59.

1977   *Yanomamö: The Fierce People*. New York: Holt, Rinehart and Winston (second edition).

1983   *Yanomamö: The Fierce People*. New York: Holt, Rinehart and Winston (third edition).

Chagnon, N. A., and R. B. Hames

1979   Protein deficiency and tribal warfare in Amazonia: new data. *Science* 203:910–13.

1980   The "protein hypothesis" and native adaptations to the Amazon Basin: a critical review of data and theory. *Interciencia* 5:346–58 (in Spanish).

Chagnon, N. A., and W. Irons (editors)

1979   *Evolutionary Biology and Human Social Behavior: An Anthropological Perspective*. North Scituate: Duxberry Press.

Clarke, W. C.

1977   The structure of permanence: the relevance of self-subsistence communities for world ecosystem management. In *Subsistence and Survival: Rural Ecology in the Pacific*, edited by T. Bayliss-Smith and R. Feachem. New York: Academic Press. Pp. 363–84.

Cowgill, G. L.

1975a   Population pressure as a non-explanation. *American Antiquity* 40(2, Pt. 2):127–31.

1975b   On causes and consequences of ancient and modern population changes. *American Anthropologist* 77:505–25.

Crist, R. E.

1982   Amazon River. *Encyclopaedia Britannica, Macropaedia* 1:652–56.

Crosby, A. W.

1972   *The Columbian Exchange: Biological and Cultural Consequences of 1492*. Westport: Greenwood Press.

Dasmann, R. E.

1976   Life styles and nature conservation. *Oryx* 13:281–86.

Davis, S. H.

1977a   The Yanomamö—ethnographic images and anthropological responsibilities. In *The Geological Imperative:*

Anthropology and Development in the Amazon Basin of South America, edited by S. H. Davis and R. O. Mathews. Irvine: University of California, Program in Comparative Culture, Occasional Papers No. 5. Pp. 6–21.

1977b *Victims of the Miracle: Development and the Indians of Brazil.* New York: Cambridge University Press.

Davis, S. H., and R. Wright
1981 *The Yanomami Indian Park: A Call for Action.* Boston: Anthropology Resource Center.

Denevan, W. M.
1976 The aboriginal population in Amazonia. In *The Native Population of the Americas in 1492,* edited by W. M. Denevan. Madison: University of Wisconsin Press. Pp. 205–34.

Diamond, S.
1974 *In Search of the Primitive: A Critique of Civilization.* New York: E. P. Dutton.

Dickinson, J. C.
1973 Protein flight from tropical Latin America: some social and ecological considerations. In *Latin American Development Issues,* edited by A. D. Hill. Syracuse: Proceedings of the Conference of Latin Americanist Geographers. Pp. 127–32.

Diener, P., and G. Avery
1979 Protein, nutritional hazard and cultural evolution: some theoretical comments. *Lambda Alpha: Journal of Man* 11(2):13–48.

Diener, P., K. Moore, and R. Mutaw
1980 Meat, markets and mechanical materialism: the great protein fiasco in anthropology. *Dialectical Anthropology* 5:171–92.

Diener, P., D. Nonini, and E. E. Robkin
1980 Ecology and evolution in cultural anthropology. *Man* 15:1–31.

Dostal, W. (editor)
1972 *The Situation of the Indians of South America.* Geneva: World Council of Churches.

Dwyer, P. D.
1974 The price of protein: five hundred hours of hunting in the New Guinea Highlands. *Oceania* 44:278–93.

Eichinger Ferro-Luzzi, G.
  1978   Remarks on D. R. McDonald's "Food Taboos." *Anthropos*
          73:593–94.
Farnworth, E. G., and F. B. Golley
  1974   *Fragile Eco-Systems: Evaluation of Research and
          Applications in the Neotropics.* New York:
          Springer-Verlag.
Ferdon, E. N.
  1959   Agricultural potential and the development of cultures.
          Southwestern Journal of Anthropology 15:1–19.
Fischer, A.
  1969   The personality and subculture of anthropologists and
          their study of U.S. Negroes. In *Concepts and Assumptions
          in Contemporary Anthropology,* edited by S. A. Tyler.
          Athens: University of Georgia Press. Pp. 12–17.
Fittkau, E. J., and H. Klinge
  1973   On biomass and trophic structure of the Central
          Amazonian rain forest ecosystem. *Biotropica* 5:2–14.
Gerbi, A.
  1973   *The Dispute of the New World: The History of a Polemic,
          1750–1900.* Pittsburgh: University of Pittsburgh Press.
Glassow, M. A.
  1978   The concept of carrying capacity in the study of cultural
          process. In *Advances in Archaeological Method and
          Theory,* edited by M. B. Schiffer. New York: Academic
          Press. Vol. I:32–48. Goldsmith, E.
  1973   Adam and Eve revisited? *The Ecologist* 3:348–55.
Goodland, R. J. A., and H. S. Irwin
  1975   *Amazon Jungle: Green Hell to Red Desert? An Ecological
          Discussion of the Environmental Impact of the Highway
          Construction Program in the Amazon Basin.* Amsterdam:
          Elsevier.
Goodman, E. J.
  1972   *The Explorers of South America.* New York: Macmillan.
Goulding, M.
  1980   *The Fishes and the Forest: Explorations in Amazonian
          Natural History.* Berkeley: University of California Press.
Grainger, A.
  1980   The state of the world's tropical forests. *The Ecologist*
          10:6–54.

Gross, D. R.
1975   Protein capture and cultural development in the Amazon Basin. *American Anthropologist* 77:526–49.
1982   Indians and the Brazilian frontier. *Journal of International Affairs* 36:1–14.
Gruber, J. W.
1970   Ethnographic salvage and the shaping of anthropology. *American Anthropologist* 72:1289–1299.
Guthrie, D. A.
1971   Primitive man's relationship to nature. *BioScience* 21:721–23.
Hames, R. B.
1979   A comparison of the efficiencies of the shotgun and the bow in neotropical forest hunting. *Human Ecology* 7:219–52.
Hames, R. B. (editor)
1980   *Studies in Hunting and Fishing in the Neotropics.* Bennington: Working Papers on South American Indians No. 2.
Hames, R. B., and W. T. Vickers (editors)
1983   *Adaptive Responses of Native Amazonians.* New York: Academic Press.
Hanc, J. R.
1981   Influences, Events, and Innovations in the Anthropology of Julian H. Steward: A Revisionist View of Multilinear Evolution. Master's Thesis, University of Chicago.
Hara, T.
1981   Yanomama: an image expected. *Japanese Journal of Ethnology* 45:360–371. (In Japanese).
Harris, M.
1974   The savage male. In *Cows, Pigs, Wars and Witches: The Riddles of Culture,* by M. Harris. New York: Random House. Pp. 83–107.
1977   Proteins and the fierce people. In *Cannibals and Kings: The Origins of Culture,* by M. Harris. New York: Random House. Pp. 45–54.
1979a   Varieties of pre-state village societies. In *Cultural Materialism: The Struggle for a Science of Culture,* by M. Harris. New York: Random House. Pp. 89–92.

1979b The Yanomamö and the causes of war in band and village societies. In *Brazil: Anthropological Perspectives: Essays in Honor of Charles Wagley,* edited by M.L. Margolis and W. E. Carter. New York: Columbia University Press. Pp. 121–32.

1979c The human strategy, our pound of flesh. *Natural History* 88(7):30–41.

Hatch, E.
1983 *Culture and Morality.* New York: Columbia University Press.

Hayden, B.
1975 The carrying capacity dilemma. *American Antiquity* 40(2, Pt. 2):11–21.

Hecht, S. B.
1980 Deforestation in the Amazon Basin: magnitude, dynamics, and soil resource effects. Where have all the flowers gone? In *Deforestation in the Third World,* edited by V. H. Sutlive et al. Studies in Third World Societies No. 13. pp. 61–108.

Heizer, R. F.
1955 Primitive man as an ecological factor. *Kroeber Anthropological Society Papers* 13:1–31.

Hemming, J.
1978 *Red Gold: The Conquest of the Brazilian Indians.* Cambridge: Harvard University Press.

Herrera, R., et al.
1978 Amazon ecosystems: their structure and functioning with particular emphasis on nutrients. *Interciencia* 3:223–31.

Hershkovitz, P.
1972 The recent mammals of the neotropical region: a zoogeographic and ecological review. In *Evolution, Mammals, and Southern Continents,* edited by A. Keast et al. Albany: State University of New York Press. Pp. 311–431.

Hirschberg, R. I., and J. F. Hirschberg
1957 Meggers' Law of Environmental Limitation of Culture. *American Anthropologist* 59:890–91.

Honigmann, J. J.
1959 *The World of Man.* New York: Harper & Row.

Janzen, D. H.
    1974   Tropical blackwater rivers, animals, and mast fruiting by the Dipterocarpaceae. *Biotropica* 6:69–103.
    1975   *Ecology of Plants in the Tropics*. London: Edward Arnold.
Jochim, M. A.
    1981   *Strategies for Survival: Cultural Behavior in an Ecological Context*. New York: Academic Press.
Johnson, A.
    1978   In search of the affluent society. *Human Nature* 1(9):50–59.
Johnson, B.
    1977   The primitive barrier to political ecology. *The Ecologist* 7:88–93.
Jordan, C. F.
    1982   Amazon rain forests. *American Scientist* 70:394–401.
Jordan, C. F., and R. Herrera
    1981   Tropical rain forests: are nutrients really critical? *American Naturalist* 117:167–80.
Kellman, S.
    1982   The Yanomamis: their battle for survival. *Journal of International Affairs* 36:15–42.
Kensinger, K. J., and W. H. Kracke (editors)
    1981   *Food Taboos in Lowland South America*. Bennington: Working Papers on South American Indians No. 3.
Kozlovsky, D. G.
    1974   *An Ecological and Evolutionary Ethic*. Englewood Cliffs, NJ: Prentice-Hall.
LaBastille, A., and L. McIntyre
    1979   Heaven, not hell. *Audubon* 81:68–103.
Lathrap, D. W.
    1970   *The Upper Amazon*. New York: Praeger.
Lieth, H.
    1975   Primary productivity of the major vegetation units of the world. In *Primary Productivity of the Biosphere*, edited by H. Lieth and R. H. Whittaker. New York: Springer-Verlag. Pp. 203–15.
Lindholm, C., and C. Lindholm
    1982   Avengers of the bloody moon. *Science Digest* 90(1):78–83, 106.

Lizot, J.
  1976   *The Yanomami in the Face of Ethnocide.* Copenhagen:
          International Work Group for Indigenous Affairs
          Document No. 22.
  1977   Population, resources and warfare among the Yanomami.
          Man 12:497–517.
  1979   On food taboos and Amazon cultural ecology. *Current
          Anthropology* 20:150–55.
Longman, K. A., and J. Jenik
  1974   *Tropical Forest and Its Environment.* London: Longman.
Lynch, T. F.
  1978   The South American paleoindians. In *Ancient Native
          Americans,* edited by J. D. Jennings. San Francisco:
          Freeman. Pp. 455–89.
Maurer, H.
  1979   The Amazon: development or destruction? *Nacla* (The
          North American Congress of Latin Americanists)
          8(3):26–37.
McDonald, D. R.
  1977   Food taboos: a primitive environmental protection agency
          (South America). *Anthropos* 72:734–48.
Meggers, B. J.
  1954   Environmental limitations on the development of culture.
          American Anthropologist 56:801–24.
  1957   Environment and culture in the Amazon Basin: an
          appraisal of the theory of environmental determinism. In
          Studies in Human Ecology, edited by A. Palerm.
          Washington, D.C.: Pan American Union.
  1971   *Amazonia: Man and Culture in a Counterfeit Paradise.*
          Chicago: Aldine-Atherton.
  1973   Some problems in cultural adaptation in Amazonia, with
          emphasis on the pre-European period. In *Tropical Forest
          Ecosystems in Africa and South America: A Comparative
          Review,* edited by B. J. Meggers et al. Washington, D.C.:
          Smithsonian Institution Press. Pp. 311–20.
  1979   *Prehistoric America: An Ecological Perspective.* Chicago:
          Aldine-Atherton.
  1982   Archeological and ethnographic evidence compatible with
          the model of forest fragmentation. In *Biological
          Diversification in the Tropics,* edited by G. T. Prance.

New York: Columbia University Press. Pp. 483–96.

Migliazza, E. C.
1972   Yanomama   Grammar   and   Intelligibility.   Ph.D.
dissertation, Indiana University.

Moran, E. F.
1979   *Human   Adaptability: An   Introduction   to   Ecological
Anthropology.* Boulder: Westview Press.
1981   *Developing the Amazon.* Bloomington: Indiana University
Press.
1982   Ecological, anthropological, and agronomic research in the
Amazon   Basin.   *Latin   American   Research   Review*
17(1):3–42.

Murdock, G. P.
1968   The current status of the world's hunting and gathering
peoples. In *Man the Hunter,* edited by R. B. Lee and I.
DeVore. Chicago: Aldine. Pp. 13–22.

Myers, N.
1979   *The   Sinking   Ark: A   New   Look   at   the   Problem   of
Disappearing Species.* Elmsford, NY: Pergamon Press.

Myers, N., and S. Ayensu
1983   Reduction of biological diversity and species loss. *Ambio*
12(2):72–74.

National Research Council
1980   *Research Priorities in Tropical Biology.* Washington, DC:
National Academy of Sciences.
1982   *Ecological Aspects of Development in the Humid Tropics.*
Washington, D.C.: National Academy of Sciences.

Neel, J. V.
1982   Infectious   disease   among   Amerindians.   *Medical
Anthropology* 6:47–55.

Newman, M. T.
1976   Aboriginal New World epidemiology and medical care,
and the impact of Old World disease imports. *American
Journal of Physical Anthropology* 45(3, Pt. II):667–72.

Nietschmann, B.
1972   Hunting and fishing focus among the Miskito Indians,
eastern Nicaragua. *Human Ecology* 1:41–68.
1973   *Between Land and Water: The Subsistence Ecology of the
Miskito Indians, Eastern Nicaragua.* New York: Seminar
Press.

Norgaard, R. B.
   1981   Sociosystem and ecosystem coevolution in the Amazon. *Journal of Environment, Economics, and Management.* 8:238–54.
Nugent, S.
   1978   Explaining Amazonia: review article. *Critique of Anthropology* 3:100–102.
   1981   Amazonia: ecosystem and social system. *Man* 16:62–74.
Odum, E. P.
   1971   *Fundamentals of Ecology.* Philadelphia: W. B. Saunders.
Orlove, B. S.
   1980   Ecological anthropology. *Annual Review of Anthropology* 9:235–73.
Parry, J. H.
   1979   *The Discovery of South America.* New York: Taplinger.
Pepper, S. C.
   1958   *The Sources of Value.* Berkeley: University of California Press.
Price, B. J.
   1982   Cultural materialism: a theoretical review. *American Antiquity* 47:709–41.
Ramos, A., and K. I. Taylor
   1979   *The Yanomama in Brazil in 1979.* London: Survival International.
Reichel-Dolmatoff, G.
   1971   *Amazonian Cosmos: The Sexual and Religious Symbolism of the Tukano Indians.* Chicago: University of Chicago Press.
   1976   Cosmology as ecological analysis: a view from the rainforest. *Man* 11:307–18.
Ribeiro, D.
   1967   Indigenous cultures and languages of Brazil. In *Indians of Brazil in the Twentieth Century*, edited by J. H. Hopper. Washington, D.C.: Institute for Cross-Cultural Research. Pp. 77–166.
Richards, P. W.
   1973   The tropical rain forest. *Scientific American* 229(6):58–67.
Roosevelt, A. C.
   1980   *Parmana: Prehistoric Maize and Manioc Subsistence along the Amazon and Orinoco.* New York: Academic Press.

Ross, E. B.
1978   Food taboos, diet, and hunting strategy: the adaptation to animals in Amazon cultural ecology. *Current Anthropology* 19:1–36.
1979   On food taboos, and Amazon cultural ecology. *Current Anthropology* 20:151–55.
1980   More on Amazon cultural ecology: reply. *Current Anthropology* 21:544–46.
Ross, E. B., and J. B. Ross
1980   Amazon warfare. *Science* 207:592.
Ross, J.B.
1980   Ecology and the problem of tribe: a critique of the Hobbesian model of preindustrial warfare. In *Beyond the Myths of Culture: Essays in Cultural Materialism*, edited by E. B. Ross. New York: Academic Press. Pp. 37–60.
Saffirio, G., and R. Scaglion
1982   Hunting efficiency in acculturated and unacculturated Yanomama. *Journal of Anthropological Research* 38:315–27.
Sauer, C. O.
1958   Man in the ecology of tropical America. *Proceedings of the Ninth Pacific Science Congress of 1957* 20:104–10.
Schefold, R.
1982   Urgent anthropology. *Survival International Review* 7(1):13–14.
Schwabe, G. H.
1969   Towards an ecological characterization of the South American continent. In *Biogeography and Ecology in South America*, edited by E. J. Fittkau et al. The Hague: Dr. W. Junk, N. V. Pp. 113–36.
Scott, G. A. J.
1977   The role of fire in the creation and maintenance of savanna in the Montana of Peru. *Journal of Biogeography* 4:143–67.
Seagraves, B. A.
1974   Ecological generalization and structural transformation of sociocultural systems. *American Anthropologist* 76:530–52.

Sioli, H.
 1975 Tropical river: the Amazon. In *River Ecology*, edited by B.
 A. Whitton. Berkeley: University of California Press. Pp.
 461-88.
Siskind, J.
 1973a Tropical forest hunters and the economy of sex. In
 Peoples and Cultures of Native South America, edited by D.
 R. Gross. New York: Doubleday. Pp. 226-40.
 1973b Village: hunting and collecting, the battle of the sexes. In
 To Hunt in the Morning, by J. Siskind. New York: Oxford
 University Press. Pp. 89-109.
Smith, N. J. H.
 1982 *Rainforest Corridors: The Transamazon Colonization
 Scheme*. Berkeley: University of California Press.
Smole, W. J.
 1976 *The Yanomama Indians: A Cultural Geography*. Austin:
 University of Texas Press.
Spielman, R. et al.
 1974 Regional linguistic and genetic differences among
 Yanomama Indians. *Science* 184:637-44.
Sponsel, L. E.
 1981 The Hunter and the Hunted in the Amazon: An
 Integrated Biological and Cultural Approach to the
 Behavioral Ecology of Human Predation. Ph.D.
 dissertation, Cornell University.
 1982 Fallacies in Criticisms of the Protein Hypothesis. Paper
 delivered to the annual meetings of the American
 Anthropological Association in Washington, D.C. on
 December 5, 1982.
 1983a The Yanomama warfare, protein capture, and cultural
 ecology: a critical analysis of the arguments of the
 opponents. *Interciencia* 8:204-10.
Steward, J. H. (editor)
 1946- *Handbook of South American Indians*. 7 vols. Washington,
 1959 D.C.: Smithsonian Institution.
 1948 Culture areas of the tropical forests. In *Handbook of
 South American Indians*, edited by J. H. Steward.
 Washington, D.C.: Smithsonian Institution. 3:883-99.

1955   *Theory of Culture Change: The Methodology of Multilinear Evolution.* Urbana: University of Illinois Press.
Steward, J. H., and L. C. Faron
1959   *Native Peoples of South America.* New York: McGraw-Hill.
Steward, J. C., and R. F. Murphy (editors)
1977   *Evolution and Ecology: Essays on Social Transformation by Julian H. Steward.* Urbana: University of Illinois Press.
Tocantins, L.
1974   The world of the Amazon region. In *Man in the Amazon,* edited by C. Wagley. Gainesville: University Presses of Florida. Pp. 21–32.
Torres-Trueba, H. E.
1968   Slash-and-burn cultivation in the tropical forest Amazon: its techno-environmental limitations and potentialities for cultural development. *Sociologus* 18:137–51.
UNESCO
1978   *Tropical Forest Ecosystems: A State-of-Knowledge Report.* Paris: UNESCO.
Vickers, W. T.
1979   Native Amazon subsistence in diverse habitats: the Siona-Secoya of Ecuador. In *Changing Agricultural Systems in Latin America,* edited by E. F. Moran. Williamsburgh, VA: Studies in Third World Societies No. 7:6–36.
Wagley, C. (editor)
1974   *Man in the Amazon.* Gainesville: University Presses of Florida.
Wagner, P. L.
1977   The concept of environmental determinism in cultural evolution. In *Origins of Agriculture,* edited by C. A. Reed. The Hague: Mouton. Pp. 49–74.
Walter, H.
1971   *Ecology of Tropical and Subtropical Vegetation.* Edinburgh: Oliver and Boyd.
Weiss, G.
1980   The aboriginal cultural areas of South America. *Anthropos* 75:405–15.
Whitten, N. E., Jr. and D. S. Whitten
1977   Report of a process linking basic science research with an action oriented program for research subjects. *Human*

Organization 36:101–5.

Wilkinson, R. G.
1973 *Poverty and Progress: An Ecological Perspective on Economic Development.* New York: Praeger.

Willey, G. R.
1971 *An Introduction to American Archaeology.* Volume 2. South America. Englewood Cliffs, NJ: Prentice-Hall.

Williams, T.
1898 Was primitive man a modern savage? *Smithsonian Institution Annual Report for 1896.* Pp. 541–48.

Wylie, P.
1971 Cultural evolution: the fatal fallacy. *BioScience* 21:729–31.

Yost, J. A., and P. M. Kelly
1983 Shotguns, blowguns, and spears: the analysis of technological efficiency. In *Adaptive Responses of Native Amazonians,* edited by R. B. Hames and W. T. Vickers. New York: Academic Press. Pp. 189–224.

Zerries, O.
1969 The South American Indians and their culture. In *Biogeography and Ecology in South America,* edited by E. J. Fittkau et al. The Hague: Dr. W. Junk, N. V. Vol. 1:329–88.

II.   CASE STUDIES
AND THEMATIC DISCUSSIONS

CHAPTER 5

STONE WALLS AND WATERFALLS: IRRIGATION AND RITUAL REGULATION IN THE CENTRAL CORDILLERA, NORTHERN PHILIPPINES

June Prill Brett

Indigenous forms of irrigation organization in the monsoon countries of Southeast Asia present social scientists with enduring testimonies to the institutional variability possible in effective environmental management. The study of time-tested methods of agricultural water distribution reveals that not only are the hydraulic characteristics of particular regions exceedingly diverse, particularly in upland areas, but also that there are equally unique values that underlie indigenous forms of water distribution.

Early studies on irrigation, especially in the anthropological literature, tended to focus primarily on large-scale irrigation works and the centralization of power in their management, or, more specifically, the role of hydraulic agriculture in the formation of the state (Steward 1955; Wittfogel 1957, 1972). More recent studies in irrigation have expanded the discussion to include a more detailed understanding of how irrigation systems work, that is, how they operate, how they are maintained, and how social conflicts are resolved.[1]

Most of the studies of indigenous Philippine irrigation organizations have concentrated on resource management in lowland regions (Coward 1979; Lewis 1980; de los Reyes 1980; Siy 1982). Analyses of small-scale mountain irrigation systems have been neglected, despite the social and ecological insights such systems offer. This bias toward lowland systems is not unique to the Philippines, but can be observed in other countries as well (Yoder and Martin 1982).

The relationship between humans and nature is a dynamic one in that both culture and environment ceaselessly adapt and readapt as each changes in response to the other's influence (Rambo 1981a, 39). On the micro level, one often finds that what first appear to be irrational, irrelevant, or random beliefs or activities surrounding irrigation control become ecologically intelligible when they are examined in the physical and social context in which they occur. As pointed out by Jamieson and Lovelace (1983), the role of ideology in human-environment interactions in general has been a field of benign scholarly neglect; most of the literature treats values and perceptions of the environment simply as a level of phenomena that is mystifyingly related to the material conditions of social existence. We have yet to agree on the exact nature of the relationship between the cultural values of a society and the way people interact with their environment.

This chapter discusses the role and effects of the Tukukan Bontok ideational system in the management and distribution of water, specifically, how ritual regulates social relationships to satisfactorily control individual access to irrigated land, water, and labor under a particular set of ecological constraints. The chapter deals with the physical characteristics of the Bontok irrigation system in Tukukan, the equities of resource distribution and forms of conflict resolution, and the way that ritual maintains a unique balance between social and environmental constraints.

The Physical Setting

The Bontok region is located in Mountain Province of the Central Cordillera, northern Luzon (fig. 5.1). Connected to the Sierra Madre in Central Luzon, the Cordillera Central is the most extensive system of highlands in the Philippines. The people of the Cordillera Central inhabit some 24,000 km^2 of northern Luzon, leaving a narrow strip of coastal flatlands and foothills to the west and adjoining the Cagayan valley to the east. The Cordillera Central extends about 270 km from north to south, and about 50–100 km from east to west. It is a broad, highly dessicated upland with peaks of more than 2,500 m in the south central area. Mount Pulag, the highest peak in Luzon at 2,930 m and the second-highest mountain in the Philippines, is located here. Most of the Cordillera Central lies

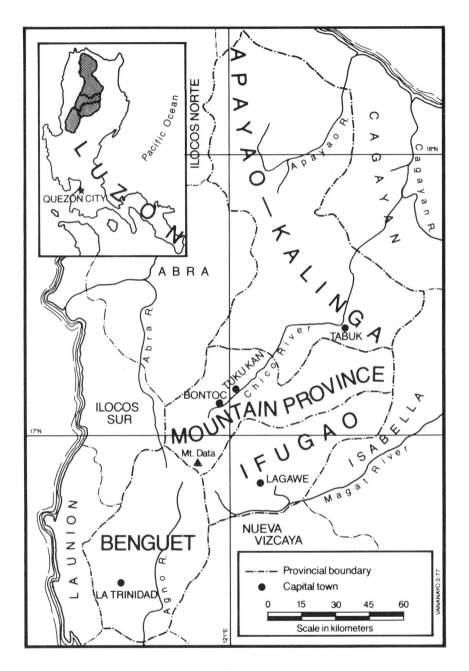

Fig. 5.1. The Bontok region.

above 1,000 m and thus forms a unique climatic, geographic, and cultural region.

Mount Data, a central peak in the mountain chain (elevation 2,289 m), shapes in large measure the regional topography. From the shoulders of its watershed flow the headwaters of four main river systems: the Agno, the Chico, the Abra, and the Magat rivers. The valley floors of the major rivers grade smoothly from an elevation of about 1,500 m in the Mount Data area to about 300 m at the edge of the Central Cordillera.

The Central Cordillera has two seasons: a dry season from November to April and a wet season from May to November. Precipitation ranges from 175 to 500 cm per year, with the maximum rainfall occurring from June to September. The dry season and low temperature are primarily responsible for the existence of pine forest and grassland (Kowal 1966, 393).

Vegetation patterns include mixed tropical montane forest and dense stands of reeds on the upper inclines, with irrigated terraces and wooded ridges dominating the lower slopes (Breeman et al 1970; Conklin 1967, 1980; Wernstedt and Spencer 1967). Agricultural terraces cover an estimated 28,000 ha of the area (Spencer 1952).

The Cordillera Central is inhabited by so-called cultural minorities composed of different ethnolinguistic groups whose life-styles today are very different from those of their Hispanized and colonized neighbors in the surrounding lowland areas (Scott 1982). Some groups are cultivators of swiddens or rainfall gardens and live scattered in small hamlets, while others are wet-rice cultivators growing crops in irrigated terraces sculptured in mountain pockets and up sheer hillsides.

The people discussed in this paper belong to the Bontok culture area and represent one of the societies in the Central Cordillera cultivating wet rice in hillside terraces. They live in compact villages with populations ranging from 800 to 3,000 persons and depend heavily for their subsistence on the cultivation of irrigated rice.

The Ili (Village)

Tukukan village is 6 km from Bontoc,[2] the capital town of Mountain Province. It is a typical compact Bontok village and has 330 nuclear families (households) with a total population of 1,839.[3] Only 975 persons, however, reside in the village permanently. The rest are engaged in wage-earning employment in the mines or other urban areas. Each household is composed of a nuclear family with an average of six members.

The village is situated on mountain shoulders along the course of the Chico River and divided into several subdivisions (*sa-ill*). Important features common to all Bontok villages are sunken, stone-walled pig pits that generate compost fertilizer for the fields (see Omengan and Sajise 1981) and numerous individual rice granaries scattered within the settlement and along the nearby mountainsides. There are 432 pig pits and 421 rice granaries in the village, where *palay* (unhusked rice) is stored after harvest. Another important structure found in all Bontok villages is the *ator*, of which there are eleven in Tukukan. The *ator* has multiple functions: it is a stone-paved platform used for ceremonial purposes, a lounging place for men during "rest" days or ceremonial occasions, and serves as a sleeping hut for boys, bachelors, and widowed men. The *ator* (*dap-ay* in Western Bontok) is also an important socioreligious and political institution (Brett 1975, 1977). Each family is affiliated with a particular *ator* through the male head of the family. Important tasks of this institution, under the direction of the council of elders, include the coordination of activities of the agricultural cycle (Reid 1972), the defense of the village, the handling of war and peace treaties, the regulation of intervillage affairs, and the coordination of rituals for *ator* members (Brett 1975, 1977). External affairs are the business of a supra-*ator* organization that directs all the *ator* in the village in carrying out certain activities (economic, religious, and political) as one body.

Food Sources

Surrounding the village for some distance (some are approximately a two-hour hike) are stone-walled rice terraces. The terrace walls may stand as high as 9.75 m on the rugged mountainsides, some of which are 60 degrees or more in slope. Stone walls are required to retain these earthworks.[4] Traditionally, only one crop of rice is grown annually in these artificially irrigated terraces.[5] The main growth period is during the dry season from January to July. The rice harvested provides roughly half of the total subsistence intake (see Drucker 1974, 11). During the rainy season, the terraces are drained and the soil is piled high and converted into sweet potato (*camote*) gardens. This second crop provides approximately one-fourth of the villagers' total subsistence.

The other one-fourth is derived from mountain and hillside *uma* (swidden gardens) that produce corn, beans, millet, sweet potatoes, and some vegetables and fruits. Meat is obtained from domesticated pigs, chickens, cows, and *carabao* (water buffalo), the latter being allowed to run half-wild in the mountain pasture lands. These food sources are supplemented by fish from the river and a variety of shells and pond vegetables found in the rice terraces. There is an increasing use of tinned foods such as sardines, dried fish, and other foodstuffs not produced in the village, available to those who have cash income.

Work in and on the rice terraces is carried out in the context of an elaborate annual cycle involving traditional practices and ritual activities. These practices are marked by rich symbolism and demand villagewide coordination and cooperation. Besides their symbolic importance, many of these practices also can be shown to be ecologically and agriculturally sound.

The Agricultural Cycle

The traditional agricultural cycle is initiated with a series of communal irrigation activities, beginning with the *khaat si arak* (literally "cutting of the grass" along the irrigation canal). The proper time for this to commence is "announced" by the arrival of the *kiwing*, a migratory bird that is usually spotted around the edge of the village. The appearance of this bird is an indication that the dry season (*chakon*) has arrived and the proper time for sowing of the rice seeds (*panar*) is imminent. In the traditional calendar, this period occurs during the eleventh moon (usually in November). Sowing is carried out at any time within this period when the moon begins to wax or wane.[6]

Prior to sowing, villagers inspect the irrigation canals (*arak*) to assess damage and labor needs. Canal maintenance begins with an announcement by the village elders (e.g., the *fukhaw nan khaat*, literally meaning "to shout" the announcement of irrigation canal cleaning). The timing of this activity is a joint decision of the elders and is guided by their observation of natural phenomena such as the climate, the migratory birds, the water level in the river, and the changing temperature. The announcement by the elders also informs villagers when each irrigation association will work, thus allowing members of more than one irrigation association to contribute their labor on separate days. Technical information on the extent of damage incurred in some sections of the waterworks is also relayed to members of each association so that they will know what tools and materials must be transported to the site.

The irrigation group usually begins work either at the receiving end of the irrigation canal or at the source of the irrigation water or diversion dam. The diversion dam traditionally has been a temporary structure made out of river rock cemented with mud and small stones. All members of irrigation associations must send a representative to help render service for a day or more depending on the kind of task required, the length of the canal, and the repairs to be done. Members furnish their own tools.

The Irrigation Canals

The longest irrigation canal in the village is 4.8 km.[7] This canal winds above the pine line through grassland until it reaches its source, a waterfall from a large mountain spring. The waterfall is temporarily dammed by rocks held together with sod, clay, twigs, and small pebbles. Only half the flowing water is diverted into the irrigation canal, which is 1.2–1.3 m wide and 0.9 m deep. The sloping areas around the irrigation canal allow detritus and rich humus from decomposed leaves, lichens, and other materials to be washed down and dumped into the main canal during a heavy downpour. These materials are transported more than 4 km down to the rice terraces.

Some portions of Bontok irrigation canals show signs of having been tunneled through solid rock.[8] The ancestors of the present farmers labored with crude tools, but they had an intimate knowledge of their environment. The intricate irrigation ditches, with their necessary wooden aqueducts and flumes to allow water to be transported across ravines, always yield to the natural contours of the mountain terrain. Sections where the soil easily erodes (*karayakay*) have been riprapped with rocks transported from nearby creeks or more distant sources.

The demands of these indigenous hydraulic engineering works in such rugged terrain would have been impossible for the Tukukan people without reciprocal labor cooperation (*ugfu*) among the original architects and builders of these waterworks. The amount of labor invested to secure and distribute water over great distances to the rice terraces is reflected in the fact that there are practically no streams or springs within the area that are not carefully diverted to even tiny plots that have been terraced for cultivation. Land that is irrigated acquires a high value immediately, regardless of the steepness of the slope on which the terrace is constructed.

In the village of Tukukan, and in some downstream Bontok villages, the Chico River is not diverted or dammed for irrigation as it is in most of the upstream villages. Whereas upstream villages in Mountain Province have fewer sources of upland water, Tukukan has ample mountain springs that can be tapped for irrigation. The town of Bontoc, for example, is land poor, and most of the rice terraces must be constructed close to the water source—the Chico River. River "islands" are converted into rice fields with intensive and continuous labor inputs since they are washed away annually by strong river currents during storms. This situation requires that the people of Bontoc rebuild not only the rice fields but also the canals and dams each year after the height of the typhoon season. As Jenks observed in 1903 during his pioneering fieldwork in Bontoc, "It is built each year during November and December, and requires the labor of fifteen or twenty men about six weeks" (1905, 91). Tukukan farmers have an apparent advantage over such upstream villages, since they are able to tap the higher mountain streams with the expectation that their irrigation works will not be washed away during annual storms. Nevertheless, their initial labor may be much greater since their water has to be channeled a greater distance.

Rice Terraces and Geographic Layout

There are fifty-one localities with irrigated rice terraces of varying sizes in the village of Tukukan.[9] The approximate total extent of irrigated land is less than 100 ha.[10] Most of the area has been worked for more than six generations, which can be traced through genealogical reckoning by informants. It is believed that many fields are much older than these, however, but genealogical reckoning has its limitations (most elders cannot remember ancestor's names beyond the sixth generation).

There are approximately seventy small springs and brooks being tapped or channeled into Tukukan rice fields, and there are four major irrigation canals. Amurong Canal, which is 4 km long, services the largest number of fields in several different localities, occupying a total of approximately 20 ha of land. This canal serves 352 rice terraces that are owned by 275 members of the indigenous irrigation association. Marupey Canal serves 144 fields owned by 77 members. It waters approximately 2.5 ha of land and covers a distance of 4.8 km from its main source. Sabfi Canal gets its irrigation water from a source more than 3 km away and services a little less than 4 ha with 242 rice fields. Amfuet, the shortest canal, is only 1.5 km in length. It services 1.5 ha of rice land encompassing 56 fields owned by 37 members of the irrigation association. The steep terrain and scattered water sources limit the number of rice-terrace systems that can be constructed in one locality.

Equity of Water Allocation and Layout of Irrigated Land

According to my census in 1982, there are approximately 4,424 rice fields of different sizes around the village, all within the territorial boundary of Tukukan. Of these, 1,358 belong to the *kakachangyan* (aristocrats)[11] and 3,066 belong to the non-*kakachang-yan*.[12] There are no villagers without at least a couple of rice fields to till; therefore, there are no landless villagers in Tukukan. Besides the rice fields, all villagers, as members of corporate groups, have ownership rights to communally held swidden land.

Individual Bontok farmers own rice fields scattered within a particular irrigated locality. Thus, a farmer may own fields located at the head (*sunga-chan*), middle (*khawa-an*), or tail end (*ucho-na*) of an irrigation system. The tail areas are the last to receive irrigation water. The implication of such a layout of parcels is perceptively pointed out by Coward regarding his Ilocos data. He writes, ". . . the fact that all members farmed parcels at the 'head' as well as at the 'tail' of the system served as the strongest incentive for cooperating to maintain the system at maximal efficency in order to adequately irrigate the entire area" (1979, 30).

It has been observed that the layout and mechanics of a canal irrigation system may differ from place to place, usually in response to variations in topography and source of water. Among the Bontok, the geographical layout of rice fields has been influenced not only by the terrain and water source, but also by inheritance patterns. Upon marriage, children inherit fields from their parents through rules of succession based on the principles of bilateral descent, homoparental transmission, and primogeniture. Fields may also be inherited from kinsmen who have no heirs, or fields may be acquired from relatives by paying the costs of mortuary ritual. Thus, an individual's fields may be scattered among different localities and also dispersed within them. This situation entails a crisscrossing of field ownership in different locations for access to main turnouts of irrigation canals. Hence, farmers have a vested interest in ensuring equitable water distribution to the lowest terraced rice fields in spite of the competition for water during critical periods in the agricultural cycle. What to an outside observer appears to be a case of severe land fragmentation, therefore, is actually the physical expression of principles designed to ensure equitability in water distribution.

Irrigation Association Membership and Recruitment

Members of irrigation associations are recruited by virtue of their being citizens of the village and descendants of original terrace-field owners. Membership in an irrigation association ranges from 37 to 275 families whose fields are scattered in different geographic locations. Rights to irrigation water are automatic for any owner of a rice field (whether inherited or purchased) and cannot be acquired in any other way.

Newly constructed fields adjacent or below older terraces are allowed to receive water from *kus-sing* turnouts (fieldspouts) originating from these older rice fields. It is a rule that no one is allowed to construct a new field above or close to the turnouts of main irrigation canals. Also, no owner of a new field is allowed to receive irrigation water before the original (older) rice-terrace owners have watered their fields, unless the new fields are extensions of old terraces belonging to descendants of original owners. This principle of exclusion and prior rights to water prevents water usurpation by newcomers who are not descendants of the original ricefield owners, who labored in the original construction of the irrigation canal.

Water Scarcity and Water Rotation

Irrigation water flows into the rice terraces continuously until the critical period when water becomes scarce, e.g., during the dry season (January to April). This is the period when rice plants need water most to ensure their growth to maturity. It is at this point that a system of water rotation is enforced to minimize conflicts caused by "grabbing" other people's water shares. Cases of water stealing do sometimes lead to altercations but rarely to physical violence.

During periods of water scarcity, villagers rely upon a method of water distribution known as *sogsog-li* (literally "rotation"), which divides and allocates water on a temporal basis. If there are five localities to be irrigated by the same canal, as in figure 5.2, the *sogsog-li* procedure would be as follows: during the first day, locality 1 starts off owning all the irrigation water shares (*chatag*) during the daylight hours. At sundown, the owners of the fields in locality 3 block (*pet-en*) all the turnouts of localities 1 and 2 and divert the water to locality 3, which then receives water during the night (sundown to sunrise). At sunrise on the second day, locality 2 farmers open up the blockade (*kwangen*) to collect (*a-mong*) and divert the water to that area. Locality 2 then owns the irrigation water for the duration of the day. At the hint of falling darkness, locality 4 representatives will go to the turnout canals and block all the subcanals leading to areas other than their own, and they receive the diverted water for the night's duration. At the next sunrise (day three), the water goes back to locality 1, and from sunset to sunrise to locality 5. The next daytime water shares belong to locality 2, and so on, with the rotation completing one full cycle before the next cycle starts.

Water rotation is controlled in two ways: (1) at the main canal level, where the owners of the scheduled share will block the turnouts to all the other localities; and (2) at the field level (fig. 5.3). The point where water is diverted to individual rice fields is blocked when it is time for one's share, as is further described below.

Within each irrigated locality in this cyclical rotation system, farmers owning rice fields are divided into two to four groups, depending on the field layout. The farmers drawing water from fields along the main canal are grouped with the terraced fields above them from which they obtain irrigation water. Each group is composed of farmers (e.g., five to eight) who till fields served by one takeoff point in the canal. Those whose fields are farther from the canal obtain irrigation through the cuts (*pakew-an*) in the embankment of the upper terrace field. When water is distributed by rotation, each group is entitled to receive a number of water shares. The allocated share is scheduled (as in the case illustrated in fig. 5.3).

During the rainy season, when water is relatively abundant, the takeoff points along the main canal and the cuts in the embankments of the upper terraces are kept open at all times to allow for drainage and continuous irrigation. The decision to enforce rotational distribution is taken whenever a group of farmers who have assessed

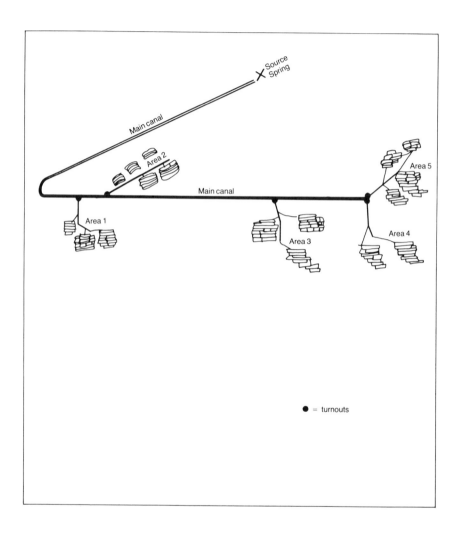

Fig.   5.2.   Water rotation at the main canal level.

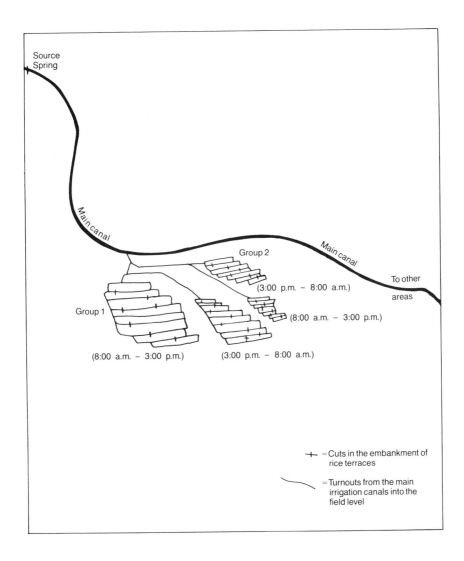

Fig.   5.3.   Water rotation on the field level.

the volume of water flowing from the main canal agree that the water is becoming too scarce to reach the other terraces situated at the tail end in sufficient quantity. They "shout" (ey-fukhaw) a call to members of the field owners who have terraces in the same locality that the water rotation (sogsog-li) will be commenced. Water rotation is strictly observed only when farmers who have the right to draw water at a particular time are personally watching the flow of water toward the fields. This system of guarding is necessary since some farmers, even though they are aware of the rotation schedule, will divert the available water into their own fields if those farmers who are supposed to obtain it are not present. Although getting the share of a farmer who is not present in the field during the rotation period is considered water stealing, it is normally thought to be the fault of the absent farmer who has allowed someone else to take his share.

When water is critical for rice plants, it is imperative that farmers who own fields in different geographical areas divide the task of watching among family members. Different members will be assigned to sleep in the vicinity of their rice fields to oversee their share of water and to divert it to their fields (mananum). The mananum activity is carried out by villagers who stay up from sunset to sunrise. Such watchfulness is necessary to make sure that fields are not being short-changed by other neighboring (cha-og) rice-field owners, who may "steal" water from their sleeping coirrigators. The activity of mananum will last for the duration of water scarcity.

The chotoken are supervisors of water distribution[13] and are recruited to this post through rotation and popular choice by comembers of the irrigation association. They are not compensated in any form, since this job is rotated among members over time. This system of managerial rotation minimizes water monopolization since, if fairness is not maintained, it will also not be reciprocated. Water distributors cannot always effectively police the canals at night, however, and a typical evening in Tukukan is one where villagers are out trying to sneak water from a dozing companion's terrace.[14]

Conflict Resolution

Water theft is considered "bad" (lawa) because it deprives the other person of his rightful share of irrigation water. When an act of water theft is discovered, it is reported to the comembers of the irrigation system. The offender is simply scolded by the owner of the water share, a verbal sanction that is then echoed by other field owners in the vicinity. To an outside observer, the lack of stricter sanctions surrounding such a critical offense as water theft may give the impression that the system is ripe for individual abuses. When informants are asked why offenders are not fined or punished by witholding their future water shares, however, the values underlying their actions emerge. Informants comment that it is not unusual to find brothers, sisters, or other close kinsmen quarreling due to water-share disputes. Should these people be fined, most of their labor time will be spent in litigation involving water theft rather than working in their fields. Since labor inputs are so crucial at particular phases of the agricultural cycle, and since petty water theft is so common, lowered production could easily result if these disputes were settled through more formal procedures. The solution, which villagers feel is the best answer, is for everyone to be physically present during the scheduling of their water shares. It is also believed that to punish an offender by withholding his water share is tantamount to "killing" him or her. The sentiment prevails that "everyone has the right to irrigation water to allow a person and his family to survive." Thus, the traditional practice of being physically present discourages villagers from the temptation to steal other people's water shares while at the same time it avoids the waste of precious agricultural time that would result from constant water-theft litigations.

This laissez-faire policy toward petty water theft between individuals, however, stands in stark contrast to the policy surrounding abuses that deprive an entire group of villagers from access to water. A case occurred in 1982 wherein the customary principles of equitable water allocation were challenged. This case highlights the hierarchical institutional control that emerges in more serious situations.

On May 19, 1982, a *te-er* (compulsory village confinement period or "rest" day) was the occasion to call the *ma-amongan nan mangili* (the gathering of villagers) to witness the litigation of two cases. The litigation was held around Parew, the central *ator*, where traditionally all important matters are discussed by the council of elders. One case concerned the burning of a mountainside, and the other case concerned the violation of water rotation rules between owners of rice fields at Otor and owners of fields at Parew. Normally, cases involving irrigation disputes are settled by the groups involved and not on the supra-*ator* level. The supra-*ator*, or village-wide council of elders, is like a "Supreme Court" that handles cases of greater magnitude than those customarily settled by the elders of individual *ator*. The supra-*ator* council handled this particular case because there was a violation of custom law regarding water rotation rules set down by the ancestors to avoid disputes due to non-equitable water sharing. Since this rule was being challenged, the case had to be heard by the council of elders on the village level. The problem arose when the Otor area field owners refused to share water with the Parew area field owners in spite of the fact that this procedure has been going on for generations. Every time the Parew group of irrigators went to get their water shares the Otor group would open the blockage and divert the water into their own area, thus depriving the Parew irrigators of their shares.

The decision of the elders, after all witnesses had their say, was to enforce the old rule allowing owners of fields at Otor to take their shares of irrigation water for one day and one night, while owners at Parew got theirs for two days. The elders emphasized that this old rule must be followed strictly. However, they also announced a new law, unanimously approved by the people present during the case hearing, that a fine (*lakaw*)[15] would in the future be imposed on anyone (from either Otor or Parew irrigation groups) who tried to deprive the others of their water rights during the *sogsog-li* water rotation schedule.

Hence, the subsistence ethic that underlies equitable water distribution in Tukukan was strictly enforced through new punitive measures designed to prevent the monopolization of water by any particular group.

Ritual Maintenance of the Irrigation System

What is most intriguing about the Tukukan irrigation system, unlike other communal irrigation systems in the Philippine lowlands today, is that much of the dispersion of water resources is regulated through the ritual system. Whereas leadership positions are inherent in lowland Philippine irrigation systems and currently required in government-sponsored irrigation projects, Tukukan irrigation associations distribute managerial responsibilities among all members. This diffusion of responsibility is quite clear about the decisions required for physically maintaining the waterworks and for dispersing the water. In the ritual system, however, leadership is held by a "priest" (an elder member of the irrigation association) who officiates during irrigation-water ceremonies (see appendices I and II). The rules surrounding irrigation and agricultural activities are the result of time-tested ecological adaptations, but they are regulated by rituals that are themselves part of a unique and complex cosmological ideology. While the cosmology of the Tukukan Bontok is not the subject of this chapter, certain aspects of their belief system must be mentioned briefly to elucidate the ideology that allows their irrigation organization to operate effectively.

The cleaning activities of the irrigation canal are done annually before the sowing of the rice seeds and also during emergencies when storms and other natural catastrophies (e.g., erosion due to landslides) cause damage to the canals. When irrigation-canal activities are called for by a consensus of the farmers, all members of the irrigation system should send a family representative to help direct water to the irrigation canal from the source, clear the canal, and cut the bushes along and around the canal slopes. In addition, eroded portions must be riprapped, debris and silt must be cleared from the clogged waterways, and sacrificial offerings must be contributed by all the rice field owners. Those who fail to fulfill their labor obligations in the maintenance of the irrigation system are fined by members who were present. Although nothing is written, everyone remembers those who did not render service and the number of days they failed to come. During the work period, the group will go to the delinquent's house to collect the fine, called *ob-ob*. This fine is five bundles of *palay* (unhusked rice) per day of absence, or the equivalent in cash (pesos), sugarcane wine, a chicken, or tobacco leaves. In general, the amount depends on the number of days the member was absent during the *khaat*. Proceeds from the collection are used by those who were present during the canal activities. If the fine is paid in the form of food, it is consumed by the people who fulfilled their labor obligations. If cash is collected, it will be used to buy tobacco, tinned food, and other needs that the comembers may decide on.

Agricultural Rest Days

All of the irrigation activities that synchronize with agricultural activities are regulated by the *te-er* institution (literally "to remain in one place"). The *te-er* refers to a compulsory village confinement period. When a *te-er* is decided on by the council of elders at the central *ator*, the young boys "shout" the announcement that no one should work in the fields (*fuk-nag*) the next day or next two days, depending on the interpretation of omens received by reading the bile sac of a chicken. *Te-er* is imposed by the elders to enable them to enforce the observance of traditional agricultural rituals, which include the important observance of omens for the welfare of the whole village. Furthermore, the *te-er* marks the end of one set of agricultural activities and the onset of another. An important function of the *te-er*, although an unconscious one, is the compulsory physical rest of farmers from heavy agricultural labor.

During the *te-er*, no one is allowed to leave or enter the village, nor are they allowed to go to their fields or to the forest to cut lumber. The task of policing the village entrances and exits to report violators is carried out by the designated *sa-ill* or geographic subdivision of the village. This group takes charge of collecting fines from those who violate the rest day. The task of policing is rotated among the different *sa-ill*. Fines for *te-er* violation are referred to as *pañgila*. There are different grades of *te-er*[16], with some considered more strict than others, depending on the type of agricultural activity concerned. *Te-er* related to planting and harvesting, for example, and village intensification rites, are more strict than other *te-er*. Breaking the *te-er* by insisting on visiting one's rice field or leaving the village to go elsewhere is believed to destroy the intended benefits of the *te-er* ritual; thus, a bad harvest is attributed to the violation of this rest day. It is believed that rice predators (rats, worms, ricebirds, and wild animals) will destroy the crop, and those who insist on going to the fields will find all their agricultural efforts fruitless due to the nonobservance of the *te-er*. The *te-er* imposition is lifted only after a ritual *patay* (sacrifice) at the *papatayan* (sacred trees "where sacrifices are offered") has taken place, and only if no bad omens are observed.[17] If no bad omens have occurred, the *te-er* will be lifted and the next day will be a work day (*fuk-nag*) in the field. The *patay* ceremony is officiated by the *puma-patay* (*patay* hereditary priests), who are the guardians of the sacred pot (the symbol of abundance and fertility of crops). During the ceremony at the sacred trees, a chicken is offered (provided by the fines from those who previously violated *te-ers*) as sacrifice to the guardian (*anitu*) spirits. A prognosis for agricultural, social, and political activities of the village is communicated by these *anitu* through the bile-sac-reading of the sacrificial chicken or pig, as interpreted by the elders.

This system of rest days sets and coordinates agricultural labor of specific kinds into restricted time periods or seasons. Such a system is advantageous because it enforces a coordinated schedule of activities to minimize risks for individual farmers. Informants narrated stories of farmers who had insisted on sowing rice much ahead of time and consequently experienced the misfortune of being overrun by rice predators, such as rats and ricebirds, that normally divide their attention and appetites between thousands of fields instead of ganging up on the few fields that were planted early. Through the ritual coordination of planting a few farmers do lose some grain, but a majority minimize the risk of total crop loss to predators. A similar situation is said to arise if farmers stagger their planting or decide to plant their rice at a time later than the rest. In these cases, the farmers are exceptionally vulnerable to losing their ripened crop to the storms that usually blast their way through northern Luzon from June to September.

Thus, this ritual coordination of agricultural activities is ecologically adaptive given the environmental pressures of

- Water scarcity during certain periods of the rice agricultural period cycle

- The threat of invasion of seasonal rice predators

- The likelihood of panicle shattering if the harvest period is unduly delayed[18]

- The certainty of heavy, destructive rains and wind at the beginning of the rice harvest season

Overall, this intricate ritual scheduling of agricultural activities serves to both maximize yields and minimize risks by proper and precise timing in preparing the fields, diverting the irrigation water, repairing the canals, green manuring, planting the rice seeds, transplanting, intensive weeding, and coordinated harvesting.

Discussions and Conclusions

The Tukukan Bontok people present us with an interesting case for analyzing how cultural values and environmental action interrelate and maintain feedback between each other. It is quite clear, for example, that there is unequal distribution of land in this community between members of the aristocratic rank and the bulk of the villagers. This inequality in resource distribution is socially recognized and accepted. Social inequality does not, however, govern the distribution of water in Tukukan, since all villagers have an equal right to water shares.

Both ecological and ideological factors lie behind social measures to prevent the inequitable allocation of water. Ecologically, the dispersed ownership of fields between the head, middle, or tail end of the irrigation canals causes all villagers, regardless of the number of fields they own, to have some vested interest in seeing that water is evenly distributed. Second, this dispersed pattern of field ownership over wide areas usually requires reciprocal labor exchange for cultivation, since (1) the optimum periods for rice planting and harvesting are short; and (2) fields located in different geographical areas cannot be efficiently cultivated within these time periods by family labor alone. People in the upper rank, or *kachangyan* group, are heavily dependent on the labor of the non-*kachangyan* group to till their extensive fields. People in the latter group, on the other hand, are in an excellent position to withhold their power from large farmers if they so desire, since they themselves rely primarily upon reciprocal labor exchanges. Hence, any attempt by the large landowners to assert greater control over water is prevented *a priori* by their dependency on the labor of many smaller landowners.

Ideologically, there are a number of cross-cutting principles that support this equitable system of water distribution. First, Tukukan Bontok place a high value on the labor expended by their ancestors in constructing both the rice terraces and the irrigation canals. It is this labor, exerted long ago, that entitles present-day descendants to their share of inherited land, whether it be large or small. At the same time, however, villagers know that the waterworks were engineered and constructed with labor inputs from all ranks of early Tukukan society. Thus, inequalities in socioeconomic rank have little relevance to water distribution in the community.

The notion that no one has the right to monopolize water is but part of a more encompassing ethic that disallows any member of the village from depriving another member of resources that ensure—and indeed are required for—survival. This subsistence ethic, which incidentally extends only to members of the village, is bolstered by strong supernatural sanctions that lessen the likelihood of violation. Tukukan people believe that supernatural retribution in the form of illness, death, crop failure, or other misfortune will be exacted on any person (or a member of his family) who violates this code of ethics.

It is perhaps the role that ritual plays in maintaining equality in water distribution that makes the Tukukan Bontok case most unique. The question arises as to why, in what is clearly a ranked society, ritual is the overarching organizational form, instead of a more secular type of managerial control vested in political authority.

There are several possible answers to this question, all closely related to the way the Tukukan have had to adapt both to their natural surroundings and to the presence of other groups in the area. Indigenous groups in this part of the Cordillera have had to adapt to a historical experience of endemic warfare between villages. Intervillage institutions such as the peace pact (Brett 1975, 1977) have evolved to allow these groups to carry on subsistence activities, but the external and ever-present threat of warfare has led to an intensification of intravillage institutions and beliefs designed to maintain internal cooperation and harmony. The structure of Tukukan society, for example, is such that individuals are members of a number of cross-cutting groups (e.g., reciprocal labor groups, irrigation assocations, and *ator*). The resulting multiple obligations and relationships with other members of the village present a unique cultural configuration that is well suited for regulating the manpower demands of irrigated rice agriculture, as well as the external situation of endemic warfare. In many contexts, these intravillage groups act with relative autonomy, but matters that concern the welfare of the whole village are controlled and decided upon by the council of elders whose members come from both upper and lower ranks of the society. Similarly, the nuclear family in Tukukan society, while maintaining great importance and symbolic value, is clearly subordinate to the community in contexts where the welfare of the group hangs in the balance.

Both the environmental and social constraints of Tukukan have led to an institutional structure that downplays the likelihood that inequalities in resource ownership can be transmitted into inequalities in power over other village members. In this particularly complex social system, group coordination and control is critical. Ritual regulation and supernatural sanctions are part of a belief system that makes events in their uncertain social and physical environment comprehensible. Given the fact of endemic warfare, as well as of climatic fluctuations that cause seasonal resource crises, coordination of communal agricultural activities through ideological values and ritual is perhaps more adaptive than coordination through organizational forms wherein leadership responsibility is concentrated and vested in specific individuals.

One issue raised by the data in this chapter remains unanswered, if not unanswerable. Is the subsistence ethic expressed by the Tukukan Bontok the real determinant of equitable water distribution in rice agriculture? Or, is the dispersed nature of rice field ownership the major factor preventing water monopolization? The real issue, it would seem, is whether it is indeed possible to separate these two viewpoints into discrete levels of explanation. For the Tukukan Bontok, there is no contradiction in accepting both ecological and ideological causation in a single explanation.

Tukukan beliefs and values regarding irrigation do not constitute autonomous systems unrelated to their overall ecological adaptation. Nor are they passive products of their physical interactions with their environment. On the contrary, their beliefs and values coordinate, as well as result, from the ongoing reciprocal interaction between their social and physical worlds. As in the example presented in the text, new rules that govern ecological behavior between or within social groups will be devised in accordance with changes in resource availability—changes that in turn prompt alterations in human behavior and belief over time.

Acknowledgments

The research on which this chapter is based was carried out from January 1982 to March 1983 in the village of Tukukan, a Bontok community. It was funded by the Ford Foundation through the Cordillera Studies Center, University of the Philippines, Baguio City. My thanks and appreciation to Dr. John Cool for all the encouragement he gave. I would also like to express my gratitude and appreciation to Dr. Susan Russell, Rockefeller Foundation, School of Economics, University of the Philippines (Quezon City), for her most helpful suggestions and critical comments on an earlier version of this paper.

Appendix I

Apey si Arak: Ritual for the Continuous Flow of Irrigation Water

On the day of the irrigation canal cleaning (*khaat*), the *apey si khina* ritual is performed. Two or three chickens, salted pork, and locally brewed sugarcane wine (*fayas*) are used during the ritual performance. A short prayer is said at the tail end of the irrigated rice fields. It is required that a descendant, preferrably a male elder, of the original owners officiate at the ceremony. This individual must be ritually "clean," i.e., none of his family members should have died as young men or women, he must not be a widower, he must have no record of having committed adultery, and he must not be presently officiating in any "polluting" ceremonies. Ritual purity is necessary so that none of these negative deeds can pollute or negate the fertility effects of the prayers and sacrifices being offered to the *anitu*/ancestors. After the *apey* (literally "to build a fire," but referred to as "warming up" the irrigation water), restrictions will be imposed during the following day (or days) on the particular irrigation-canal members. No one is allowed to go to the rice fields, as it is believed that the ritual will then be invalidated or nullified (*maswak*). Thus, the rice plants and their yield will be poor, the effects of the ritual cancelled out. A free translation of the prayer goes:

> *Nay apeyan me sigah ay arak, et entat-aranak nan chanum sina ya en fafaru nan sinamar sin-a.*
>
> Now we "warm" you up irrigation canal, may you (water) flow continuously so that we will have a bountiful harvest.

After the short prayer, all the irrigation canal members eat the food that each has brought from his or her home. A *pu-chung* (*runo* reeds with topmost leaves tied into a knot to denote a sign of ritual observance) is stuck into the ground and the long tail feathers of a chicken are arranged around it. When the traditional priest arrives home, he must refrain from attending wakes or other feasts, as this act is also believed to negate the ritual for the increase of water. The priest is traditionally expected to avoid all village feasts until the moon has completed a full cycle, which constitutes in Bontok cosmology a calendrical month. Due to the hardship this entails, however, the ritual is usually carried out before the first quarter of the moon so that the next quarter waning moon can conveniently be considered "a month." It is believed by the Bontok that certain characteristics of the sacrificial animals (as well as of the officiators of the sacrifice) have some influence upon the intended effects of their prayers. For example, the color of the sacrificial chicken is thought to effect the quality of the rice crop, and chickens selected for the ritual must not be white because it is believed that this will result in "white," "sickly," or "fruitless" rice plants. The sacrificial chicken's sex is also important and only hens, a symbol of fertility due to their ability to produce offspring that multiply quickly, are used in the ritual. The reproductive ability of hens is believed to induce an increase in the amount of water available for irrigation.

Appendix II

Apey nan Lumayaw ay Chanum: Ritual for the Recapturing of a Water Spirit That Transferred to Another Village

Sometimes reliable water sources dry up or irrigation water levels decrease for reasons that are explained by informants as due to their having not received the required amount of respect.

When the irrigation water is low or dries up, it is said to have transferred to another village (*lumayaw ay chanum* or "runaway water"). The spirit (*anitu*) of the water sometimes communicates with the people in the new area through a medium who becomes possessed by the spirit. The *anitu* says, "I am the water spirit from _____ (name of the original village), and I have transferred to this place." The Bontok believe that the water spirit transferred because it was not "warmed up" (*maapeyan*) through the *apey* ceremony (described in appendix I).

The members of the irrigation canal then select a priest (a male elder) and a descendant of an original field owner. These individuals, together with some assistants dressed in their newest *chinangtar* (special g-strings), make a trip to the village where the water spirit allegedly now resides. The new attire worn in this context symbolizes the fertility of crops and water. When the group arrives at the water spirit's new home, they try to "persuade" the spirit to come back with them. This is done by the elder putting water from the new water site into a small gourd bottle while saying a prayer of persuasion. The water spirit is then brought back with the group when it returns to the original village. The party must be careful not to let anyone from the water spirit's new village prevent them from taking the water spirit away. People from the new village may try to stop them, especially if they believe that their own water spirit is also being enticed to leave. Should anyone from this village catch the original group leaving, the ritual just performed would be negated.

The gourd with water is kept in a *pas-king* (rattan woven back-basket) during the trip back to the home village. Some members of this group have stayed behind outside the home village to form a reception committee to escort the water spirit back to its original source. They proceed to the dried-up water source and pour the captured water spirit into the spring site, again reciting a short prayer to persuade the spirit to stay.

After this procedure, the men go directly to the house of the member of the irrigation association who owns the largest number of fields. This person will contribute a pig to be butchered for a *chao-es* ceremony, to be held in connection with the activities of "recapturing" the spirit of the irrigation water. During this ceremony the *ayyeng* (men's song) occurs. The words of the song concern the water's fertility and abundance. It is believed this song will make the rice crop bountiful and prevent the water spirit from leaving them again. Irrigation-canal members will be restricted from visiting their rice fields the next day to ensure the ritual performance's effect. The persons who "brought back" the water will have to refrain from attending "unclean" ceremonies and wakes for the duration of a "month." A wake is perceived to be polluting because of its association with death, while the rice crop symbolizes fertility and a "life-giving" essence. Mixing the opposed symbol of death with life is believed to negate the power of the latter, and therefore is dangerous.

If followed faithfully, these procedures are believed to result usually in the return of the irrigation water. Ritual performances such as these appear to lessen stress and allow farmers to go about their daily work with greater confidence. In this sense, ritual performs an essential stress-reducing function in a society where environmental crises, such as marked variations in precipitation, create continued insecurity among farmers; ritual provides them with some kind of "control" when faced with the unpredictable forces of the natural environment.

Notes

1. See Geertz (1967); Lewis (1971, 1980); Hunt and Hunt (1976); Mitchell (1976); Coward (1979, 1980); De los Reyes (1980); Leach (1980); and Siy (1982).
2. It is generally accepted practice to distinguish between the town name of Bontoc and the name Bontok (referring to a cultural group in the Mountain Province of northern Luzon) through a different spelling of the last consonant.
3. 1981-82 house-to-house census was conducted by the researcher.
4. See Conklin (1980) for more detail on terrace-stonewall technology.
5. After the turn of this century, planting of a second crop was introduced. However, roughly less than 50 percent of the villagers plant this second crop, which does not entail any rituals.
6. Agricultural activities related to seeding are never carried out during the absence of the moon or during the time of the full moon. It is believed that sowing the rice seeds during the full moon will make the rice plants lush but the fruitheads less than full. Sowing seeds during the absence of the moon is unfruitful as well since the rice plants will not bear many grains on the panicle.
7. The longest irrigation canal in Mountain Province is close to 25 km long. This canal extends from the village of Tanulong to Besao in western Bontok, which is the source of irrigation water (see Bacdayan 1980).
8. The traditional method of cracking rock is by firing it (building a strong fire around the rock) and then pouring cold water over the heated area. When the rock cracks, the broken fragments are chipped away and the procedure is repeated until the desired result is obtained (refer to Conklin [1980, 18] for a detailed description of Ifugao methods of building stone walls).
9. The following are the terms for culturally recognized size differences in Tukukan rice fields:

    Mar-wa—the largest and widest fields in size. They are 18–25 m wide and 30–45 m long.
    Faneng—the second largest field category with

measurements approximately 3–4.5 m wide and 12–24 m long.

Fag-liw—a narrow strip of rice terrace, usually sandwiched between two larger fields. Measurements are usually from 0.6–1.2 m wide to 3–9 m long.

Sinkong—approximately 0.6–1.2 m wide and 1.5–3 m long. These fields may have various shapes such as rectangular, round, or oval.

Sak-kob —smaller than the sinkong, these fields occupy any small space between any rice terrace. They are approximately 0.6 m wide and 0.6–1 m long.

When a family head has to divide one large rice field among his children, an equal division is established by an elder and a boundary marker called *ked-cheng* is ritually erected. Rice wine or sugarcane wine is poured over the monument.

10.   It is difficult to accurately survey the total hectarage of rice land in Tukukan, or for that matter anywhere in the Mountain Province. The data are almost impossible to obtain, since not even the Bureau of Agricultural Extension, the National Irrigation Administration (Bontoc Office), or the Bureau of Census have exact data but only estimates. The people will not allow their fields or areas to be measured by "outsiders"; thus, only estimates have been made by the concerned offices of the area under irrigation in Bontok.

11.   Those people who occupy the highest ascribed rank in the village are referred to as *kakachangyan*. These are the descendants of earlier generations of *kachangyans* and own the largest number of inherited rice fields, animals, and titles in the village. They own exclusive rights to rituals and marry individuals of similar rank.

12.   The non-*kachangyan* are generally referred to as *lawa* (literally "bad," since "to be poor is bad"). These families cannot trace their ancestry to any *kachangyan* lines. Their birth, marriage, and death practices also differ from those of the *kachangyan*.

13.   These water distributors are usually males, a practice that dates back to past intervillage warfare when women were not allowed to go out to check on the water flow in main irrigation canals. There was the fear that enemy headhunters may have intentionally diverted the irrigation

water in order to lure people to be ambushed.

14. The different strategies of water theft are: (1) siphoning the water from another field with the use of a rubber or plastic hose (not readily detected during the night); and (2) *loñgog*, the insertion of bamboo or reed sticks at the bottom of the terrace embankment or laying bamboo reeds across the rice field to channel water into one's own fields.

15. *Lakaw* is one of the graded fines that involves the payment of a pig, carabao, or huge iron vats and other big items. The offender's rice fields may have to be transferred to enable him or her to pay this fine (*lakaw*). These fines do not usually go to the offended party directly but rather to the village council of elders. In Bontok custom law, the fine is considered "unclean" and polluting; only the elders are exempted from its polluting effect.

16. *Te-er* are carried out during the whole agricultural cycle. Other *te-er* are related to litigation cases being heard at the central *ator*, or to political affairs between two villages (such as those related to armed conflict or peacemaking). There are also *te-er* that are carried out exclusively by particular households who are performing ceremonies, or who have just performed ceremonies.

17. Examples of bad omens include rainbows, hawks squawking above the village, deaths, and accidental house burnings.

18. This problem is particularly acute during the dry-season crop (*chinakon*) due to the photoperiod sensitivity of the traditional varieties of rice.

References

Bacdayan, A. S.
  1980   Mountain irrigators in the Philippines. In *Irrigation and Agricultural Development in Asia: Perspectives from the Social Sciences*, edited by W. E. Coward, Jr. Ithaca, NY: Cornell University Press. Pp. 172–85.
Breeman, N. Van, L. R. Oldeman, W. J. Plantinga, and W. G. Wielemaker
  1970   The Ifugao rice terraces. In *Aspects of Rice Growing in Asia and the Americas*, edited by N. Van Breeman, A. J. F. Hydendael, D. Hille Ris Lambers, H. C. Moster, L. R.

Oldeman, W. J. Plantinga, and W. E. Wielemaker. Miscellaneous Papers No. 7. Wagneingen: Landbouw-hogeschool. Pp. 39–73.

Brett, J. Prill
1975 Bontok Warfare. Master's thesis, Department of Anthropology, University of the Philippines, Diliman, Quezon City.
1977 The Bontok Peace Pact Institution. Research paper submitted to the Social Sciences and Humanities Research Committee, Office of Research Coordination, University of the Philippines, Quezon Hall.

Conklin, H. C.
1967 Some aspects of ethnographic research in Ifugao. *Transactions of the New York Academy of Sciences*, Ser. II. 30:99–121.
1980 *Ethographic Atlas of Ifugao.* New Haven, CT: Yale University Press.

Coward, W. E., Jr.
1979 Principles of social organization in indigenous irrigation systems. *Human Organization* 38:28–36.

Coward, W. E., Jr. (editor)
1980 *Irrigation and Agricultural Development in Asia: Perspectives from the Social Sciences.* Ithaca, NY: Cornell University Press.

de los Reyes, R. P.
1980 *Managing Communal Gravity Systems: Farmers Approaches and Implications for Program Planning.* Quezon City, Philippines: Institute of Philippine Culture.

Drucker, C. B.
1974 Economic and Social Organization in the Philippine Highlands. Ph.D. dissertation, Department of Anthropology, Stanford University.

Geertz, C.
1967 Tihingan: A Balinese village. In *Villages in Indonesia,* edited by Koentjaraningrat. Ithaca, NY: Cornell University Press. Pp. 210–43.

Hunt, R. C., and E. Hunt
1976 Canal irrigation and local social organization. *Current Anthropology* 17:389–411.

Jamieson, N., and G. Lovelace
  1983   Cultural Values and Human Ecology: Some Initial
          Considerations. Working Paper. Honolulu: East-West
          Environment and Policy Institute, East-West Center.
Jenks, A. E.
  1905   *The Bontoc Igorot*. Ethnological Survey Publications Vol.
          I. Manila: Bureau of Printing.
Kowal, N. L.
  1966   Shifting cultivation, fire and pine forest in the Cordillera
          Central, Luzon, Philippines. *Ecological Monograph* vol.
          36, no. 4.
Leach, E. R.
  1980   Village irrigation in the dry zone of Sri Lanka. In
          *Irrigation and Agricultural Development in Asia:
          Perspectives from the Social Sciences*, edited by W. E.
          Coward, Jr. Ithaca, NY: Cornell University Press. Pp.
          91–26.
Lewis, H. T.
  1971   *Ilocano Rice Farmers: A Comparative Study of Two
          Philippine Barrios*. Honolulu: University of Hawaii Press.
  1980   Irrigation societies in Northern Philippines. In *Irrigation
          and Agricultural Development in Asia: Perspectives from
          the Social Sciences*, edited by W. E. Coward, Jr. Ithaca,
          NY: Cornell University Press. Pp. 153–71.
Mitchell, W. P.
  1976   Irrigation farming in the Andes: evolutionary
          implications. In *Studies in Peasant Livelihood*, edited by R.
          Halperin and J. Dow. New York: St. Martin's Press. Pp.
          36–59.
Omengan, E., and P. Sajise
  1981   Ecological Study of the Bontoc Rice Paddy System: A
          Case of Human-Environment Interaction. Paper
          presented at the IRRI Thursday Seminar, March 26,
          1981.
Rambo, A. T.
  1981a   Introductory essay: the conceptual development of human
          ecology. In Conceptual Approaches to Human Ecology: A
          Sourcebook on Alternative Paradigms for the Study of
          Human Interactions with the Environment, edited by A.
          T. Rambo. Unpublished. Honolulu: East-West Center,

Environment and Policy Institute. Pp. 1–49.

1981b   Introductory essay: what is human ecology? In The Scope of Human Ecology: A Sourcebook on The Application of The Ecological Perspective in the Social Sciences, edited by A. T. Rambo. Unpublished. Honolulu: Environment and Policy Institute, East-West Center. Pp. 1–9.

Reid, L.
1972   Wards and working groups in Guinaang, Bontoc, Luzon. *Anthropos* 67:530–51.

Scott, W. H.
1982   The creation of a cultural minority. In *Cracks in the Parchment Curtain*, edited by W. H. Scott. Quezon City, Philippines: New Day Publishers. Pp. 18–27.

Siy, R. Y., Jr.
1982   *Community Resource Management: Lessons from a Zangera*. Diliman, Quezon City, Philippines: University of the Philippines Press.

Spencer, J. E.
1952   *Land and People in the Philippines*. Berkeley, CA: University of California Press.

Steward, J. H.
1955   *Irrigation Civilization: A Comparative Study*. Social Sciences Monograph No. 1. Washington, D.C.: Pan American Union.

Wernstedt, F. L., and J. E. Spencer
1967   *The Philippine Island World: A Physical, Cultural, and Regional Geography*. Berkeley, CA: University of California Press.

Wittfogel, K. A.
1957   *Oriental Despotism*. New Haven, CT: Yale University Press.

1972   The hydraulic approach to pre-Hispanic Mesoamerica. In *The Prehistory of the Tehuacan Valley*, edited by R. S. McNeish et al. Vol. IV:59–80. Austin: University of Texas Press.

Yoder, R. D., and E. D. Martin
1982   Water Resource Management for Rural Development in Nepal. A joint research proposal presented during a meeting of the Canadian Council for Southeast Asian Studies, held at the Institute of Southeast Asian Studies, Singapore, June 21–24, 1982.

CHAPTER 6

MEMORY, MYTH, AND HISTORY: TRADITIONAL AGRICULTURE AND STRUCTURE IN MANDAYA SOCIETY

Aram A. Yengoyan

The organization and structure of human societies, especially nonliterate ones, depend on how myth and history are combined to provide cultural coherence, which, when reproduced through time and space, articulates the past with the present. All human societies possess a number of ways in which history is made meaningful in everyday activities and actions. In most small-scale, nonliterate societies, however, the basic means by which the past is reproduced—either as history or as myth—may depend on one of two forms: either a well-developed oral tradition culminating in a classical orality expressed by only a few individuals, or mnemonic structures or devices that act as cues in individual retention and as storage of past information and its use in the present. Oral traditions in many societies, especially in Southeast Asia, are usually associated with the presence of orators, most of whom are adult males who possess a keen sense of how past events relate to present activity.

Oral tradition is not simply the transfer of information. It also maintains an aesthetic quality in terms of speech usage, style of delivery, and the ability to use profound words and abstract concepts. At the same time, this tradition combines and recombines words and phrases in rhythmic and poetic constructions. Thus, the use of speech in a unique and powerful delivery is the basis for gaining respect and esteem as an orator. These talents, combined in one person, are unusual; yet, in many societies oratory is the central means by which knowledge is transmitted. Moreover, the power of speech as control of information is also a form of social control. Power rests in the possession and use of knowledge. Since many forms of knowledge are highly privatized, the control of other individuals by orators (in many cases professional orators) is relatively widespread as a form of political development.

Mnemonics, namely the use of memory devices for the storage and use of cultural knowledge, has its roots in early Greece. Its development in Western culture has been admirably dealt with by Yates (1966). In the West, mnemonics is commonly associated with written literacy, either as an expression of literacy or as a vehicle for cultural transmission. Outside the West there are many nonliterate forms of mnemonic structure that are used in different societies for information storage and retrieval. In a more recent paper, Hage (1978) describes the differences between the Puluwatese of Micronesia and the Iatmul of New Guinea in terms of how memory devices are constituted, information is coded, and the kinds of cues employed for transmission of knowledge. As we would expect, spatial references act in many cases as images for the storage and maintenance of knowledge. The ethnography of New Guinea, Aboriginal Australia, and the Pacific abound with cases in which space is highly categorized. The number of relationships that hold spatial categories together promote a complex means by which all forms of cultural data, be they highly concrete or abstract, are dealt with as instant cultural recall.

Spatial axioms are usually based on concrete cues such as material objects or particular localities where an event took place. Mnemonic devices are verbal cues directing how an individual is socialized in a particular culture. The ability to recite stories, legends, and accounts is not only a form of verbal art. Most often, there is a tendency to combine concrete and verbal cues into a system in which every item of cultural knowledge or event is seen as possessing a physical and a symbolic referent. How physical and symbolic referents coordinate with one another may vary from one culture to another, but it is imperative for us to understand that the connection between them is neither linear nor causal. Thus, mnemonic structures are not simply linear registers of spatial and temporal events; they must be interpreted as a means by which all cultural information is maintained and transmitted in a society. Legends, myths, history, and ritualized speech are cultural codes of the past, which need to be brought into the present as an expression of cultural coherence.

By focusing on mnemonic devices as a means of articulating time and knowledge through the present and into the future, we assess how memory structures may or may not operate as metaphorical signification. Although one could conclude that all mnemonic structures and devices are by necessity metaphorical, if they are to serve their purpose, such a conclusion is not very interesting nor is it theoretically powerful. The metaphorical component may be either at a deep and abstract level (which is the way most anthropologists understand it) or it may be overt and highly literal. The question of deep-abstract as opposed to literal-overt is essentially an empirical one that must be understood as being highly relative as one moves from culture to culture.

By analyzing mnemonics as a memory bank, we can feel relatively confident that what a society thinks of as being imagery or mystical might also be interpreted by members of that society as highly literal thoughts and actions. This would allow the investigator a means of breaking from the assumption that all inquiries that are mystical, metaphysical, or imaginative must by their very nature be interpreted as deep, covert, and abstract. Memory devices relate imaginative thought, myth, and legend to a web of specific spatial or contextual referents that provide the means for individuals to recall and store cultural information.

The expression of cultural information would also vary from society to society. As Hage (1978, 91) notes, "memorized knowledge is corporate and inherited (patrilineally)," but its display depends in part on the social structure of the society under investigation. Among the Puluwatese of Micronesia, memory knowledge usually was used in highly competitive situations, either between heads of navigation groups or as part of the ongoing tension and interaction between different titled groups. In Iatmul society, memory is a more collective activity, and it has a broad social expression. In both cases, the ability of certain individuals to enact the role of better speakers is highly variable. Thus the desired quality of erudition can be consciously cultivated. As Bateson (1958, 227) notes, erudition is linked to knowledge and wisdom that is broad gauged and ranges over a number of issues and subjects. To gain respect and generate an entourage, an orator or skillful speaker must combine the ability to speak well, through the use of repetition, rhyme, and verse, with an unusually keen sense of wisdom and profound thought.

In comparing Puluwat and Iatmul, Hage has been able to detect cultural variations in mnemonic devices that are in part a product of different social structures. In both cases, however, one is comparing cultures that have marked geographical constraints for the arena in which memory must operate. For Micronesians, the sea and the land are markers for one another through which myth, legends, whaling accounts, fishing expeditions, and long-distance travel are contextualized in a meaningful way for retelling stories or retrieving cultural information. With a system of ranked titles and complex rules of land tenure, Puluwat social structure may be interpreted as a series of complex and highly involuted groupings and categories that overlap both in social structure and economic organization. As land and population pressures increase, one would expect these influences to be recorded in the knowledge system and content of mnemonic structures.

For the Iatmul, the internal complexity of the society is not found in titles or land-tenure arrangements but in a highly detailed totemic system that is elaborated as a series of permutations of several personal names for any individual. Through these personal names and the corresponding clan designations, an individual may have more than thirty names, and, in turn, every clan has hundreds of ancestral names that refer to myths and ancestral beings (Bateson 1958, 127). How this complex of names and categories is ordered and conveyed depends on a set of concrete and verbal cues that make up the mnemonic structures. This kind of knowledge is learned by people of erudition.

A basic isomorphism exists between the content and structure of the mnemonic system and the social and cultural elaboration of any society. We should not conclude, however, that this isomorphism is simply a lineal process, nor is it a case of mirroring, in which one component is a reflection of the other. In many cases, memory devices have a special kind of complexity for the retention of historical details or the recording of a minute geographical variability. This minute knowledge is used both for pragmatic effects and for symbolic reckoning, and one simply cannot argue that all memory knowledge exists solely for utilitarian ends. In fact, most memory devices refer to events that may have occurred only once or twice. Although it might have no contemporary pragmatic purpose, this kind of knowledge is not dismissed and over time becomes part of the cultural lore of a society. While memory devices for history and myth are related and regulated by the structure of society, the stored cultural and personal content and the structure of these devices may be variable and complex. Stored knowledge may be retrieved for personal and social needs, yet at the same time knowledge is also utilized in ways which need not be pragmatic.

In this chapter, I illustrate some of these thoughts with ethnographic observations made among the Mandaya, a group in the southern Philippines (fig. 6.1).

Upland Mandaya Economic and Social Structure

The Mandaya reside on the coastal and upland areas of southeast Mindanao, Philippines, in the provinces of Davao Oriental and Surigao. Early accounts of this group come from the Jesuit *cartas* of the 1850s to the 1880s. The Catholic Church and the Spanish colonial administration were moving southward in eastern Mindanao at this time with the aim of establishing missions and military settlements to curtail the Moro expansion along this coast.

With the coming of the Americans after 1900, the missions became municipalities. Today the towns of Mati, Manay, Caraga, Baganga, Cateel, and Lingig are major coastal settlements. Most of these towns are populated by Mandaya and by Visayans who moved into southeast Mindanao after the early 1930s.

The upland Mandaya occupy the eastern cordillera of Mindanao. This mountain range runs north and south from Surigao, continuing south through Davao and to Cape San Agustin. Interior settlements are generally located up to an elevation of 4,000 feet (see fig. 6.1). In some cases, Mandaya have moved beyond this elevation, though the effects on swiddening are highly negative. The early ethnographic accounts by Cole (1913) and Garvan (1931) indicate that formerly the Mandaya were seldom found higher than 750 m. The push into the higher interior is a result of the extensive nature of Mandaya slash-and-burn cultivation.

Like many other non-Christian, non-Islamic groups practicing upland swidden cultivation, the Mandaya have only recently become involved in a market economy. Abaca (hemp) production in swiddens that are no longer useful for upland rice is now found scattered throughout the foothills from the Mati-Tarragona area north to Cateel. Most of these small-scale abaca farms are still maintained by the Mandaya, but in certain localities Cebuano farmers have taken

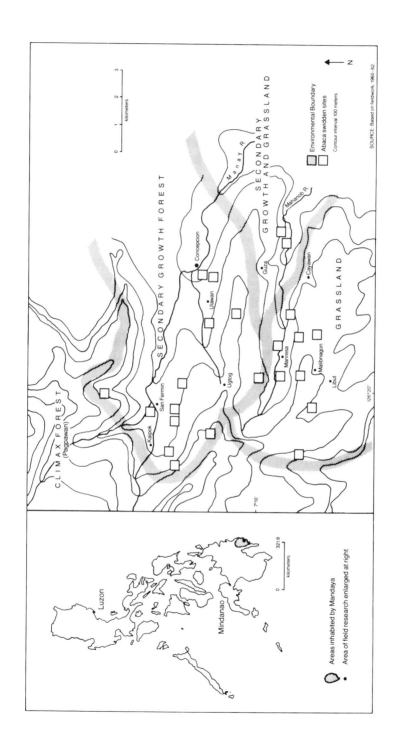

Fig. 6.1.   Environmental zones and abaca swidden sites among the Mandaya, southeastern Mindanao.

land from the upland Mandaya. The Visayans are also committed to a mixed farm system of rice, corn, and vegetables for household consumption, while abaca is planted as a cash crop.

Rice, along with a large variety of root crops, is the staple food of all upland Mandaya, especially those living at swidden sites at elevations of more than 750 m. The system of cultivation is similar in many ways to what Conklin (1957) has described for the Hanunoo of Mindoro. However, in eastern Mindanao, the burning of cut and dried forest growth and secondary growth seldom occurs above an elevation of 750 m. This is due to a high annual rainfall of 4,195 mm per year, with an average of 224 days of rain per year. Because the longest periods without rain are seldom more than five days, burning is seldom practiced, and with the absence of a pronounced dry season most burning results in smoky fires. Seldom does the vegetation burn off in a way that adds nutrients to the soil. The net result of the lack of sufficient burning is that rice yields are markedly lower in comparison with other systems in which swidden burning is practiced. Also, most Mandaya cultivators place their swiddens on high terrain slopes, which allow them to clear the forest by rolling cut trees and vegetation to the lower edges of the plot. This practice, although labor saving, results in much of the topsoil being washed off; bare areas are created that cannot be planted in rice. A more complete description of the cultivation cycle is found in Yengoyan (1964, 1965, 1966a, 1966b, 1966c, 1970a, 1970b, 1971, 1975a, 1975b, 1977).

Whereas among Mandaya communities who cultivate abaca village nucleation has started, upland rice cultivators still maintain a dispersed settlement pattern. This is characterized by a single field that is cultivated for one year or possibly a second year, with each household being located adjacent to the swidden. Households are moved as often as new swiddens are planted. Further, each household is composed of a single nuclear family or in a few cases a polygynous household consisting of an adult male, two or more spouses, and unmarried children. Neolocality is the general pattern of residence following marriage, a pattern considered to be the cultural ideal by the Mandaya. In rare cases, virilocality and uxorilocality as well as geriatric filialocality occur but most of these cases are the results of special circumstances.

Swiddens are generally located from 2 to 10 km apart. There is no settlement nucleation, but new swiddens are never located at sites from which one cannot see another household, although such neighboring households might be located across a valley or on a distant mountainside. Visual contact with neighbors is important since this may be the only means by which one can obtain information. Reciprocal labor exchanges emerge from these dispersed households. Most work in the fields is done by family and household labor, but during rice planting and harvesting work groups are larger and are composed of a number of individuals from different households. One of the other reasons for locating households within sight of each other goes back to the early days of warfare. At the turn of the century and up until the 1920s or early 1930s, warfare was endemic among the uplanders. Most warfare revolved around the *bagani*, political leaders whose fame and power was in part generated and maintained by their prowess in war and armed conflict. Each *bagani* had an entourage, which came from closely knit settlements and which, in turn, raided distant communities. Attacks usually took place in the predawn hours when a household was burned, some of the elders killed, or young children taken as slaves. It was only through the burning of the house and the screams of the victims that neighbors could hear or see what was happening and could thus either flee or arm themselves. The Mandaya still claim that the distances between households are limited by the need to be able to communicate the occurrence of any dangerous event.

The largest kin group is the kindred, which normally includes all consanguineal kin plus affines up to fifth cousins. This kin group, or kin category, is called *akong kalumonan*, roughly defined as "all my relatives." *Kalumonan* comes from the root *lumon*, which means "siblings." The composition of the kindred varies over time and most members of one's kindred are dispersed. There are few, if any, occasions in which all members of an individual's kindred might come together for common action. However, the kindred is the group from which one selects individuals for particular kinds of assistance or the reciprocal exchange of labor and services. Furthermore, the kindred is not strictly an exogamous group. One can marry within his kindred, but the closer the kin connection in terms of marriage, the greater the amount of the fine a male must pay to his future wife's parents. Thus marrying a first cousin might mean that a man would give a cow or a carabao to his in-laws, while marriage with a fifth cousin would involve only a few days of labor service. A more detailed description of the kindred is found in Yengoyan (1964, 1973).

The community is the largest spatial unit. Among uplanders living in Pagpawan, the community encompasses a number of small settlements. Here, settlements are composed of dispersed households held together by the exchange of labor and other forms of assistance. Although the spatial arrangement of individual settlements depends upon the immediate terrain—the nearby valleys and mountains—settlements normally include all households within a walking distance of up to two hours. Since households are moved yearly, an individual might find himself in a number of different settlements over the course of his lifetime. A number of settlements compose a community; communities were historically linked to the presence of a *bagani* warrior. With the demise of warfare and the end of the *bagani* era, communities were characterized by a relatively high rate of marriage endogamy. In 1965, more than 80 percent of all marriages were outside of the settlement but within the same community.

Oratory in Myth and History

Mandaya myth and history may be understood by how they transect and combine with certain other aspects of Mandaya culture, especially with religious knowledge and belief systems on the one hand and with historic and geographic knowledge on the other. The oratorical tradition is based on the manner in which certain kinds of knowledge are retrieved and utilized to maintain the continuity of religious thought. This section, and the following one on mnemonic structures, discusses how knowledge is stored and retrieved. The utilization of structures of oratory and memory depends not only on the vehicle of expression, but, more importantly on the kind and source of knowledge.

Oratory and the overall ability to use and control verbal art is a highly desired quality among the Mandaya. A good speaker not only has a fine sense for turning a phrase, using a metaphor, or stressing repetition, but also possesses the wisdom and experience for animating knowledge and thought in a meaningful way. This combination is rare among individuals. It is difficult (if not impossible) to learn, and it is even more difficult to transmit. Almost all of the warrior leaders (*bagani*) seem to have this ability, but it is clear that one need not be a *bagani* to be a great orator. In fact, during the 1960s when I was doing fieldwork among the Mandaya, many claimed that the *bagani* tried to learn this talent and that in some cases they had advisers (*angtutukay*) who were selected partly on the basis of speech ability.

Given the changes that have occurred over the past thirty years, it is difficult to assess exactly how the system of oratory formerly worked. From the living memories of older people, however, we do have some idea of how it was structured. First, it appears that the vehicle of expression in oratory ranged from riddles and proverbs to the lengthy recitation of epic poetry (*dawot*), both in its classical and nonclassical forms. Oratory also included a variety of oral traditions, such as ritual songs, folk songs, folktales, morality songs, certain kinds of love songs, and some puns. Within this range of genres, no single orator could master the complete inventory of all forms of expression. Certain orators were called upon for particular forms of oration, especially those regarded as classical. A characteristic feature of classical oratory, particularly when it related to obscure accounts, was that very few listeners could provide a precise translation. In many cases, the whole effect was based on sound and lyrical composition. The reaction of the audience was that of a mesmerized body.

A second feature of oratorical structure raises a question about the common "demoninators," or themes, that embraced all these different forms of expression, and why some features were part of the oratorical tradition while others were not. In many of the more lengthy accounts, such as the Mandaya *tagadiwatanan* or the *gambong*, which are almost like epics or sagas, the sky is always the major point of reference. In attempting to understand which aspect of the oral tradition falls within oratory, the role of the sky as the source of meaning and action appears to be crucial. For the Mandaya, questions of morality, immorality and amorality, truthfulness and wrongdoing, and evil and goodness—all ethical issues—posited the sky as the initiator of action as well as the source from which the recited story or epic derived. In turn, the sky was also seen as being responsible for the way in which the orator interpreted the account to the audience. The function of oratory and erudition was not only to store and retrieve cultural information, but also to convey values to people who were earthbound. By contrast, many puns, riddles, folk stories, and ritual songs took as their source other referents, such as the ground, the forest, a rice field, and occasionally the sea. Relatively few of these accounts could be included as part of the realm of oratory, and their absence in the corpus of oratorical forms indicates that the sky was primary and special in explaining the basic values and cultural codes the Mandaya used to promote and judge action and thought. It appears that stories, poetry, and epics, which had their existence in the sky, were given to humans as a form of revelation that was not to be debated, nor could it be understood in any physical expression. At best, these accounts were accepted as cultural givens. In one sense, oral traditions that use the sky as the only referent of existence are comparable to what in Greek tradition is referred to as *The Word*. Most of these oral accounts, either in folk songs or lengthy epics, start with the phrase *long naan* ("it is said").

Mandaya oratory was performed by both men and women; both sexes could achieve a high level of oratory. Men controlled political and military functions, but most of the religious life and ceremonial activities were limited to part-time specialists or practitioners who were commonly women known as *ballyan* or *bailan*. Although most religious activities pertained to agricultural and medical matters, many female *ballyan* also had a well-developed sense of oratory. Oratory and high levels of erudition among the *ballyan* occurred frequently during trance-like states commonly linked to ritually charged activities. Yet there were times when women recited and interpreted epics in conjunction with men who were renowned as expert orators. Gatherings in which both sexes participated commonly occurred after the last rice harvest. These gatherings did not appear to be competitive in the sense that one orator would try to outrival another. In most cases, the quality of one orator as compared with another was not measured by the general ability to recite a legend, but by the emphatic and creative manner in which the moral message was conveyed. This required the use of both informal and colloquial expression and of highly formalized speech.

Oratory is the primary means of stating and conveying what myth and religion are, how they are to be interpreted, and how morality and cultural truths are related to all human situations. Mandaya accounts of *magbabaya* (which is vaguely conceived of as a supreme being), or of *tagamaling* (the fairy god who promotes goodness in individuals), require recitation that will allow all members of society to realize what inspires individuals to strive for certain goals. At the same time, the question of wrongdoing is regarded as an enduring form of evil that must be controlled and conquered.

Knowledge promoted via oratory deals primarily with questions of human conduct—actions and thought—on the basis of an understanding of values and morals, in the context of how the individual relates to family, to kinsmen, and to others. Almost all knowledge contained in myths and epics and conveyed through oratory is based on what it means to be a social being and how each individual fits into the social fabric. How the individual and society are constituted is the vital message in Mandaya myth.

Oratory is not the only major expression of the Mandaya concept of humanity; another is mnemonic structures. However, knowledge transmitted through oratory is markedly more generally accessible than knowledge vested in mnemonic structure. In defining these differences, we might conclude that myth is regarded as a revelation, an account accepted at face value and pertaining only to human conduct and thought. Just as myth is something special, oratory is special as well. Even though only a few individuals have oratorical ability, the message of myth is not privatized. It represents a form of collective knowledge that all individuals must comprehend. Even in folktales, where the verbal art of raconteurs makes the message special and unique, it must have meaning for all individuals who consider themselves Mandaya.

If oratory is the expression of myth, where does history fit into Mandaya cosmology? For the Mandaya, historical statements are events that sometimes have specific referents in the land and the ground, and more often in the forest and the cultigens the Mandaya have planted. History is the form of events that have transpired. All of these events are recalled or can be recalled through memory.

Memory and History

Mandaya oratorical traditions have their source and existence in myth, which is the primary means for understanding how the individual and society are created and how values are shared. The realm of history deals primarily with what humans have done with their environment, how they have developed through particular cultural events, and how these events relate to one another through genealogy and recent time. Mnemonic devices are probably the most important ways in which time and history converge, while genealogy —either factual or imputed—links individuals and groups to one another.

Nearly all mnemonic devices are related to plants, either cultigens or wild plants, or to localities in which significant cultural events have taken place. Most cultigens, such as rice, other grains, and vegetables, have a growing period of a year or less. Since these crops are planted annually in new swiddens, they are not used as memory devices. Cultigens which last for many years, however, such as fruit trees (durian, mango, banana, *atis*), are used as a primary means of tracking the sequential events attendant to the movement of families from one settlement to another. Each bearing fruit tree still gives the original planter a residual right over the swidden in which it was planted, even many years after the swidden has been abandoned. In cases where rights are given up, for instance through an exchange, memory still records the initial planter of that cultigen as well as the one who later obtained rights to that swidden. Thus the link between individuals who succeeded one another in any swidden is maintained through history by means of the cultigen itself. In one sense property determines the relationship between and among a succession of owners. My data on this subject, which were recorded in 1965, indicate that memory of particular swiddens goes back as far as twenty to thirty years, perhaps a little longer, depending on the longevity of particular fruit trees. Individuals move from site to site because of the extensive nature of upland swidden cultivation. Each new "owner" of a swidden notes all past "owners" of a site and how and when "ownership" was transmitted. This knowledge is orally conveyed as genealogy, while information about what happened—when and why—is part of the broader cultural lore. Thus, over the years stories and accounts of happenings, disputes, encounters, and other events are "registered" in terms of the original and still-living cultigen.

Tree growth, especially of those trees that are feared, is another memory device. In this regard, the *bud-bud*, or banyan, is the most important tree, one to which almost all Mandaya relate certain events. In turn, they avoid contact with it. *Bud-bud* trees are characterized by long branches that extend outward and by aerial roots that grow from branches toward the ground, take root, and eventually provide additional support. In some cases, the crown of a single banyan tree may cover an area ten to fifteen m or more in diameter. The dense crown provides a heavy shade so that it is usually very dark underneath these trees. The Mandaya claim that malevolent spirits such as the *asuang* are invariably found in such surroundings. Swiddens are never planted near banyan trees, and Mandaya always avoid walking within shouting distance of them. However, each of these trees signifies a long story of past events that brought harm to someone. Furthermore, the banyan is usually the nesting and resting place for the *limokon*, a white dove that plays an important role in Mandaya concepts of fear and death. Whenever the call of the *limokon* is heard, all activity is stopped and a Mandaya retires to his residence. If the call is unheeded, one is believed to be vulnerable to sickness and death, if not immediately at least relatively soon.

The *bud-bud* tree, the call of the *limokon*, and other frightful signals are not simply warnings for avoidance. They cue the recounting of events in the past and thus the transmitting of history. History—structured by memory—provides cultural continuity. In some cases observed during fieldwork, events that occurred as long ago as eight or ten generations back were discussed as if they had just happened. Memory is verbally transmitted, but each physical and geographical cue is used to describe events that have occurred and will recur. The concrete cues support and reaffirm the impact of the verbal account.

Offerings to spirits, both beneficial and malevolent, also involve the use of memory. One of the better examples is the offering of areca nuts to spirits who must be fed to maintain their beneficial qualities. Mandaya claim that areca nuts have a soothing effect on spirits so that their malevolent qualities are curtailed. On gathering examples of how this works, I was told that in the past these offerings were neglected only once, and at that time pestilence ravaged the rice fields.

The roots of the areca-nut plant are used as a form of divination in ascertaining how much must be given to the spirits. Abnormal or stunted roots are viewed as a sign that the spirits must be provided with nuts. The use of plant indicators to guide one's actions suggests that, within memory, some kind of connection was perceived between the state of the plant world and human behavior and then conceptualized in terms of the relationship of the spirits of the forest and of rice to human problems.

Memory, as it is conveyed orally and through physical cues, is also important in denoting localities considered dangerous. Not only are parts of the forest regarded with caution, but deep pools of water are also considered places in which dangers lurk. Deep pools as well as the tracks linking pools are avoided since evil spirits move along particular trajectories from pool to pool. The means of linkage need not be water; in many cases, spirits will move through the forest, over the land, and even from rice field to rice field. Paths of evil spirits are noted and this information is conveyed through mnemonic means. Directions of spirit movement between pools indicate that spirits travel in a linear direction. Thus, human paths are commonly curved and criss-crossed to avoid contact. Throughout the forest and open rice fields, markers indicate the direction in which humans should move. Each locality also has a story about past disasters that befell individuals who did not heed the advice concerning the movement of spirits. Not only is history expressed through spatial and physical referents, but the continuity of action indicates that past events may be used as indicators of what might happen if violations and trespassing provoke malevolent spirits.

Physical localities and certain cultigens, especially fruit trees, are the major means by which individuals relate to their environment. The use of memory to link individuals and groups with one another, on the other hand, is expressed in two ways. One is the role of past exploits or conflicts and feuds that were led by a specific *bagani* and his warriors; the other is the use of genealogy as history and as memory.

Almost all conflicts and raids that resulted in significant warfare are noted in the living memory of individuals, especially among elder males. In discussing the long-past exploits of a leading *bagani*, the elders recall such activities as if they had only recently occurred. However, when one tries to reconstruct the time-frame of these attacks, it appears that they cluster in the late Spanish period (1880–1900). Although some battles are noted after 1910, the fact of well-established pacification and the gradual demise of the *bagani* by the 1930s indicates that the high point of armed conflict ceased more than three generations ago.

The exploits of past warfare live on not only in the memory of some individuals; in some cases mock battles are held to demonstrate how the conflict took place. In the upper portions of Malibnagon, a number of clearings are remembered as the locations where battles occurred. When elder males travel through these clearings, the chants and songs of war are sung and a mock enactment is staged to show who was where, what happened, and who was killed at what spot. Bravery and heroism are noted and stressed as the exploits are repeatedly recalled in a manner meant to evoke all the spirits of the warriors and emphasize their meaning to the living generation. Clearings and localities of war are semisacred and future swiddens cannot be located near them. As expected, it is difficult to establish what really happened, since each succeeding generation embellishes the list of heroic deeds. These stories are passed on to children as a means of maintaining a memory of historical events which are not simply locked in time but become timeless. Memory is thus used as a means of moving events through time. Since the meaningfulness of these activites is not vested in the fact that they occurred in the past, memory creates a form of history that collapses the immediate past into the present. This feature does not occur with myth, which is related to questions of cosmogony, the origin of things, and how Mandaya humanity is constructed.

Genealogy as a mnemonic device is critical in relating history and culture to and through the present. Most Mandaya genealogies are characterized by their shallowness: they go back two generations, occasionally three, and seldom four. Although kinship terms for cousins extend to fifth cousins, the link between fifth cousins cannot usually be demarcated by tracing genealogical connections. Thus, an individual learns that another is a fifth cousin only by being told. Although there is a lack of genealogical depth for the average person, most genealogical connections spanning three generations relate to how individuals draw kinship ties to a famous *bagani*. Some of these ties, most of which are imputed and fictional, go back six to nine generations and are combined into a conical type of relationship, or what can be called "linking to the axial line." Cultural events relating to large ceremonies, warfare and conflict, pestilence, and population movement are "tacked on" to each axial line. The memories of past events are sequentially linked through conical and axial genealogical connections to founders or particular warriors.

Because the Mandaya are nonliterate, the historical dimension of their lives is invoked through memory devices, of which genealogical connections are one of the major means to project past events into the present. The use of geographic, environmental, and sociological idioms in the constitution of memory structures permits us to understand how history is transmitted through a number of cultural vehicles, the content of which is markedly variable.

Conclusion

The structuring of memory devices varies across cultures, yet we may note some parallels in how mnemonic devices are constituted and how they relate to the total environmental and cultural context. The use of physical referents as markers in oral tradition has also been noted among the Trobrianders by Harwood (1976), while Colby and Cole (1973) and Cole and Gay (1972) have shown how memory connects with narrative.

Among the Mandaya, memory devices are associated primarily with historical events; spatial and environmental cues are used in this context to relate the past to the present and future. Most recent events are incorporated into the mnemonic repertoire as a means of registering their existence for future discussion. Memory provides the basis for an ongoing dialogue in which individuals continuously discuss what happened, when, by whom, to whom, and what it meant. Over time, the memory of events changes, with reinterpretation adding new dimensions to the events and thus creating a continuous remetaphorization of the context.

The tradition of oratory is the major vehicle for expressing how cosmological and ontological forces emerged and what they mean in terms of Mandaya life. However, oratory does not only consist of the transmission of this cultural knowledge. It also involves the ability of the orator to state convincingly and forcefully how these values and philosophical concepts relate to behavior and everyday reality. The moral basis of human action is founded on a close and harmonious connection between the pragmatics of everyday life and the symbolic underpinnings that provide meaning and emotional sustenance to what people are doing.

Structure (as a set of rules, models, or patterns) and behavior (as the actions of individuals and groups) are among the central issues in anthropological theory. How they relate to one another in terms of linear causal models or dialectics or reflective theory is central to what anthropolgists conceptualize as the nature of society and the meaning of culture. Most anthropological theories going back to Radcliffe-Brown, to Malinowski, Murdock, Barth, and British structural-functional approaches have held that rules and patterns are expressions, explicit or implicit, etic or emic, of what is happening on the ground and of how individuals and groups cooperate and behave. The philosophical basis for these views is best exemplified by the rampantly behavioristic social-science theories anchored in Western thought that hold that rules, models, and patterns can only be analyzed through their behavioral expression. Furthermore, it is assumed that, if rules and patterns cannot be related to behavior, they must be in the realm of metaphysics. Epistemologies prevailing over the past two hundred years have left no place in science for metaphysics.

The major theoretical point of this chapter is that in many societies rules as structures are anterior to behavior. Behavior is structured. The existence and source of the structure are to be determined, not in terms of behavior, but in the moral and ontological principles that provide content to structure. For the Mandaya, myth is the critical source of all rules that pertain to Mandaya humanity. Whereas myth relates to how moral behavior should occur and what it means, the system of mnemonic devices acts as a set of coded structures that allow individuals to deal with the contingencies of everyday life in a historically meaningful way.

Acknowledgments

Fieldwork among the Mandaya was conducted for twenty-one months during 1960–62, for four months in 1965, and for three weeks in 1970. I wish to thank the following foundations for their support of this work: the Foreign Area Training Fellowhsip Program of the Ford Foundation (1960–62) and the Agricultural Development Council, Inc. (The Rockefeller Foundation) for the fieldwork in 1965.

References

Bateson, G.
1958   *Naven.* Stanford: Stanford University Press. 2d ed.
Colby, B., and M. Cole
1973   Culture, memory and narrative. In *Modes of Thought,* edited by R. Horton and R. Finnegan. London: Faber and Faber.
Cole, F. C.
1913   *The Wild Tribes of Davao District, Mindanao.* Chicago:

Field Museum of Natural History.

Cole, M., and J. Gay
1972   Culture and memory. *American Anthropologist* 174:1066–84.

Conklin, H. C.
1957   *Hanunoo Agriculture.* Forestry Development Paper, No. 2. Rome: Food and Agriculture Organization of the United Nations.

Garvan, J. M.
1931   *The Manobos of Mindanao.* Memoirs of the National Academy of Sciences Vol. 23. Washington, D.C.: Government Printing Office.

Hage, P.
1978   Speculations on Puluwatese mnemonic structure. *Oceania* 49:81–95.

Harwood, F.
1976   Myth, memory, and the oral tradition: Cicero in the Trobriands. *American Anthropologist* 78:783–96.

Yates, F. A.
1966   *The Art of Memory.* Chicago: University of Chicago Press.

Yengoyan, A. A.
1964   Environment, Shifting Cultivation, and Social Organization among the Mandaya of Eastern Mindanao, Philippines. Ph.D. dissertation, Department of Anthropology, University of Chicago.
1965   Aspects of ecological succession among Mandaya populations in Eastern Davao Province, Philippines. *Papers of the Michigan Academy of Science, Arts and Letters* 50:437–43.
1966a   Baptism and 'Bisayanization' among the Mandaya of eastern Mindanao, Philippines. *Asian Studies* 4:324–27.
1966b   Marketing networks and economic processes among the abaca cultivating Mandaya of eastern Mindanao, Philippines. Report to the Agricultural Development Council, New York.
1966c   Marketing networks and economic processes among the abaca cultivating Mandaya of eastern Mindanao, Philippines. Reprinted from report and abstracted in *Selected Readings to Accompany Getting Agriculture Moving,* edited by R. E. Borton. New York: The

Agricultural Development Council, Inc. Vol 2:689–701.

1970a   Open networks and native formalism: the Mandaya and Pitjandjara cases. In *Marginal Natives: Anthropologists at Work*, edited by M. Freilich. New York: Harper and Row. Pp. 403–39.

1970b   Man and environment in the rural Philippines. *Philippine Sociological Review* 18:199–202.

1971   The Philippines: the effects of cash cropping on Mandaya land tenure. In *Land Tenure in the Pacific*, edited by R. Crocombe. Melbourne: Oxford University Press. Pp. 362–76.

1973   Kindreds and task groups in Mandaya social organization. *Ethnology* 12:163–77.

1975a   Introductory statement: Davao Gulf. In *Ethnic Groups of Insular Southeast Asia: Volume 2, Philippines and Formosa*, edited and compiled by F. M. LeBar. New Haven: Human Relations Area Files Press. Pp. 50–51.

1975b   Mandaya. In *Ethnic Groups of Insular Southeast Asia: Volume 2, Philippines and Formosa*, edited and compiled by F. M. LeBar. New Haven: Human Relations Area Files Press. Pp. 51–55.

1977   Southeast Mindanao. In *Insular Southeast Asia: Ethnographic Studies, Section 4: Volume 1, Philippines*, compiled by F. M. LeBar. New Haven: Human Relations Area Files. Pp. 79–116.

CHAPTER 7

BOUNDARY AND BATIK: A STUDY IN AMBIGUOUS CATEGORIES

Alice G. Dewey

For many years I have studied the peasant markets in Java, and the longer I have done so the less satisfied I have been with the usual Western economic view that they consist of small individualistic competitive units perennially short of capital and unable to mount cooperative efforts that would enlarge the scale of operations. In some sense these things are perfectly true, yet they fail to describe some very fundamental principles of market organization that I have found difficult to express in abstract terms, but which I have observed in personal behavioral terms. For example, traders who were overtly competitive were sometimes close kinsmen who commonly cooperated; in fact, such cooperation was not restricted to kin. While I was puzzling over this problem, I came upon an article by a biologist discussing the ecology of mussels. The researcher had come to a conclusion that puzzled him. He summarized his findings by saying: "Yet cooperation and competition are generally viewed as being virtually opposite extremes of organism interactions. My results suggest the reverse. In certain cases, interspecific competition may provide the very selective pressures that lead to the evolution of cooperation" (Buss 1981, 1014). Ecologists, then, who look at organisms in their natural context seem to be seeing phenomena that parallel those I was observing, and yet their findings were felt to be logically contradictory.

I believe this sense of incongruity arises from a tendency of Western cultures to put things in a cognitive framework that polarizes things and conceptualizes things as having hard-and-fast boundaries. This predisposition has been crystallized in scientific procedures in which the item under examination is abstracted from the complexity of its natural setting and is purified and rigorously controlled in such a way as to render the results of experiments as unambiguous as possible. This has been an enormously powerful approach to understanding our world, but I believe that at the same time it systematically masks our ability to perceive some aspects of the situation. In time these will force themselves on the observer, as can be seen in the case of the biologist studying mussels. As an anthropologist, I began to wonder if other cognitive models existing in other cultures might be more amenable to seeing things in the context of participating in multiple relationships of mixed quality. The example of the cooperative and competitive relationships of mussels showed they not only coexisted but were mutually productive of a more successful adaptation than either alone would have produced. Were Javanese market traders more successful because they mixed cooperation with competition? Since over the years I have begun to understand the way the Javanese look at the world, their "eidos," to use Bateson's (1958, 25) term, and because I sense that this is rather different than the eidos most common in the West, I turned to Javanese culture to seek an alternative model. As I did so, I realized that my interest in various aesthetic domains of Javanese culture had already begun to lead me to a different world view, some aspects of which I will try to sketch in this chapter. After analyzing batik patterns and a number of other cultural domains, I will attempt to apply the approaches derived in this way to other, more mundane aspects of Javanese culture, most especially the behavior of people in the market. I start by looking at traditional Javanese batik patterns.

The aspect that serves as a focus is the handling of the definition of design elements, their bounding, and the interrelationships among elements. If one views these patterns with a more typically Western frame of reference, there is often ambiguity in the definition of elements. From a Javanese point of view, I believe, a greater complexity of boundary relationships is possible and shifting identities are not felt to be ambiguous. Theirs is a view that allows the world to keep more of its holistic integrity. Things are seen in context with first one element or pattern highlighted, then another. Foreground and background interchange and boundaries shift so that elements bond or break up and interdefine each other.

Batik

One of the first things to strike the eye in Java is the prevalence of negative images. They are found not only in the shadow-puppet plays but also in all of the major cloth-decorating techniques of batik, ikat, tie-and-dye, and tritik, which are all dye-resist techniques. That is to say, the cloth is waxed or tightly bound in various ways so that the areas treated resist the dye; when the process is finished the treated areas will appear in the "negative" or uncolored state. In positive decorative techniques, wherever the cloth (or other surface) is touched, a mark is made and untouched areas remain unaltered. In negative techniques, the cloth is altered everywhere that it is not touched. The habit of producing designs in this way may possibly lead those who do them to be more aware of the structure of both the colored and uncolored areas simultaneously.

If one examines batiks, it can be seen that they commonly display characteristics that suggest this is true. First, many patterns exist in both light-on-dark and dark-on-light versions, rather like a photographic negative and the print produced from it. The pattern, then, has been conceived as focusing first on the undyed and then on the dyed portions. Second, there are patterns for which there are two ways of conceptualizing the structure, one by focusing on the dark areas and the other by focusing on the light areas created by the dark areas. A simplified version of this phenomenon is met with in the Chinese yin-yang symbol. A circle is divided into light and dark areas that meet in an S-curve such that the light area defines, and is defined by, an identically shaped, but inverted, dark area. The pattern is so structured that at the interface of light and dark areas one can conceptualize it equally as the dark area forming the background of the light area, or the light area forming the background of the dark area. In the case of the yin-yang symbol, the two shapes are identical; in a batik they would usually be different, but each would be capable of standing visually on its own as an aesthetically satisfying motif and be equally worthy of attention. It is as if a craftsman, in making a stencil, paid as much attention to the aesthetics of the form of the stencil as to the pattern of the colored areas it would produce, so that when the process was completed the blank areas would create a pattern of equal importance to the colored areas. In such a pattern, figure and background are reversible and each area stands in turn as the focus.

Another aspect of pattern manipulation can be seen in batiks where the elements can be grouped into a variety of units, the boundaries of which shift in complex ways. One of the patterns in which this can be seen most clearly is the classic *kawung* (fig. 7.1). Most simply, it consists of diagonal rows of ovals set at right angles to each other with diagonal rows of diamonds set into the spaces between the ovals, alternating with a motif consisting of four tiny circles set in a cluster. Inside the curve at each end of each oval is a small cross. (There are many variants of this pattern; for example, the crosses may be replaced by dots.) Westerners commonly assume that the ovals form the petals of a four-petalled flower with the tiny circles being the stamen in the center and the diamonds acting as background filler. In fact, one sees examples of the *kawung* elaborated into just such a flower. Javanese, however, say that the basic meaning of the pattern, the thing it represents, is a fruit cut in half, with the ovals being four seed chambers lying within the fruit, the diamond forming the core, and the crosses signifying the seeds within the chambers. The same basic elements can thus be conceptualized in two quite different ways. In one, the ovals group around a central diamond and the clusters of circles lie at the periphery. In its other interpretation, the small circle motif moves into the center, with the ovals radiating out and the diamonds fitting into the space between the outer ends of the petals and forming a border for the flower pattern. The conceptualizations crosscut one over the other as the eye singles out first one and then the other possible interpretation. This same sort of shifting conceptualization characterizes many batik patterns, and in some cases more than two constructs can be made. There are interesting parallels here to mathematical set theory.

In a recent article, Edmund Carpenter (1980) discusses this same phenomenon. He points out that what he calls visual punning is common in ancient (and non-Western) art. He includes plates of such things as New Guinea masks and a Pre-Columbian gold plaque from Panama vividly demonstrating dual images. They are designed in such a way that two quite separate figures can be seen, one when the piece is viewed one way up, and the other when it is inverted. In some cases, the duality does not depend on the orientation of the piece but on the shift in the viewer's perception. The batiks described here fall into this class. Carpenter noted that this technique appears occasionally in Western art and cites Dali's "Paranoiac Face" and

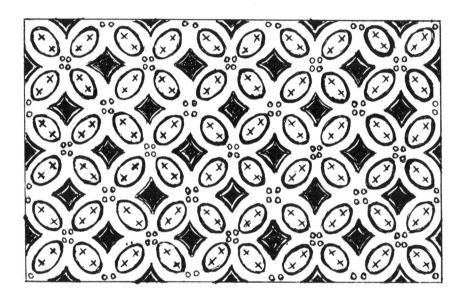

Fig.   7.1.   The classic *kawung* Javanese batik.

the work of some surrealistic painters. I must admit that I find these works clumsy and contrived. I think it is also important to note, as does Carpenter, that these paintings are considered to lie outside the normal as the very title of Dali's work reveals. Carpenter goes on to say that dual images have been considered in the West to be the work of madmen and children, but his research indicates that neither children nor psychotics produce true dual images. For some years he has asked his students to attempt to produce drawings of such images and has discovered that they find the process frustrating and difficult; he further notes that, as art, the results are awful. He quotes M. C. Escher, the one Western artist who has mastered the technique, as saying that the process of drawing a line that has a double function with a meaningful shape on each side is an complex business. Thus, Westerners have trouble producing dual images, and Carpenter attests that they often have trouble seeing them in non-Western art. When they do meet them, they are often disturbed and tend to think of them as the product of disturbed or childish minds. And yet Javanese and other non-Westerners obviously produce them with ease and elegance, finding them aesthetically rewarding. The need of the Westerner to deal with what Carpenter calls "one-thing-at-a-time" is clearly not felt by people from many other cultures, Javanese among them. However, there may be non-Western cultures that have yet other views of the world. Hindu culture, with its caste system and fear of pollution by anything that crosses boundaries, may be an example of such a variant.

Gamelan

In the domain of Javanese classical gamelan music, there are similarities to the patterns I have described for batik. Judith Becker (1981, 176) has described this so elegantly that I will quote her at length. She writes:

> One of the difficulties of recording gamelan music is that there is no absolute ideal of balance in the sense of that for a Western symphony orchestra. All the parts of a gamelan are there to be backgrounded or foregrounded according to the interest or focus of the listener, unlike a symphony in which one is invited by the composer to notice certain instruments at certain

times. The best way to hear gamelan music is from inside the gamelan, in which case one never hears the parts balanced. Those instruments nearest are prominent, some may be only heard faintly. The center of focus can be anywhere. The ideal Javanese listener will follow one part for a while, then another, and another as his or her own interest leads.

As I have argued, there is no single true meaning, no one focus; rather, one is faced with a multifaceted richness.

If we look at the instruments of the gamelan orchestra, another interesting and characteristically Javanese feature emerges. In contrast to gamelan, Western music has been created and played on a fixed scale of notes so that when a score calls for C# it is expected that any instrument playing that note will be tuned so as to play a note as close as possible to the ideal sound-frequency of that note, and a singer is praised for having "absolute pitch." Each Javanese gamelan orchestra, however, is a unique set of instruments, and (with the exception of a few, mainly tunable stringed instruments) they cannot be moved from one gamelan to another, for each gamelan is permanently tuned to a somewhat different set of pitches, and in fact within a set of instruments one may even find a slight difference between one octave and the next. The cast bronze gongs and keys, which constitute the majority of the instruments, cannot be altered in pitch without great difficulty. Rather than finding fault with this variation, the Javanese value it, for it gives variety—some gamelans being bright and lively in texture, for example, and others being solemn. The ideal scale, or scales (there are two sets of instruments in gamelan: one tuned to a seven- and one to a five-tone scale), were never traditionally given precise definition and even at an abstract level do not have single correct forms. There are, of course, limits beyond which the notes cannot properly vary, but within those fairly flexible limits there is no meaning to the statement that a particular gong is "off key" in an absolute sense; it can only refer to its relationship to the other gongs, etc., in that ensemble. Singers, on the other hand, do sing with more than one gamelan, but rather than trying to achieve some "absolute pitch" they develop their own version of the two scales, which they may have to modify somewhat depending on the gamelan with which they are performing at the moment. Perfection, then, lies not in a rigorous matching of the instruments to exact pitches, but rather in the creative use of the unique qualities of a particular set of instruments related to the sung parts to produce a pleasing version of a piece.

The difference in the musical domain between a focus on an absolute standard and ensembles that vary from one to another but are internally matched is a matter of cultural preference, not of better or worse. If we examine a similar situation in a more mundane domain we find an interesting parallel. This example concerns the production of ready-made clothing. Western buyers are often distressed by the fact that Javanese seamstresses do not seem to pay sufficient attention to producing garments in exact sizes and tend to assume the problem lies in lack of skill, laziness, or a lack of business acumen. In my experience, Javanese craftsmen and women are skilled, hardworking, and knowledgeable in business, and I felt the situation needed further exploration, especially as the concept of *cocok* (the match or fit of things to each other) is very important in Javanese thought. I found that Javanese usually make clothing to order for a particular customer, and, since people come in an endless variety of sizes and shapes, a good fit can only be achieved by making the garment match the unique form of the customer. This would only rarely be identical to a standard size. Javanese cannot believe that foreigners really come in perfect size 10s, 12s, 14s, etc., so they have trouble understanding why standard-size garments should be produced since they will not fit any better than, or even as well as, garments of more varied sizes. There is a good deal to be said for the Javanese attitude which focuses on the fit between the garment and the wearer. Western commercial sizes were designed for the convenience of mechanization, not for the comfort of people. Javanese judge the fit of things not in terms of rigid absolute standards, which attempt to force them all into identical molds, but rather in terms of how well they fit with each other—how *cocok* they are. Moreover, the better a thing is fitted to its function the more efficiently it will perform. Diversity, then, is felt to be both pleasing and practical. The idea of uniformity in itself as desirable and good does not, I think, exist in traditional Javanese culture. Where uniformity is needed it is suitable and good, where it is not needed it is irrelevant, and in a world not dominated by machines it is not usually needed. Put another way, a jacket that does not fit you is not a "bad" jacket; it is merely unsuitable. It may fit someone else perfectly and for that person be a "suitable" jacket. *Cocok* is a judgment not of single things against an ideal but of the way two or more things interlock, so that it does not really make sense to say the jacket itself is "good." It is the relationship that is the focus of judgment. The fit of the jacket to the wearer is good, not the jacket itself. A "bad" jacket is one that is poorly made, one that will never function properly for anyone.

Wayang

If we look to the stories portrayed in the dance dramas and the shadow plays (*wayang orang* and *wayang kulit*) we find another domain where variety is needed in order to fit various functions and where there are many admired characters, not just a few stylized heroes and heroines. There is a great variety of characters in *wayang*, ranging from those who are extremely refined and restrained (*alus*) to those who are very coarse (*kasar*). The qualities of *alus* and *kasar* are not, as some Western scholars have suggested, a set of polar opposites with *alus* considered good and *kasar* viewed as bad. A better description of them is that they are contrasting qualities like salt and pepper, each good in its appropriate place, the particular mixture depending on the demands of the recipe. Similarly, the balance of *alus* and *kasar* in *wayang* characters is *cocok* with his or her nature. Some, like Prince Yudistira, are extremely *alus*; others, like King Rahwana, are markedly less *alus* and have a strong element of *kasar* in their makeup as is suitable to a demon king. Others, like the ogre Terong, are completely *kasar* and lack any touch of *alus*. Then, there is the complex clown-god Semar who is very *kasar* externally but wise and powerful internally. The particular mixture of *alus* and *kasar* in each character is essential if that character is to play its proper part in the story.

It is also important to note that there are characters of all types among both friends and enemies who are much admired by Javanese. One of the most respected figures is Kumbakarna, the gigantic, ugly, immensely powerful brother of the demon king. He fulfills his role to perfection, being the essence of a giant in size and strength, the compassionate elder brother protecting a younger sibling, and a loyal subject who bravely gives his life for his country in battle against the hero, Prince Rama, who mourns him after his

death as a noble and honorable warrior. In fact, when one shifts the
focus of perception away from Rama as hero and sees the pattern
from the other side, Kumbakarna becomes a hero.

If one turns to the *wayang kulit*, or shadow-puppet plays, one sees
another instance of the use of the negative image and
complementary patterns. The puppets are designed to present
images that relate to each other in quite complex ways. They are
made of leather scraped thin so as to be slightly translucent. The
leather is cut into the outline of the figure and internal details are
defined by cutting pieces out so that light shines through in lines and
points indicating eyes, the texture of hair, the pattern of clothing,
etc. On an unpainted puppet these cuts are clear and it can be seen
that the puppet and its shadow relate to each other much as do a
photographic negative and its print (an interesting parallel to the
first class of batiks noted previously). A finished puppet, however, is
painted in bright colors and intricate patterns with the result that the
pattern of the cuts is largely obscured. This means that the
appearance of the puppet is quite different from the appearance of
the shadow it throws on the screen. When a play is performed, the
audience is free to move around, and it is quite common for an
individual to spend some time viewing it from the shadow side and
some time watching the puppeteer manipulate the puppets
themselves. Both faces of the puppet are meant to be seen, and its
quality is judged by the aesthetics of both aspects. Here, truly, the
stencil is as important as the image it produces. In a real sense, a
puppet is both itself and its shadow. The reversal is not that of figure
and ground set side by side, as the interfaces of light and dark zones
reciprocally define different images in batik, but rather that of a
figure and its hidden obverse, the subtly different shadow. The
complexity of this relationship is enhanced by the manipulation of
the puppets as they are brought to lie flat against the screen then are
moved back and tilted and twirled so that the shadow grows larger or
smaller and becomes lopsided or fades away at one edge. In the days
prior to electricity and before Coleman lanterns became common,
light was provided by an oil lamp whose flickering flame gave the
shadow image even more movement and flow. The various shapes
that the shadow can take are constrained and given continuity
through the puppet that creates them, despite the fact that they
have no substance in and of themselves, and thus no stable shape or
precise boundaries. The shadow, then, has a crisp central form that

expands fluidly until it becomes so diffuse that it eventually loses definition. Though it is still linked to the focal shape, it no longer can be comprehended as a representation of that shape.

It is interesting that Westerners often have difficulty grasping the implications of this dual nature of the *wayang kulit*, where the shadow side represents a kind of inversion of reality, where shadow represents substance and normal rules reverse themselves. One realizes the inflexibility of the Western view of the world when one sees tourists at a shadow play (which takes place at night) attempting to photograph the shadow side using a flashbulb. They seem to assume that anything they view will be more clearly photographed if it is flooded with light. They cannot cope with the idea that in *wayang kulit* they must learn to see darkness as well as light.

In the *wayang kulit*, the audience takes a more active part than does the audience of a Western play. As they move from one side of the screen to the other, they alternatively perceive the image as puppet and as shadow and give continuity and unity to the multidimensional experience of the *wayang kulit* in much the same way as the puppet gives continuity to the many-shaped shadow. The viewer of batik and a member of the audience at a gamelan performance also participate in creating shifting, overlapping patterns as he or she moves closer or farther away and focuses first on one conceptualization and then on another. Javanese viewers fully comprehend a phenomenon only by shifting their point of focus repeatedly so that they can perceive the various potential patterns. Meaning, then, is sought in the interaction between the phenomenon and its various contexts, and there are as many meanings as there are relationships. To isolate a phenomenon from its context, to put a sharp boundary around it, is to destroy the full complexity of its meanings and hinder understanding by singling out a partial, distorted meaning.

Javanese Peasant Markets

To test the utility of using a Javanese model to comprehend Javanese behavior, I have chosen the peasant market system since I have been studying it for some years. One problem immediately attracted my attention: the question of the supply of capital and the size of the units in the market system. The usual conception of

Javanese markets, as of other peasant markets, is of a multitude of very small, competing, independent units consisting often of single individuals or family units supplementing family labor with two or three hired workers at most. Generally, it is further assumed that the capital resources available to these trading units are very low and that this constrains their commercial activities. Despite the overt accuracy of this characterization, my impression during my initial study of a market in East Java during 1953–54 was that, when there was an opportunity for profitable transactions, sufficient capital could usually be found to finance them. This impression was given support some years later when Professor Mubyarto of the Economics Department of the Universitas Gadjah Mada told me that in a survey administered by members of his department only the wealthiest traders complained that a shortage of capital seriously hindered their operations.

The question then becomes, why are the perceptions of the traders themselves in regard to their capital resources so different from the appraisal of the trained economists? At least part of the answer, I believe, lies in the nature of the units of trade. On one level they are, as I have said, small, self-sufficient firms, each individual or small family group managing the business, making decisions, bearing risks, and taking profits and suffering losses in competition with all similar units and independent of all other units of like or unlike type. Nevertheless, because of their very independence, they are free to break their boundaries and shift their functions as convenience dictates. The patterns are not of static isolated units of fixed form, but rather are similar to a batik in which the assemblages can shift from moment to moment. There are several ways, for example, in which the barrier of competition becomes a tie of cooperation, shifting the competitive boundary outward so that opposed units fall within a larger grouping. One of the most straightforward ways in which this happens has been described in my first work (Dewey 1962) on the market system. In that work, it was seen that during the height of the harvest season for onions, the most important local cash crop, there was a desire for greater amounts of capital to handle the increase in goods flowing through the market since dealing on a larger scale could bring economies of scale and wholesaling advantages. At this time small groups of traders—sometimes as few as two or three, sometimes as many as seven or eight—would pool their capital and become a unit instead of competing. Not uncommonly, a trader would be involved in several deals at once. The personnel of any two partnerships might be identical, or they might overlap while including other partners in either or both of the groups. Or the two partnerships might be completely separate except for the presence of one linking individual who might at the same time be operating independently for some transactions. The smallest units, the single individuals or families, therefore, at one level retained their separate identities as competitive units, while at the same time they entered into shifting cooperative arrangements with their competitors. The definition of the relations between the units is thus situational, and the boundaries that separate them can become internal partitions bonding the units into a larger whole.

Capital and supplementary labor can also be made available by a different set of strategies, that is, by the conversion of competitive relations between two traders into an arrangement where one supplies the other. This is seen commonly among traders who handle batik and other types of traditional cloth. There is a seemingly endless variety of patterns and combinations of patterns, and no trader can hope to begin to stock the whole range. When a customer comes into the market, unless there is an existing special relationship with a trader, all of the traders make lively efforts to attract his or her attention. However, once negotiations have started with a particular trader, the others fall quiet until the customer has finished dealing with the first trader and indicates a desire to view the stock of another trader. In fact, the other traders may help a rival in making a sale by extolling the merits of a piece being offered. At this stage, the relationships have shifted from overt competition to a pattern in which the primary trader takes the focus and the other traders play neutral or supportive roles. Should the customer ask for an item that the primary trader does not have, she (most of the traders are women) can go to other traders and procure the cloth requested. There are various systems used, but in all of them the primary trader buys the cloth from the second one at a price below that which she hopes to get from the customer. The relationship between the two traders has again shifted from that of competitors occupying coordinate positions in a market system to supplier and buyer in complementary positions.

If we step back and look at the implications of the flexibility of relationships between traders, it becomes clear that, from this perspective, other interesting facts emerge. Each trader has an inventory at her command far larger than her own holdings, and, conversely, she has a pool of labor available far beyond her own efforts. Seen in this way, a pattern of small, separate, competing units transforms itself, and the cloth traders can now be viewed as forming an unfocused, vaguely bounded unit in which the total inventory and all of the labor represented by all of the traders in the market (and in fact other traders outside the market) are linked. Within this larger aggregation, the capital and labor resources that are potentially mobilizable in a large market are formidable, and even in a small market they are substantial and are far more than any one trader could possibly need unless he or she is operating on a very large scale. The shift in perspective has allowed us to see why the traders do not see themselves as hindered in their operations by a lack of capital.

This collectivity has no set boundaries, no fixed structure, and many potential foci. Like a batik pattern, it is capable of being conceptualized in complex ways as the smaller units are bonded to each other in constantly shifting groupings altering their focus or boundary relationships. These multicentric entities are difficult to describe or categorize, but any attempt to force them into a mold that would freeze the process shifts and tidy up their boundaries would so distort their essential nature as to make it difficult for us to understand them.

If we turn to the approach used in ecology, we find a number of parallels between it and the model of the Javanese market as I have described it. Most obvious is the fact that individual organisms, like individual traders, are not removed to a controlled setting in a laboratory, but rather are examined in their context. Interest is not just on the individual units but on the interrelationships through which their position within the system is defined. The web of interaction is the primary and characteristic domain of study. A particular study focuses on one species or a linked set, such as a predator and its prey, and the web is traced outward as far as is convenient or until the web becomes so interwoven with other relationships that its relationship to the focus becomes inconsequential. Another study may choose a different focus and look at the web of interrelationships from that point of view. Thus, various constellations and arrangements are highlighted in much the same way as I have dealt with the traders, or as Javanese look at a batik pattern, or listen to a piece of gamelan music. And, as in this music, importance lies not so much the absolute number of individuals, or the quantity of calories flowing through the system, as in the ways in which individuals, species, and calories fit into the pattern of relationships established in a particular system.

Since ecology examines behavior in natural settings, it seems reasonable that the concepts used would have parallels to the context-sensitive models of Javanese thought. Since social anthropology also works primarily with behavior in field settings, Javanese models may be of use here too. As a tentative test of their utility, I suggest a brief look at a common problem, namely, the relationship between a society's rules and the people's actual behavior. A common procedure is to see rules as generating behavior, but behavior does not follow rules in any straightforward way. Lévi-Strauss seeks to explain this by positing the existence of a level of deep structural rules, of which the people are not aware but which actually control their behavior. He likens behavior to a picture puzzle and the cutting of it to folk-structural models with the deep structure as the thing that drives the cutting saw, saying that " . . . its key lies in the mathematical formula expressing the shape of the cams and their speed of rotation; something very remote from the puzzle as it appears to the players although it 'explains' the puzzle in the one and only intelligible way" (Lévi-Strauss 1960, 52).

There seem to be serious problems with this metaphor, and a look at Javanese cognitive models, which focus on the way elements interact to create more complex entities synergistically, suggests a better description of picture puzzles and of the relationship of social-structural rules and behavior. Let us first look at the nature of picture puzzles. Their essence lies in the randomness of the pieces as they relate to the picture reproduced on their surface. The pieces are produced by cutting the picture up in some purposeful way, possibly by a camshaft-driven saw. The cams would produce a regular, comprehensible pattern of the cuts, but this pattern must have no relationship to the picture or else the puzzle would be very simple (as some made for very young children are). The goal of the cutting is to destroy the picture and reduce it to incomprehensible bits. Thus, we have a comprehensible picture and a regularly patterned set of cuts, which, through the fact that they are irrelevant to each other, create a picture puzzle. The interface between the two is the essence of the puzzleness, and it is at this synergistic level that we must seek explanation, not at the level of the mathematical formula, which determines that the particular curve of the cuts is incidental. There is, after all, a virtually endless series of ways in which a picture could be cut up that would effectively reduce it to meaningless pieces. The real meaning of the puzzle lies in the mental process of the doer who sees through the random relationship of cutting and picture and manages to reconstitute the picture.

If the picture-puzzle metaphor does not work very well, can a better one be drawn from the Javanese models I have discussed? I suggest that an interesting one can be found in the shadow-puppet plays. Here a puppet interacts with the light as the puppeteer moves it in such a way as to throw a variably shaped shadow on the screen. The puppet, like social rules, is fixed. The light and the screen can be likened to environmental resources and constraints, and the puppeteer to the human actors. The shadow is then the resulting behavior in all of its variety. And just as the shadow is the fundamental focus of the *wayang kulit*, so is behavior the fundamental data of anthropology. Metaphors should never be pushed too far, but I submit that this one is more thought-provoking than that of the picture puzzle, because it more closely parallels the phenomenon it represents.

Acknowledgments

This article is the result of some thirty years of interest in Java. The people of Java who have helped me understand their culture during these years are, of course, too numerous to recount. Their unending courtesy and patience with my strivings to understand will, I am sure, be extended to my failings in this latest effort. Certain people, including some non-Javanese, should be mentioned because of their help in the writing of this article. First, Professor Hardjo Susilo of the Music Department of the University of Hawaii has for years let me be a guest of his gamelan group. He (and all of the other participants) have been a rich source of insight and a continual pleasure for me. Among these, Joan Suyenaga, anthropology student and gamelan musician, deserves special mention. It was she who pointed out that both *batik* and *wayang kulit* deal with negative images and who explained the problems of pitch faced by those who sing with a gamelan. Another important person in the development of my understanding is Dr. Roger Long of the Department of Drama and Theatre of the University of Hawaii. He has helped me in many ways to understand the *wayang kulit*. And it was Dr. Ruth Sando, formerly of the Department of Anthropology of the University of Hawaii, who showed me the importance of active participation by the audience at a *wayang kulit* performance in unifying the puppet and shadow images.

From a rather different point of view, I must thank Dr. Otto Soemarwoto of the Institute of Ecology of Padjadjaran University who heard this chapter presented as a paper. When he, a Javanese ecologist, found it worthy of thought I was much encouraged. Finally, it was Dr. Deverne Reed Smith, a fellow anthropologist and lover of gamelan music, who pointed out that the underlying theme of this chapter is a description of various synergistic processes.

References

Bateson, G.
   1958   *Naven*. Stanford: Stanford University Press. 2d ed.
Becker, J.
   1981   Review of "Javanese Court Gamelan, Volume III" (recordings by Robert E. Brown in the Kraton Yogyakarta, 1976 and 1978), Nonesuch Label. *Ethnomusicology* 25(1):175–76.
Buss, L. W.
   1981   Group living, competition, and the evolution of cooperation in a sessile invertebrate. *Science* 213 (4511):1012–1019.
Carpenter, E.
   1980   If Wittgenstein had been an Eskimo. *Natural History* 89(2):72–77.
Dewey, A. G.
   1962   *Peasant Marketing in Java*. New York: Free Press.
Lévi-Strauss, C.
   1960   On manipulated sociological models. *Bijdragen tot de Taal-, Land- en Volkenkunde* 116:45–54.

CHAPTER 8

PEOPLE AND NATURE IN JAVANESE SHADOW PLAYS

Roger Long

This chapter is written from the perspective of a performer and teacher of theater who has been impressed by the depth and range of conceptual frameworks relating to the topic of human values and tropical ecology. While my observations may appear simplistic in the eyes of the social scientist concerned with the relationship between ecology and culture, I am hopeful that a decidedly different perspective may provoke discussion and provide insights into the relationship of the arts and the environment.

It is evident that ecology and environment have a significant impact not only on symbolic forms employed in theater and the meanings they may carry, but also on the early evolution of theater as an art form itself. Among nomadic tribes such as the Bakhtiari, tribes that spend virtually all of their time and energy moving their herds from one grazing spot to another, there is virtually no aesthetic or artistic expression of cultural values through the medium of theater. The environment dominates and controls them, demanding vast amounts of time and energy for mere survival. In fishing villages throughout Southeast Asia, a similar situation is frequently encountered. Although there may be substantial amounts of ritual and ritual-related performance activity, there is typically little sustained or highly developed theater in these societies.

It is in communities of village agriculture that the arts begin to flourish and where expressions of cultural values are readily seen in dance, drama, and music. Agriculture, especially the type of wet-rice farming found in much of Southeast Asia, provides periods of intense work followed by periods of relative leisure. And without leisure, it is difficult for humans to develop art forms into complex expressions of themselves and their society.

The unique environment of the royal court offers even more opportunity for the development of richly textured and refined literature, poetry, and the performing arts, for it provides not only substantial leisure time, but also the subsidized wealth that permits specialization, dedicated training, and elaborate support. As a political center, the court also exposes its artists to new, often foreign, ideas and visitors that can stimulate the development or adaptation of new art forms.

Although I have referred almost exclusively to traditional art forms, it should be noted that the environment of urbanized societies continues to have an impact on modern theater and the cultural values it expresses. Corresponding with contemporary life-styles and social pressures, modern urban theater and literature are typically more specifically single-issue focused than the traditional arts. They also tend to reflect the values of a rather small, urban, intellectual elite, although this is rapidly changing as artists increasingly concentrate on the plight of the common people in their societies. Modern Southeast Asian theater may well provide interesting insights into the relationships between urban ecology and human values.

Value of the Humanities

The importance of the humanities in reflecting human values is mentioned frequently in this volume. Aram Yengoyan's comments on cultural storage and historical memory are useful because they imply that, as windows to the past and mirrors of the present, the humanities (and here, again, I refer to the traditional arts) almost always attempt to maintain the status quo of the society in which they are embedded. A brief excerpt from the opening narration in a standard Javanese shadow play illustrates this point.

> The King of Dwarawati is endowed with the characteristics of sage, judge, warrior, and noble. The most excellent of kings, he is powerful yet humble, wise yet shares his wisdom. He cares for his subjects; he clothes the naked and feeds the hungry, aids the infirm and offers shelter from the sun and rain, cheers the mournful and heals the sick. To

> recount all the King's virtues and the greatness of his
> Kingdom would consume the long night. What has
> been said must be sufficient. (Brandon 1970, 87)

Even the clown scenes in *wayang kulit*, scenes that are often
commented upon because of their social and political satire and their
references to contemporary issues, do not seriously challenge
authority. Although perhaps irritating to some segments of the
audience, the scenes are more cathartic than subversive.

As a cultural mirror, however, the Javanese *wayang* is perhaps
unparalleled in Southeast Asia. It serves as a constant reflection of
traditional values and ideals of Javanese society while at the same
time incorporating current social and political events into its content.

An excellent example of this occurred after the bloody trauma
following the attempted Communist coup in August, 1965. One of
President Sukarno's chief advisors was the leftist-leaning Subandrio,
a man with close ties to Chinese and Indonesian Communist leaders,
who was the target for much abuse and blame for the country's ills
after the failed coup. Subandrio's position as an advisor to Sukarno
paralleled that of Durna, one of the major advisors of the powerful
but corrupt Kurawa clan in Javanese *wayang*. Although traditionally
Durna has been seen as a highly respected teacher of weaponry to
both warring clans in the Mahabharata epic from which the *wayang*
stories are often drawn, the similarity of his position in government
to that of Subandrio effected a pronounced change in his dramatic
characterization. Less and less often was Durna depicted as an
outstanding teacher who out of loyalty had chosen the less-admirable
faction. Instead he assumed the negative traits of a conspiring,
Machiavellian prime minister—traits that mirrored the popular
attitude toward Subandrio. Ten years later, the character of Durna
had reverted to type and was once again depicted as a misguided but
essentially honorable and loyal servant of his government.

Thus, the *wayang* molded itself to reflect the values and attitudes
of the Javanese viewers and the current political mentality,
emphasizing and exaggerating the negative aspects of Durna in
order to comment on important issues of the day. Perhaps more
important, accusing someone of being a "Durna" during that volatile
period was tantamount to calling him a Communist—a charge that
could result in very serious consequences. *Wayang* reacted to social
and political change and was then used to reinforce current attitudes

and ideology.

Modern theater is much more likely to challenge and threaten institutionalized power structures and traditional cultural values than the older folk and court forms. Controversial modern Indonesian writers such as W. S. Rendra and Arifin C. Noer have had works banned, and Rendra was placed under house arrest for an extended period. A comparison of cultural values reflected in modern and traditional art forms would provide an interesting measure of changing values and perspectives, although I suspect there would be more similarities than one might expect.

Traditional Javanese Theater, Nature, and Cultural Values

Nature images abound in both Western and Asian literature and drama. They are especially common in Javanese *wayang kulit*. One particular puppet, the *kayon* or *gunungan*, is an excellent example (fig. 8.1). It represents both the world tree and the holy mountain. Filled with animals, flowers, birds, and branches, it also combines the four basic elements of fire, wind, earth, and water—all surrounding the palace gates. Thus, the palace is seen embraced by nature, represented by a dense web of symbols—a surprising image given the amount of open, uncluttered space around most remaining Javanese palace complexes. The tree of life serves as the boundary for the dramatic action of the play. It demarcates the beginning and the end of every performance, marks every major scene transition, and generally surrounds the actions of humans and the gods. It can also serve as a symbolic prop for wind, forests, volcanoes, magical and spiritual powers, and the blood of fallen heroes.

Three of the most important scenes in any *wayang* performance take place in the midst of an ecologically rich environment and directly concern the relationship of people and nature. These three scenes, the "world upheaval" (*gara-gara*), the "hermitage scene" (*jejer pandita*), and the "flower battle" (*perang kembang*), reflect much of what many Javanese consider to be the essence of *wayang*. These three scenes deserve some discussion.

The Gara-gara

In almost every *wayang* performance, at about midnight, a scene of cosmic turmoil occurs. The hero of the play, who has yet to make an appearance, begins a deep meditation in order to gather himself

Fig. 8.1.   The *kayon* or *gunungan*.

for the tasks ahead. So powerful is his concentration that the entire universe begins to tremble. Volcanoes erupt, and hurricanes and tornadoes sweep across the oceans and continents. Winds rip trees out of the ground and send them flying through the air. Earthquakes split the earth and fire leaps from the crevices. All the world is split asunder because of the hero's meditation. Man's turmoil and the energy he exerts to control himself are powerful enough to disrupt the harmony of the universe.

Here, humans and nature are linked by a powerful bond. But what is important is that the humans are pictured as the controllers of nature. It is nature that reacts to human actions, not humans reacting to create harmony between themselves and the environment. If a cultural value is present here, it is that nature is subservient to humans and is thrown out of balance by the power of human concentration.

The Jejer Pandita

The *jejer pandita,* or hermitage scene, the next major scene in a *wayang* performance, occurs when the hero is introduced for the first time as he seeks counsel from an aged and respected seer. Typically, the hero wants to stay in the isolated peace of the mountaintop where, away from the intrigues and turmoil of the court, he can absorb the wisdom and meditative power of the seer. But he is never allowed to do so. He is told that his duty commands him to go back into society to confront and solve the problems that have driven him to ask for advice in the first place. And this he does. Armed with new knowledge and perhaps the boon of a magical power or weapon, he leaves the isolation of the mountaintop.

Here, again, humans and nature are linked. It is to nature that one turns to contemplate one's worldly obligations—back to the isolated summit, nearer the abode of the gods, and closer to the unity of humans and nature. But nature is not depicted as the concern of the active and committed solver of societal ills. It is the seer, not the heroic knight, who is allowed to remain isolated in nature. Peace in and with nature is a possible reward for services rendered to society, but nature offers no escape from a conflict-plagued world.

The Perang Kembang

The "flower battle," so named because of the beauty and complexity of the movements used in this scene, is the third major scene that traditionally takes place in a "nature" environment. Upon leaving the hermitage, the hero descends the mountain and enters a dark and foreboding forest frequented by numerous giants, ogres, and demons. These frightful, though often comic figures engage him in battle and are killed one by one. These are the first of a series of deaths that become more frequent as the play progresses. The forest is presented as a place of danger, not one associated with tranquil, leisurely walks among the vines and flowers. It is significant that love scenes in *wayang* characteristically occur in the luxury of the court, not in the seclusion of the woods.

These three scenes are especially impressive, not merely because they are more concerned with nature than other standard scenes, but because of their apparent significance, as indicated by their ritual position and role. The scenes begin in the dead of the night, usually around midnight—one of the most dangerous times to be up and about. It is the witching hour. And it is in these three scenes that the hero is introduced, given wisdom, and subjected to possible death. It is possible that these scenes constitute the primordial essence of the shadow play itself—an initiation rite for the young adolescent male who must go alone into nature to meditate, acquire the confidence and mark of manhood, and successfully defeat the enemies (both real and symbolic) that men encounter.

The Importance of Structure

Several aspects of Alice Dewey's chapter deserve special attention, for they provide fascinating insights into Javanese culture and values. Yet, in some cases they may be slightly misleading as well. I was especially taken with her remarks about how things in Java so often interdefine each other. Her references to reversal images in batik design and *wayang kulit*, the constantly shifting patterns of complementary images, were graphic and well supported, especially the wonderful example she gave of the tourist taking flash pictures of the shadows in a *wayang* performance, "looking at what isn't there." And yet I see this somewhat differently, for the tourist isn't taking a picture of what is not there. The shadow is indeed there, but it is being eradicated by inappropriate "enlightenment." The misapplication of light, just like the misapplication of knowledge, may obscure what one is seeking to clarify.

Professor Dewey also uses the example that in *wayang* left and right seem freely interchangeable depending on which side of the screen one sits. Yet, I have no doubt that all Javanese have a firm concept of the "left" and "right" sides and factions in a *wayang* performance. Such knowledge implies, not a capacity for perceiving altered states of reality, as Dewey suggests, but rather a tolerance for changing perspectives based on a fixed and mutually accepted reality.

A kind of filtration system is at work within this idea of image or situation reversal that attempts to achieve maximum harmony within a potentially chaotic world. I must disagree with her concept of coexisting equal relationships in favor of a belief in firm, culturally perceived structures that permit or tolerate almost endless improvisation *within* those boundaries. This certainly exists in gamelan music with its colotomic structure that is always present regardless of whether it is explicitly invoked in a given piece, and in *wayang kulit* dramatic structure and characterization, where variation is always permitted *within culturally acceptable boundaries.*

The *punakawan* (clown servants) in *wayang* provide excellent examples of the form's complementary aspects of rigidity and flexibility. The clowns appear to have great latitude within the *wayang* performance. They are the only characters who can speak in foreign languages, who can directly address the audience or members of the gamelan orchestra, and who can refer to contemporary issues that pertain to village or national life. They are renowned for their social, political, and religious satire and playful mockery of their masters. Yet the clowns never seriously present a radical challenge to the status quo or to the feudal context of their master-servant relationship. Their flexibility, so apparent and admired, is bound in an iron-clad mold of tradition.

A system of cultural checks and balances that is almost existential in nature seems to be at work. Since little is exactly what it seems in *wayang kulit*, the only way to understand the essence of an argument or situation is to filter the meaning through a number of images or interpretations that in combination allow one to perceive a *good* solution—not the *best* solution. Given the vast number of variables, it is difficult (and, perhaps more important, useless) to determine an absolute solution to any problem. If a good solution is achieved, that should suffice.

This process of filtering meaning toward one's perception of an acceptable solution to the play's problem (or, for that matter, a character's interpretation) through all of the complexities inherent in a lengthy performance (dozens of scenes, scores of characters, and between eight and nine hours of nonstop action) finds another parallel in the activities of the clown servants. In Bali, Malaysia, and especially Thailand, clown servants act, in effect, as interpreters and translators for their higher-ranking masters who use such elevated or elaborate language that the average audience member understands little of it. The clowns discuss, repeat, and interpret their masters' words in the vernacular language, thus giving the audience a perception of what has transpired, but a perception nevertheless filtered through and subtly transformed by the words and actions of the common person.

What we see is a highly flexible system of expressing and interpreting cultural values—much like that of the Bontoc village elders June Prill Brett mentioned in chapter 4. These mountain farmers reach important decisions only after prolonged discussion filters through their collective memory and permits a consensus viewpoint. The solutions to problems that arise in Javanese *wayang* are entirely predictable. It is the skill and ingenuity of the process of arriving at the solution that is so intriguing and admirable.

The Problem of Measuring the Strength and Legitimacy of Cultural Values

This is the area that disturbs me greatly, for it is so easy for outsiders to bring into their observation and evaluation a decided, if well-meaning, bias. As modern technologies and life forms intrude, traditional forms inevitably undergo change, albeit at different rates. Technology also stimulates changes in performing style. Radio and television reach vast audiences, and popular performers are often widely and faithfully imitated by younger or less talented-puppeteers. In an extreme example, cassette recorders have come to replace live performers entirely. This raises the broad question: how do we evaluate, rather than merely discern, the values we perceive to exist in a culture—be they directed at theater or tropical ecology? Perhaps this is an area where the scientific expertise of the anthropologist will prevail.

Reference

Brandon, James R. (editor)
    1970   The reincarnation of Rama, *Wahju Purba Sedjati*. In *On Thrones of Gold: Three Javanese Shadow Plays*. Cambridge, MA: Harvard University Press. Pp. 81–170.

CHAPTER 9

CONSTANCY AND CHANGE IN AGROECOSYSTEMS

Otto Soemarwoto

Environmental concern as a major public issue originated in the developed countries in the 1950s and culminated in the early 1970s with the holding of the Stockholm Conference in 1972. Presumably, it would have continued to rise if in the following year the Arab-Israeli war had not erupted again and if OPEC had not used oil as a diplomatic weapon, raising oil prices to very high levels. This precipitated the energy crisis, and, as a consequence of the economic difficulties that followed, the environmental movement lost a great deal of its appeal, although it continued to occupy the minds of many people. In a way, this development was fortunate, since this loss of appeal also was accompanied by a decrease in sensationalism and emotionalism. This has allowed for some soul-searching and some time to look at environmental issues in a more logical and rational way.

In the 1960s and early 1970s, particularly prior to the Stockholm Conference, development and the use of advanced technology were by many simply considered bad for the environment. "Zero growth" was advocated (e.g., Meadows et al. 1972) and environmentalists agitated against large-scale development projects such as high dams, industrial complexes, and harbors. Modern life was considered to be dysfunctional, and such concepts as "ecodevelopment" and "intermediate technology" and the slogan "small is beautiful" were offered as alternatives. Traditional technology as practiced by less-developed or even primitive peoples was considered good for the environment. Hence, to save the world from ecological disaster it was proposed that solutions should be sought in the traditional life-styles of these peoples. There was always an inner conflict, however, in that the very people who promoted this life-style did not want to relinquish the so-called modern way of life themselves.

The fact is, of course, that while traditional life-styles may not be as polluting and as wasteful of resources as modern industrial life, they can offer only minimum standards of living. On the other hand, while modern ways of life are polluting and wasteful of resources, they offer many conveniences and pleasures that are now taken for granted. Therefore, for practical reasons alone, a return to traditional life-styles cannot avert "doomsday." Also, whether we like it or not, traditional peoples will continue to adopt modern ways of life. This can be seen most vividly in the cities of developing countries such as Jakarta, Manila, and Bangkok, which are swamped by cars and consumer goods, but this change in life-style is spreading to the countryside as well. As these changes occur, the ways in which people exploit their environment also change to suit their new needs and wants. Sociobiophysical ecosystems, even so-called traditional ones, should therefore not be considered as homeostatic, but rather dynamic. They are in constant flux, sometimes slow and sometimes rapid, sometimes leading to successful new adaptations, but not seldom also to maladaptations. In a world of changing environmental conditions, changes are necessary for the sake of survival, although there is never an assurance of survival in the long term. To remain constant, however, would assure extinction.

Traditional Agroecosystems in Indonesia: Evolution and Change

The changing relationships between people and their environment can be observed clearly in the various traditional agroecosystems that are being maintained by human groups. The examples to be discussed here are from Indonesia, but in principle they are applicable to a wider region.

To provide a background for a discussion of traditional agroecosystems in Indonesia, the evolutionary development of these cultivation systems is schematically presented in figure 9.1. Shifting cultivation can be considered the oldest form of agriculture. It is still practiced in many areas of the country, and even in Java, the most agriculturally developed part of Indonesia, remnants of this technique are found. Two major types of sedentary agricultural

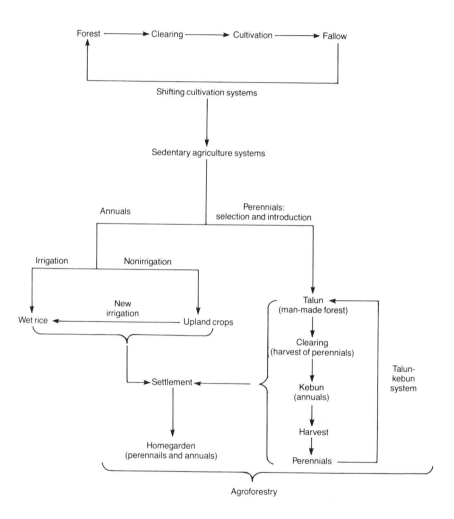

Fig.   9.1.   The evolution of agroecosystems in Indonesia.

systems evolved from shifting cultivation. One system involves the cultivation of annuals, either wet rice where irrigation is possible or upland crops where it is not, while the other system involves perennials. The cultivated perennials are selected endemic forest species that the people consider useful and a number of introduced species. The combined cultivation of selected endemics and introduced species takes the form of a man-made forest which in West Java is called *talun*. In this system, the people continue to practice a "shifting" pattern of cultivation, which is called the *talun-kebun* system, *talun* being the equivalent to the fallow period (i.e., the forest stage) and *kebun* being equivalent to the garden in cultivation. The clearing of the *talun* entails a harvest of perennials, for example, bamboo and *Albizzia falcataria*. It also occurs in various other forms in other parts of the country. When the need arises, lands that are used for crop cultivation, such as rice fields or *talun*, may also be used for settlement. In these instances, perennials are grown in home gardens surrounding the houses.

As shown in figure 9.1, agroecosystems are not constant. They have evolved from one form into others and continue to change, although change today tends to occur at more rapid rates.

Wet-Rice Agriculture

Wet-rice fields are a prominent feature in Indonesian life. Not only are they a source of food and income, but they are also an important social symbol. The possession of a rice field confers a certain social status upon the owner. Closely connected with this is the fact that rice itself is also a symbol of high status. To eat rice means to belong to a higher social class. Nonrice staple foods such as corn and cassava are considered suitable only for the poor.

Because rice needs a layer of water for its growth, its cultivation requires flat land with a very slight slope so that water can slowly and evenly flow through it. In practice such land is divided into plots surrounded by dikes. In mountain areas, this technique results in beautifully terraced mountain slopes, a field configuration that is completely in accordance with the prescription of contour planting. The Javanese call it *nyabuk gunung*, "like belts around a mountain," and the Sundanese *ngais gunung*, "like carrying the mountain with a batik cloth." From the environmental point of view, wet-rice cultivation entails an excellent technique of soil conservation, particularly in the tropical wet climate of Indonesia with its high rainfall and high rain intensity. Consequently, we may consider it as an adaptation related to subsistence, the abundance of water, and the need to protect the soil from erosion. Can we deduce, then, that our forefathers had the wisdom to place a high social value on rice, and hence on rice fields, in order to motivate the people to grow rice and by doing so protect the soil from potential erosion caused by the high rainfall and high rain intensity of the humid tropics? An alternative ecological explanation would be that the high social value of rice is based on its excellent nutritional qualities, which were discovered empirically by the people, while the effect of its cultivation on soil conservation constitutes an accidental by-product.

The social value of rice is not only an asset, however; in certain ways and under certain conditions it is also a liability. In the monsoonal climate of Indonesia, rice is traditionally grown in the wet season. During the dry season, the land is planted with second crops such as corn, cassava, or peanuts if enough soil moisture is available, or is left fallow if it is too dry. This pattern of crop rotation or fallowing has a beneficial effect on pest control in that the long interval between plantings of the same crop makes it difficult for many pests to maintain large populations. The people do not seem to be aware of this ecological benefit, however, and when water becomes available, for example, through development of irrigation facilities, they invariably grow a second crop of rice in order to enhance their social and economic status. This trend has been promoted by the government, which, as a response to population growth, has constructed expensive dams and irrigation systems in areas formerly unsuitable for rice cultivation. The continuous cultivation of rice throughout the year, however, has become an important factor in outbreaks of major pests, among them the brown planthopper (Soemarwoto 1981a).

To boost rice production further, the "miracle" rice varieties developed by the International Rice Research Institute at Los Banos, Philippines, began to be introduced to Indonesia in the early 1960s. However, while the extension of rice-producing areas was spontaneously undertaken by the farmers since it agreed with their social values, the introduction of these high-yielding varieties of rice met with many difficulties. A major obstacle was that, since rice means not only food for the people but also has social value, quality was a major consideration in the acceptance of the new varieties. Since the new strains of miracle rice were considered to be of inferior quality, government persuasion, and in many cases coercion, were used in their promotion in order to meet target yields. The resulting spread of genetically homogeneous rice varieties over very large areas had the ecological effect of increasing the vulnerability of the rice-cultivation systems. Major outbreaks of diseases and pests became more frequent, which in turn necessitated the increased use of pesticides.

Another, somewhat ironic, effect of the government's rice program has been a strengthening of the social value of rice. In parts of Indonesia where rice formerly had not been a staple food (for instance, in Madura) it has now become established as such and the nation as a whole has become more dependent on a single staple food (Birowo et al. 1978; Dixon 1979).

We see in this development an example of a process in which the ecological factor of population growth has stimulated a governmental intervention that is so permeated with the social value of rice that it has caused its cultivation to develop by leaps and bounds at the expense of ecological wisdom. The important question is whether the highly productive new system of cultivation will be sustainable in the long run. However, we should also consider the other side of the coin. Had there been no government intervention as a response to population growth, would the traditional pattern of wet-rice farming have been sustainable? Presumably not, because the traditional pattern could not provide the increased yields that were necessary for the expanding population. Without these additional yields, the pressure of this population upon the land would have continued to increase and vast areas of uplands would have been subject to deforestation for the purposes of agricultural production of dryland crops, without proper attention to soil and water conservation. This would undoubtedly have led to soil erosion; siltation of reservoirs, irrigation canals, and rivers; flooding during the wet season; and shortages of water during the dry season. Under such conditions it would have become increasingly difficult to manage wet-rice cultivation. In this sense, constancy in the wet-rice agroecosystem under conditions of increasing population density would in the long run also lead to its collapse.

The Home Garden

The home garden is an agroecosystem that differs from rice fields in many respects. It is a block of land with definite boundaries within which a house is built. Hence, a home garden is part of a settlement. A home garden represents an agroforestry system in the sense that it resembles a forest in structure and combines the natural functions of a forest with those fulfilling the social and economic needs of the people.

The structure of a home garden is based on the presence of many plant species, both annuals and perennials. They are of different ages and heights, so that a home garden normally has a layered structure (Karyono 1981). The intensity of sunlight gradually diminishes as it penetrates through successive layers of canopy (Christanty et al. 1980). Accordingly, plants making up these layers belong to sun-loving, half-shade-loving, and shade-loving species, respectively. From experimental studies, observations, and interviews we have come to the conclusion that the villagers know the light requirements of the plant species and plant them in the garden in accordance with these requirements. Undoubtedly this knowledge has been acquired by experience, presumably over a long period of time.

Because of its layered structure and the fact that in most parts of the garden plant litter remains on the ground, the home garden can effectively protect the soil against erosion. It also creates a microclimate that is quite pleasant for the people, because temperature and glare are reduced. The high diversity of plant species also means that the home garden constitutes a rich genetic resource. In addition, the home garden is an integral component of an effective system of nutrient recycling that is characteristic of West Javanese agroecosystems, as shown in figure 9.2, because plant, animal, and human residues are treated as nutrient sources for production rather than as wastes. Broadly speaking, the home garden simulates the ecological characteristics of forests in terms of soil and water conservation, microclimatic effects, nutrient cycling, and conservation of the diversity of genetic resources. Thus, the maintenance of traditional home gardens can be considered to represent ecological wisdom. But do the people acquire this wisdom through conscious effort or do they obtain it accidentally? More important, are the people conscious of it, and do they cherish it, so that they can defend it in times of change or stress? To arrive at an answer to this question, let us examine the functions of the home garden as they pertain to the daily needs of the people.

From analyses of the plant species found in home gardens, as well as from observations and interviews, it has become apparent that the home garden has multiple functions in providing the owner with food and cash income and in serving the social needs of the people (Soemarwoto and Soemarwoto 1982). The home garden is considered to have high social value; it serves as a place for social gatherings, for children to play, and for other social activities. Home gardens in rural areas are typically open, i.e., people can freely enter them or walk through them. Ornamentals are planted to provide an aesthetic environment. The subsistence crops raised in home gardens can provide significant supplements to the daily diet of the people. This is particularly true in remote areas where the market economy is not well developed. In such areas, ornamental plants and cash crops play a minor role, while the subsistence production and social function assume a prominent place.

As we move closer to cities, however, subsistence production and social functions become less important. Commercial production, on the other hand, increases in importance. This can be seen clearly near large cities. Here, the most important products from home gardens are usually fruits. Home gardens, then, take the form of orchards, planted with only a few, or just one, plant species. Plant diversity decreases and the value of home gardens as a genetic resource diminishes considerably. The layered structure also disappears and the positive effect of home gardens on soil conservation is largely lost.

At the same time, more ornamental plants are cultivated and the aesthetic function of home gardens increases in importance. In cities, the function of the home garden in the production of plants for subsistence largely ceases to exist, although there is now a new trend toward the use of home gardens for the commercial production of orchids and chickens. With this, however, the structure changes completely from a forestlike plot of layered canopies to an orderly and neatly planted garden, although the genetic diversity may still be high. Under such conditions, the home garden is usually surrounded by a fence; its "openness" has disappeared. People cannot freely enter the garden, except by permission or invitation, and its social function is thus lost as well (Soemarwoto and Soemarwoto 1982). These changes in structure and function are schematically presented in figure 9.3.

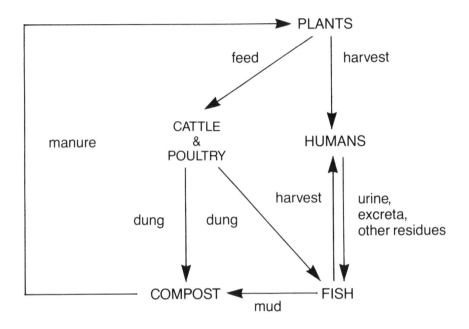

Fig.   9.2.   The traditional recycling system in West Javanese villages.

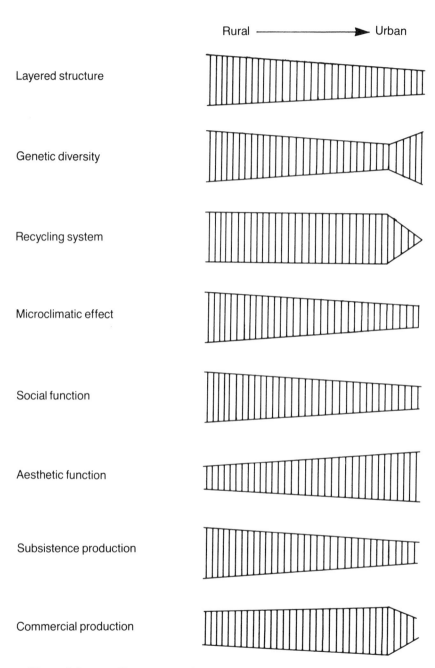

Rural ——————→ Urban

Layered structure

Genetic diversity

Recycling system

Microclimatic effect

Social function

Aesthetic function

Subsistence production

Commercial production

Fig. 9.3. Changes in the structure and function of the home garden.

Since economic development is proceeding rapidly and spreading from cities to villages, we may expect that changes in the home garden in rural settings will generally follow the direction of changes observed between home gardens in rural and urban contexts, respectively. These changes seem to indicate that the people are not consciously aware of the beneficial ecological functions of the home garden. The form of the home garden, and thus the realization of its ecologically desirable functions, is subject to, and follows, changes in the social and economic values as perceived by the people and as promoted by the social context in which they live. Unlike the case of rice, however, the changes in the attitudes toward home gardens are not directly caused by governmental interference, except in a few cases. Yet these changes are no less dramatic.

In rural West Java, one element of the traditional home garden, the fishpond, has remained relatively constant, perhaps because it is economically productive and at the same time serves a number of social functions, for example, as a community lavatory. The fishpond is an essential element in the human food chain (plant-animal-man) of West Java. Nutrients enter the pond through human excreta and in one way or another contribute to the production of fish that are in turn consumed by humans (fig. 9.2). Since the people do not eat raw fish, this food chain constitutes only a minor danger for the spread of parasitic diseases. The people do not use the water from fishponds directly for their daily use but instead bring in water from upstream sources. When population density was low, the distance between use and re-use was great. This allowed time for the water to undergo a natural purification process. Fishponds did not play an important role in cholera outbreaks and such epidemics were rare. As population density has increased, however, the distance between water use and re-use has become shorter, and there is not sufficient time for natural purification to be effective. As a consequence, cholera outbreaks have become more frequent and often take a large toll. It is apparent that the people are not aware of the ecological consequences of increasing population density, and more specifically of the link between the use of polluted water and the outbreak of cholera. The use of fishponds in West Java is, therefore, an example of constancy in a changing environment and represents an element that has become dysfunctional. However, since recycling of wastes is an ecologically sound principle, particularly in a world of decreasing resources, the solution to the problem should not be sought in abandoning the recycling system but in making it sanitary. Thus, the question is not "How do we dispose of wastes in a hygenic manner?", but rather, "How do we use residues as a resource in a hygenic manner?" The latter question is also in accordance with the cultural and economic values of the villagers.

Conclusion

More examples can be cited of changing agroecosystems. Even shifting cultivation systems in remote areas are changing because of logging, transmigration, and other development activities (Kartawinata and Vayda 1983). Shifting cultivators, for example, have adopted the use of chain saws and have learned that selling logs brings more immediate financial rewards than burning them to fertilize their crops. Their traditional values are radically changing. Many of them have changed their practices from careful husbandry of the land to careless opportunistic exploitation. Indeed, economic and political pressures are powerful forces that can push people to abandon their ecological wisdom and to exploit new opportunities without regard for the consequences. Change, rather than constancy, is the rule. In fact, development implies change, and development is a *sine qua non* for improving the living standard of the peoples in developing countries.

In these countries, issues of development versus environment have arisen and caused many difficulties. In many instances, these difficulties have occurred because both environmentalists and proponents of development such as engineers and economists have often fallen into the trap of "tunnel vision." Environmentalists tend to think in terms of absolute "good" or "bad." Reforestation and traditional technology, for example, are considered good. Logging, dams, and pesticides, on the other hand, are believed to be bad. Yet, ecological change cannot be assessed in absolutes but instead always involves a mixture of good and bad. Something is considered good if the beneficial effects outweigh the bad and vice-versa. However, good and bad are relative in space and time. What is good for one person may be bad for another, or even bad for the same person at other times and under different conditions.

Consider malaria and DDT, for example. When a large proportion of the population was stricken with malaria, DDT was considered good because it reduced the risk of being bitten by malaria-carrying mosquitoes and of contracting the disease. The risk of absorbing DDT into the body and suffering ill effects sometime in the future was accepted. In other words, at that time the "good" of DDT outweighed the "bad." When the population of malaria mosquitoes was sufficiently suppressed, the risk of getting malaria became negligible. At that time, however, the perceived risk of suffering a DDT-related disease in the future began to loom larger. As a result, DDT's status changed from good to bad. This situation may be likened to the ambiguous categories so clearly described by Dewey in this volume. As the threat of malaria vanished, our focus shifted from DDT as a protector of health to DDT as a health hazard.

The case of malaria and DDT shows that in the real world environmental benefits are always accompanied by environmental risks, the potential costs we have to pay for obtaining the benefits. In development, both the environmental benefits and the risks must be considered and weighed. Focusing attention only on potential benefits usually will result in action leading to environmental degradation of one sort or another. Focusing only upon the environmental risks, on the other hand, precludes or at least greatly inhibits the derivation of any benefits. For this reason, environmental benefit and risk analysis is a more useful tool in development and planning than is the better known and more widely used approach of environmental impact assessment (Soemarwoto 1981b) since its objective is to optimize the ratio of environmental benefits to environmental risks while at the same time allowing for the preservation of the essential processes of the global life-supporting system.

Since change is an integral part of the world in which we live and of the ways in which we perceive it, our emphasis in ecological research should be more directed at changing ecosystems. Changes are occurring at a fast rate and time is running short. If we intend to play a role in guiding these changes for the better, we must work fast and be willing to face the realities of the world.

References

Birowo, B. T., W. Sugiyanto, and H. Tedjokosoemo
 1978   *Supply and Demand of Food in Indonesia.* Bogor, Indonesia: National Workshop on Food and Nutrition (in Indonesian).

Christanty, L., Hadyana, M. Sigit, and Priyono
  1980   *Distribution and Interception of Light in Homegardens.*
          Seminar on Ecology of Homegardens III. Bandung,
          Indonesia: Institute of Ecology (in Indonesian).
Dixon, J.A.
  1979   Diversity in the diet and staple food consumption patterns
          in Indonesia (1969/1970 to 1976). *Kajian Economi
          Malaysia, Journal of the Malaysian Economic Association*
          16(1 & 2):137–52.
Karyono
  1981   Structure of Homegardens in Villages of Citarum River
          Basin. Ph.D. dissertation, Padjadjaran University,
          Bandung, Indonesia (in Indonesian, mimeographed copy).
Kartawinata, K. and A. P. Vayda
  1983   Forest conversion in East Kalimantan, Indonesia: the
          activities and impacts of timber companies, shifting
          cultivators, migrant pepper farmers, and others. In
          Ecology in Practice, Proceedings of the UNESCO-ICSU
          Conference, Paris, September 1981. Dublin: Tycooly.
Meadows, D. H., D. L. Meadows, J. Randers, and W. W. Behrens
III
  1972   *The Limits to Growth.* New York: Universe Books.
Soemarwoto, O.
  1981a   Interrelations among population, resources, environment
          and development. *Economic Bulletin for Asia and the
          Pacific* 32(1):1–33.
  1981b   Management of Environmental Benefit and Risk. First
          National Congress of Biology, Semarang, Indonesia (in
          Indonesian, mimeographed copy).
Soemarwoto, O., and I. Soemarwoto
  1982   Homegarden: its nature, origin and future development.
          In *Proceedings of the Workshop on the Ecological Basis for
          Rational Resource Utilization in the Humid Tropics of
          Southeast Asia.* Serdang, Malaysia: University Pertanian
          Malaysia. Pp. 130–39.

CHAPTER 10

IN THE LONG TERM:
THREE THEMES IN MALAYAN
CULTURAL ECOLOGY

Geoffrey Benjamin

My title is not as vague as it may sound: there are indeed identifiable and distinct, value-laden, organizational themes embedded in the materials of Malayan prehistory and ethnography. The available prehistoric data are still very thin (cf. Peacock 1979), but enough is known for me to draw some conclusions as to the kinds of choices between different modes of environmental exploitation that have been made in the past. The ethnographic data on the recent indigenous cultures—by which is meant both the Orang Asli ("Aboriginal") and Malay (Melayu) cultures—are, on the other hand, now quite rich. We possess a growing body of data on at least two of the domains of culture that are everywhere crucial to the organizing of environmental (or any!) values, namely, kinship and religion. (The data on the third such cultural domain—language—are regrettably not so rich as yet.) For some years I have been working to pull this material together, in the attempt both to understand and to explain the rather subtle patterns of social and cultural variation present in the Malay Peninsula. In this essay I have taken the opportunity to look again at the same material—the kinship material in particular—and re-examine it in the light of culture-borne ecological and environmental values. Before proceeding with the Malayan materials, let me first say something about the general theoretical issues involved in the study of values and of ecology. In particular I want to urge, in opposition to what appears to be the current received opinion, (1) that cultural "values" have more to do with power and politics than with morality or aesthetics; and (2) that the study of human ecology should be undertaken more as a kind of biologically aware, historical enterprise than as a systems-analysis or evolutionary one.

Cultural Ecology as Dealing with Power and History

Cultural Values

Like the constructs labelled "cultures," "societies," and "peoples," cultural values (as opposed, perhaps, to purely individual values) originate as preeminently political constructs. Such values do not emerge willy-nilly out of the flux of people's lives: they are, rather, imposed on those lives, but usually in such a subtle form as to disguise their political nature.[1] Values, indeed, are at their most political when so thoroughly diffused into the wider culture that they have been made to merge with the taken-for-granted "reality" of the people. This degree of diffusedness can be achieved only by continuing action on the part of some people to block the social communicating of alternative ways of viewing the world and to make the preferred way seem commonsensical. But, by the same token, only those values are truly values that have become condensed into an individual's taken-for-granted personal knowledge: the so-called attitudinal "values" beloved of questionnaire-wielding social scientists, being so readily expressible in words, are usually mere dogmas or stock answers rather than the principles on which the respondents actually live their lives. To the extent that values come to be shared, then, they pertain to the nondiscursive, unsaid, parts of culture: they have been "mystified."

Contrary to opinions commonly held by social scientists, values (which should be understood in this essay as an abbreviation for "cultural values") do not work by providing people with explicit or even implicit rules, instructions, or programs for behavior. Values work, rather, by confronting people with choices between various courses of action—but the resultant action is still always constructed by the individual agent, not by the values. It is the ideographic properties of values that matter, not any supposedly "rule"-like character they may exhibit. They are held in the mind as (often fragmentary) notions or pictures of the individual's own life-world, making certain ways of acting feel "right" (i.e., coherent) while leaving other, alternative, ways of acting relatively unconsidered by the individual.

At least three further points arise from these assertions. First, values cannot simply be gathered by ethnographic observations or just by asking questions—they must be teased interpretatively out of the available data. Second, cultural values never merely exist statically in an individual's actions or in a "culture"—they are, on the contrary, constantly being propagated, changed, and symbolically condensed (mystified) as part and parcel of the sheer politics of human social life (regardless of whether this politics relates to simple interpersonal interaction or, more conventionally, to imposed administrative rule). In other words, they are historical and artifactual products. Third, values bear at most an implicational, type-token, relation to action, and not a rule-like cause-and-effect or stimulus-response one. The study of cultural values should accordingly be much more akin to semantic, literary, or even theological analysis than to either empirical observation or formal rule-gathering ethnoscientific analysis. In this essay, for instance, I shall be looking at various kinship rules more for the kinds of mental pictures of social life that they generate than for any direct overt shaping they may exert on actual interpersonal interaction or on processes of group-formation. Only then will I attempt to show how kinship "rules" might relate to the political and ecological contexts from which they derive their meaning.

Ecology

The term "ecology" has come to replace the much older term "natural history," but even though the focus of ecological research has justifiably shifted somewhat in recent decades from individual species to the study of interrelations between species, ecology should to my mind still be built on a solid natural-historical base. Ecologically minded anthropologists, committed as they are to making their ethnographic data as richly contextualized as possible, are hardly likely to argue with this contention, however.

What the ethnographic approach does leave out, more often than not, is time-depth. In some cases, admittedly, the environmental consequences of a population's actions come about so rapidly as to be open to investigation by conventional ethnographic fieldwork. But at the other end of the scale, even ten thousand years has hardly yet been long enough for us to fully comprehend the consequences of food producing. Most ecological anthropologists are, of course, concerned with time-frames somewhere between these extremes. It should be noted, however, that people everywhere can at best only guess at the environmental consequences of their actions, for those consequences usually take far too long to manifest themselves. In many cases, it is true, the people in question may be able to observe a neighboring population whose way of life has already come to manifest the consequences of some much earlier choice of environmental action. But even then, long after the event, it will not usually be easy to decide just which of the possible antecedent factors were the ones responsible for the present-day consequences. This suggests that throughout most of history people's actions with regard to the environment have usually been aimed at holding on to what they already think of as the advantages of their current way of life, rather than at changing their way of life so as to produce some hoped for, but as yet unexperienced, advantage.

The uncovering of orientational themes of this kind is an essential part of explaining the why's of people's environmental actions, but from an ecological point of view this is insufficient. While individuals may see particular advantages to themselves in following one or other mode of subsistence, organized according to the terms of one or other environmental orientation, what they usually will not see is how their actions produce changes in the environment that might in the longer term seriously modify the chances that they or their descendents could continue in that particular way of life. In this regard, it is not without significance that the named sociocultural formations that constitute the "societies" or "cultures" of most ethnological work have tended to display very short historical life-spans as distinguishable entities. Indeed, Wallerstein has suggested (1978, 156) that they tend to remain recognizable for no more that about six generations.

People's own ideas about what they are doing are, then, of restricted relevance to narrowly ecological studies; in any case, cultural values (as we have seen) do not translate directly into specific individual actions. The ecology of human life has to be analyzed in its own terms—with, as far as we can see, one crucial difference from the ecology of other living species: in the tropics, at least, and in the time-spans usual in social enquiry, there is a disjuncture between the environmental consequences of people's actions and the continued performance or nonperformance of those actions. Therefore, the reasons adduced in understanding why and how people do things bear little direct relation to the reasons adduced in explaining the environmental effects of those actions. This is not to deny the possibility that, as Marvin Harris (1977, 1978) continues to assert, there is often a notable causal-mechanical fit between the material consequences of certain actions and the maintenance of the circumstances in which it is possible to continue performing those actions. But in the short time-spans of most anthropological analyses, such degrees of fit are neither analogous to nor homologous with processes of natural selection, for the simple reason that people's actions with respect to the environment are not reactions to some external force but the active constructs of their own individual and collective choices. Human choices, especially in the relatively forgiving (if not overabundant) circumstances of the Asian tropics, could until recently be variously good, bad, or indifferent from an adaptive point of view without coming under any marked environmental constraint during the lifetimes of the people who made those choices or during the lifetimes of their known ancestors and descendents.

"System" and "History"

These considerations change, however, if we shift our attention to the much longer time-scales of prehistoric and ethnohistorical anthropology. While "system" and "function" are to my mind quite misleading when applied to ordinary ethnographic field-data, something like these concepts seems, paradoxically, to become more relevant as the time-scale is increased. I say "paradoxically" because "system" and "function" are analytical concepts that were developed primarily in relation to supposedly synchronic issues. Synchrony, however, is an illusion—and with it the concepts of "system" and "function" as commonly understood. But clearly, it is in the long time-spans considered by prehistoric anthropologists that the environmental consequences of, and constraints upon, different ways of life show themselves.

We must still beware, however, of assuming that even a ten-thousand-year time-span will allow us to talk entirely in systemic, functional, or evolutionary terms. For one thing, the consequences of unpredictable extra-"systemic" accidents—that is, historical factors—also become more apparent after long periods of time. For another, the units of cultural analysis must now become fundamentally different from the kinds of decision-making and culture-bearing units normally studied on ethnographic time-scales:

we are dealing now not with identifiable sociopolitical groupings or formations but with shadowy organizational themes or clusters of ideas, the tokens of which are manifested in different concrete sociocultural forms at different times and places. Furthermore, the evidence for active choice, far from disappearing, often becomes clearer at longer time-depths, and the presence of choice becomes closer to being confirmed if it can be shown that several more or less equally viable alternative ways of life have been held to in a particular area for long periods of time.

The distinction I am making here between "system" and "history" may appear somewhat tendentious (and indeed it may in many instances not make much difference in practice). I believe, however, that the distinction does at least have an important emblematic or heuristic value. Those who work under the banner of "system" are likely to assume that things are the way they are because they could not have been otherwise. But those who regard their work as a fundamentally historical exercise (as I do here) take it for granted that things could always have turned out otherwise, and therefore that, while it is possible in principle to explain what happened, it is not in principle possible to have predicted it with any certainty, if at all.[2]

The Malayan Setting

The Malay Peninsula provides an excellent field for studies within the framework just outlined. The indigenous cultures have developed *in situ*, in response both to each other and to certain specifiable exogenous influences. Malayan populations have shared the same basic pool of knowledge for so many millenia that they can safely be considered to have differentiated from within the same cultural matrix—even if the textbooks and local newspapers continue to hold to outmoded "migratory-wave" theories of the extant cultural differences.[3] In recent years, there has been a rapid increase in empirical research on the cultures, languages, archaeology, and ecology of the Peninsula; although much of the material is as yet unpublished, enough of it is in circulation to provide the basis for the comparative work undertaken in this paper. On looking through this growing body of data, I have been increasingly impressed at the long-term persistence in the Peninsula of certain orientational themes embedded in the shifting details of sociocultural organization.[4]

Semang, Senoi, Malay

In the present essay I point to some of the ecology-related issues encountered in attempting to bridge the gap between the more "objective" long-term evidence embedded in the linguistic, archaeological, and ethnological data and the more "subjective" short-term evidence uncovered in recent ethnographic fieldwork. I shall try to demonstrate, with special reference to the data of kinship and social organization in the Malay Peninsula, the existence of the "themes" of my title as a set of persistent, but different, patterns of linkage between at least the following sets of phenomena:

1. The various ways of distributing people with regard to the land and to each other, as evidenced by their settlement histories and communication patterns.
2. The various preferred, or dominant, modes of environmental appropriation and modes of production—the Peninsular "ways of life."
3. The various dominant orientational modes of interpersonal and societal consciousness embedded symbolically in each of the different societal and religious traditions.

I shall distinguish three such themes in the Malay Peninsula, the "Semang," "Senoi," and the "Malay." These labels are chosen deliberately so as to highlight the intimate connection between the organizational themes discussed in this paper and the tripartite social and ethnological division recognized in studies on the Malay Peninsula since before the publication of the major compendium on the subject by Skeat and Blagden in 1906.

The Semang pattern is typified by the one thousand or so foragers in the northern part of the Peninsula, who mostly speak languages of the Northern Aslian subfamily of Austroasiatic. These people are usually referred to as "Negritos" in the anthropological literature and in Malaysian administrative practice, but I shall keep to the term Semang so as to emphasize that I am referring to a type of culture rather than to a group of people—least of all to a group defined biologically.

The Senoi pattern is best exemplified by the Temiars and upland Semais, swiddening populations currently numbering around 13,000 and 17,000 respectively, who live in the central portions of the Peninsula and speak languages belonging to the Central Aslian subfamily of Austroasiatic. (The terms "Semang" and "Senoi" are conventionalized Malay and English versions of words meaning "human being" in various Aslian languages.)

The Malay pattern is expressed in both centralized and uncentralized ("tribal") social forms, but the choice of an appropriate terminology is complicated by conventional usage and by modern political norms. In contemporary Malaysia the term "Malay" (*Melayu*), if used without further qualification, refers only to the politically dominant, lowland, Muslim population now numbering several millions, who are regarded as the citizenry of the various kingdoms whose courts have been situated historically at strategic points, mostly around the Peninsular coastline. These kingdoms have for the most part survived as constituent sultanates or "states" of the Malaysian Federation. The qualified term "Aboriginal Malay" (*orang Melayu Asli* in official, if not colloquial, usage) is employed, however, for a variety of Southern Aslian- and Malay-speaking tribal (or recently tribal) groups numbering in the hundreds or thousands, and distributed all over the southern parts of the Peninsula. But since the ethnological and historical links between these two divisions of the Malay people are actually much closer than present-day Malaysian ideology will allow, it is necessary for me to treat their cultures together. I shall therefore refer to each of the "Aboriginal Malay" groups separately by the names used in the contemporary ethnographic literature, such as Temuan and Orang Hulu, and I shall use the Malay-language term "Melayu" to refer specifically to the culture more conventionally known in English as "Malay."

The approximate locations of the various tribal or recently tribal non-Melayu groups are shown on the accompanying map (Fig. 10.1); these peoples are collectively known as Orang Asli, a modern Malay-language term created to replace the English term "Aborigines." The Melayu, not indicated on the map, live predominantly on the coasts, in the lowlands, and along the major rivers, but in some districts, such as Negri Sembilan and Ulu Kelantan, they are also found in upland areas.[5] The Malay societal pattern has historically been associated with a way of life in which sedentary farming or fishing has been combined with such outward-oriented activities as trade, piracy, and hiring oneself out for labor.

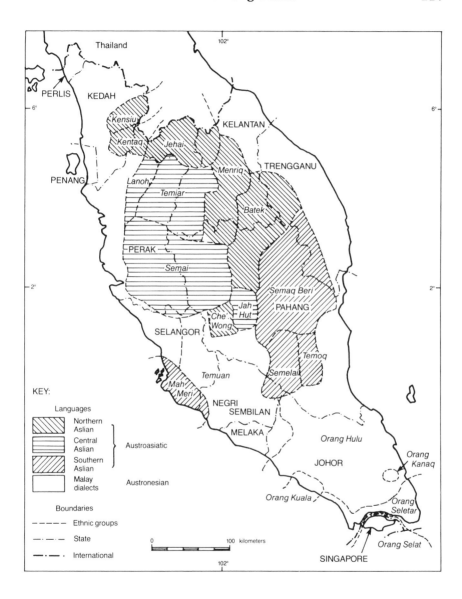

Fig. 10.1. Orang Asli groups of the Malay Peninsula (approximate distribution).

What, then, are the characteristic features of the Semang, Senoi, and Malay societal themes as they affect ecological issues? The greatest degree of contrast falls between the Semang and Malay patterns; the Senoi pattern shares some of the properties of both, but is best not thought of as forming part of any continuum between them. I must emphasize that the following characterizations are attempts to spell out the nature of the three themes as culturally embedded ideal types—as values, in a word. The empirically discoverable, on-the-ground alignments of social and spatial relations are, of course, another matter, but it is part of my argument that those alignments result in large part from people's struggles to put their values into operation.[6]

The main characteristic of the Semang pattern is the way in which it militates against permanent sedentism by encouraging constant dispersal in time, place, and consocation. Any persistent aggregation (whether of kinspeople with each other or of social groupings with respect to territory), such as might result from a long-term commitment to the agricultural way of life, is blocked by values embedded in kinship and other cultural domains. The ideal societal form as pictured by the Semang theme is an inclusive network, the nodes of which are individuals or conjugal families (nothing bigger), constantly realigning their relations with each other so that no bounded or corporate groupings are formed and the focus of attention remains on the detachable here-and-now of the present moment. Local aggregations are constantly in flux, and no core-groups are found, because no one person or group is felt to have a better claim of attachment to a particular place than anyone else. The Semang land-to-people relation is essentially random in character: local groups (insofar as they exist) consist entirely of "peripheral" members, for the formation of persistent core-groups is blocked.

The Malay pattern demonstrates the reverse tendencies. The emphasis here is on the formation of temporally persistent but spatially exclusive consocations. Each such consocation expresses its continuity by engaging in a linear progression through time and space, which manifests itself in a sedentism punctuated by periodic shifts of the whole group to a new site, coupled with a reluctance ever to return to the old site. As far as possible, the local group is organized so as to consist only of core-group members, thereby avoiding the accretion of a social periphery to the community.

Relations with other settlements are deliberately restricted and outsiders are not generally welcome to settle down. But this exclusiveness with regard to settlements of otherwise similar people is associated with a turning-outwards to the wider world beyond the home area, for the purposes of trade especially. The Malay land-to-people relation is thus very clearly determined, but it carries with it an inherently linear-historic consciousness. Each group's progress through time and territory forms a kind of temporal chain, moving ever outwards and forwards while continuing to keep the past in focus through what Wee (1985) has referred to as a cultural rear-view mirror. The Malay cultural theme is very there-and-then-and other-orientated.[7]

The Senoi pattern is not so easily characterized. Where the Semang and Malay patterns are, in their way, singleminded, the Senoi pattern is highly dialectical. While Senoi sedentism is punctuated by movements to new settlements according to the demands of swiddening and of certain religious observances concerned with death and burial, their conscious expectation is that in the long run they or their descendents will make repeated returns to previously occupied sites. The Senoi pattern of consociation is something of a fudge between the Semang preference for completely random inclusiveness and the Malay preference for a highly ordered exclusiveness. The Senoi picture their settlements as consisting in roughly equal proportions of a core-group (usually organized on cognatic-descent principles) and of peripheral people. The core-group members are essentially those who move their place of residence only when the whole settlement moves; the peripheral people move from settlement to settlement as the whim takes them, even if there are one or two such settlements in which they could have claimed primary core-membership. Thus, the Senoi land-to-people relation is cyclical in character: there is a short-term concentration, lasting for a season or more; a mid-term diffusedness, persisting for a generation or two as settlements are founded and then abandoned within the group's territorial range of exploitation; and a long-term concentration again as the cycle starts over. In the long run, the picture this subtends is that nothing much will change—though in the short term and beneath the surface pattern of interlocking circles, much intense interaction is expected to take place, both within and between local groups.

The characterizations just presented are an attempt to sketch something of the different modes of understanding and organizing social relations that appear to have been propagated and maintained in different parts of the Malay Peninsula for considerable periods of time. Relatively abstract though these formulations are, they relate to very concrete ecological and sociocultural phenomena: the themes under discussion are, after all, representations of the ways in which the people themselves have lent coherence to their own experience of the real-world circumstances under which they live. Even though the relationship between the cultural themes and the relevant concrete circumstances is not of a simple one-to-one kind, it certainly is not arbitrary either. From an ecological point of view the three different themes relate fairly obviously to the differing degrees of concentration of the particular resource(s) most favored by each category of people. I shall say more about the environmental circumstances later, but here let me first give an outline of the normal pattern of relations between resource-concentration and societal organization in each of the three traditions.

Generally, the foraging Semang meet with *no* notable degree of concentration in plant and animal resources in the forest depths (although some resources are ecotonally concentrated, as we shall see later). Semang-type society thus needs to maintain a wide network of relations, for localized social concentration would usually be ill adapted to their normally foraging mode of subsistence. In Malay-type societies farming has throughout most of their history usually been subsidiary to trading and forest collecting as the favored activities, and Malay trading relations have usually been concentrated rather than diffused, primarily at coastal ports or at upstream entrepot points.[8] This pattern of "resource"-concentration has implied a fair degree of competition between different groups; so, in a sense, the fewer intrasocietal links and the more extrasocietal links there have been, the better. The archetypical Senoi resources, on the other hand, are swidden fields and the stands of long-lasting seasonal fruit trees that succeed them. In forest conditions, fields are characteristically spatially concentrated in the short-term context of village and swidden. But, as they turn into increasingly overgrown orchards, the erstwhile fields become diffuse again in the mid-term context of the wider territorial range that each village-community regards as its own domain of productive activities. Eventually, the former fields, now thought of as the overall collectivity of past and present "home" sites, become somewhat concentrated again in the long-term context of the river-valley (the population of which tends to correspond to an in-marrying deme or dialect-group). In this way the Senoi have developed a nice dialectical balance of in-out relations at the local-group and river-valley levels (cf. Benjamin 1966, 13–22).

Socially and culturally, the thematic differences I have just outlined devolve onto kinship, the different patternings of which, I shall argue, serve as ideographic picturings of the preferred spatial and social alignments between people in the different traditions. At the most general pan-Peninsular level we find that Semang kinship contrasts with Malay kinship in respect of certain features that hold together as bundles at each pole in a relationship of mutual implication, thus:

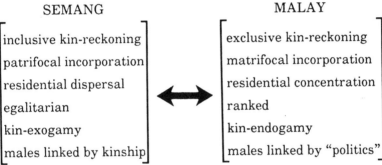

SEMANG

- inclusive kin-reckoning
- patrifocal incorporation
- residential dispersal
- egalitarian
- kin-exogamy
- males linked by kinship

MALAY

- exclusive kin-reckoning
- matrifocal incorporation
- residential concentration
- ranked
- kin-endogamy
- males linked by "politics"

Senoi kinship is both inclusive and egalitarian, like Semang kinship, but it differs from both the Semang and Malay patterns in being quite bilateral in preference, eschewing both patri- and matri-filiative bias. The Senoi residence pattern combines both dispersal and concentration (insofar as there is a core-periphery distinction in each settlement), and the marriage pattern combines prescriptive consanguineal exogamy (like the Semang pattern) with preferential affinal endogamy (somewhat similar in effect to the Malay pattern).

Semang, Senoi, and Malay kinship and social organization differ in many more specific respects too, as a glance at tables 10.1 and 10.2 will show. Among the more interesting and puzzling features are the various avoidance and joking relations typical of the different traditions: Malay kinship has neither avoidance nor joking relations; Semang kinship has a full set of cross-sex-avoidance relations; Senoi kinship has some avoidance relations and some joking relations. I shall touch on these and other aspects of social organization later; first let us look at the ecological, ethnological, and prehistoric contexts within which, and in response to which, these variables emerged.

Malayan Culture History

It is implicit in what I have been saying that the indigenous cultures of the Malay Peninsula in premodern times were dominated from the Neolithic onwards by three modes of environmental appropriation: foraging (nomadic or seminomadic hunting and gathering), horticulture (sedentary or semisedentary swidden-farming), and collecting (the gathering of natural products for trade with outsiders). These, separately or in combination, have been the major factors in the emergence of the Peninsular forms of social organization. More modern modes of appropriation, such as permanent-field agriculture, mining, piracy (which covers a wide range of economic adaptations in the Malay world), and administrative service, have not yet obliterated the earlier patterns—to some extent, indeed, they have had the effect of further institutionalizing them.

Most of the physical opening-up of the Malayan interior has taken place during the last few decades. While the changes this has induced are of great importance in themselves and probably irreversible, I shall say relatively little about them in this essay, which is more concerned with centuries and millenia than with decades. At the present time, the so-called "Aboriginal peoples" of Peninsular Malaysia (the Orang Asli) constitute a small and politically powerless minority. But the sociocultural and political realignments by which this situation was brought about are of a relatively recent date, having more to do with the effective redefinition of Melayu-ness consequent on the 1874 Treaty of Pangkor and the resultant spread of the concept of "Aborigines," than with any dying away of the Aboriginal populations themselves. The massive in-migration of nonindigenous populations over the past century has, of course, only made for a further apparent diminution in the Aboriginal presence. The term "Aborigines" used in this restricted, capital-A, sense (along with its Malay translation *Orang Asli* "original people") is misleading, for it implies that only the Aborigines are aboriginal to the Peninsula, when there is no reason to doubt that the core of the Melayu community is just as aboriginal. It is twentieth-century politics rather than scholarly research that has led to the current, but clumsy, Malaysian distinction between "Aboriginal" (*Orang Asli*) and "Indigenous" (*Bumiputera* "soil-son") populations (as if the differences were parallel to the differences between Whites and Aborigines in Australia). In this essay I shall mostly disregard these issues in favor of an approach that emphasizes what might reasonably be considered to have been the indigenous (i.e., "small-a" aboriginal) sociocultural patterns of the Malay Peninsula throughout most of the last few millenia.

While foraging, horticulture, and collecting-for-trade have been reported from almost all parts of the Peninsula, their overall geographical distribution as the dominant appropriative modes in the different interior areas in which they occur is not random. In the north, near the Thai border, foraging has been dominant; horticulture has been dominant in the mountains and hills of the Peninsular center; and collecting—almost always combined with horticulture in a distinctive way—has been dominant in the lowlands of the south.[9] The three main modes of appropriation are undoubtedly ancient in the Peninsula. Foraging, probably part-sedentary, must have been present from the earliest, paleolithic, times. Horticulture, it is now known, developed several thousands of years earlier in Southeast Asia than most sourcebooks yet acknowledge. Collecting for trade, though well known from historical and protohistorical studies (see, for example, Wang 1958; Wolters 1967; Wheatley 1961, 1964), is probably thousands of years older than usually realized (Dunn 1975). Nevertheless, as we shall see, these modes probably differentiated out of an earlier ("Hoabinhian") sociocultural matrix which, though variable, had not yet come to display the institutionalized differences discussed in this essay.[10]

The Evidence of Language

Since each of these modes of appropriation when dominant in its area tends to generate a distinctive pattern of social communication, it is not surprising that the overall pattern of language differentiation has, over time, come to reflect those differences. This, allied with glottochronological techniques, has made it possible to suggest probable dates for the first arrival or development of each of the appropriative modes in the Malay Peninsula among the earlier speakers of the Aslian (i.e., Peninsular Mon-Khmer) languages (for the full argument, see Benjamin 1976, 81 ff.). Assuming that foraging was present from the earliest times, horticulture and/or sedentary fishing and gathering appear to have been established in the Peninsula by around 6,500 years ago; forest collecting for trade (probably with the Austronesian-speaking coastal people to be mentioned later) was underway by around 5,000 years ago; permanent farming, associated with rudimentary state-formation involving linkages with civilized polities elsewhere in Asia, had started by around 1,000 years ago. These dates are, of course, little better than reasonable guesses, but they do accord satisfactorily with what is now being discovered through the more rigorous methods of archaeology.

As the patterns of language differentiation typical of each of the three Aslian subdivisions (Blagden 1906, 470; Benjamin 1976, 74 ff.) have a direct bearing on issues taken up in more detail below, I shall discuss them briefly here. The Southern Aslian languages (Semaq Beri, Semelai, and Besisi or Mah Meri) appear to have split apart and moved away from each other much more cleanly than the Central and Northern Aslian languages, and there is little secondary lexical borrowing between them. This suggests that the ancestral Southern Aslian-speaking populations became more interested in contacts with outsiders than with each other. Among the ways of life then available, such a pattern would have been characteristic only of collecting-and-trading. The associated preference for living in autonomous communities that are relatively closed off from each other but adapted primarily to relations with members of other societies persists among the Melayu and the Malay-type southern Orang Asli to the present day, as we have seen.

The Northern Aslian languages (Kensiu, Kintaq, Jehai, Mendriq, Batek, Mintil, and Che' Wong) present a quite different picture. Simple vocabulary counts provide an insufficient basis on which to decide their mutual relationships, or even to decide how many languages there are, and special statistical techniques are necessary to sort them out. This, and the high degree of lexical borrowing between the languages, shows that the Northern Aslian speakers have from early times maintained a continuous mesh of communication with each other extending from Isthmian Thailand right down to central Pahang. The only interactional pattern likely to have generated such a result is small-group nomadism coupled with wide-ranging intermarriage—which is exactly what has been ethnographically reported for the majority of the Semang up to the present day. Today most Semang camps contain members of several different locality- or ethnicity-based subgroups, and each individual appears to draw on a slightly different lexicon than everyone else.

The Central Aslian languages (Temiar, Lanoh, Semai, Jah Hĕt) display a third distinctive pattern. These, like the Southern Aslian languages, broke away from each other at various times in the past, but (as is also true of the Northern Aslian languages) this did not lead to a cessation of intercommunication between the respective populations, for there remains a high rate of secondary lexical borrowing between them. This third pattern must have been the result of continued but relatively restricted communication, mediated largely through intermarriage, between sizable sedentary populations separated by natural barriers. The Central Aslian speakers still exhibit such a pattern today, being swidden-cultivators who, as we have seen, move every few years between village sites the locations of which are narrowly circumscribed by the valley walls and the locations of the neighboring villages upstream and downstream.

It thus appears that each major subdivision of the Aslian languages encapsulates within its pattern of differentiation the decisions of its earliest speakers to follow distinct ways of life: the Northern Aslians preferred to remain foragers, the Central Aslians to become horticulturalists, and Southern Aslians to become collectors-for-trade.[11]

In contrast to the various patterns displayed by the Aslian languages, the still-extant Austronesian ("Malayo-Polynesian") languages of the Peninsula display a relatively unvarying pattern, almost all of them being dialects of the Malay language. Today there appears to be nothing left of the distinctively non-Malay varieties of Austronesian that, comparative evidence suggests, must once have been spoken in the Peninsula.[12] This suggests that the extant Austronesian languages of the Peninsula result less from differentiation than from replacement. What interests us here is not so much that it was Malay that swept the other languages before it, but that the speakers of those languages were so ready to accept a new language in what must have been a short time. The currently recognized "standard" literary variety of the Malay language is probably the result of creative processes dating back some five or six hundred years, even though inscriptional records in other varieties of

Malay date back a thousand years earlier than that. In my view, the dialects of the southernmost Orang Asli, which are found also in the islands of the Straits and on the Sumatran mainland (Kähler 1960), provided the matrix out of which "Johor-Riau Malay" was created.

The speakers of the Southern Aslian languages are at the present time shifting to Malay; the speakers of Northern and Central Aslian, however, still resist the loss of their languages although they mostly speak excellent Malay when communicating with outsiders. Institutional factors, such as the choice of Temiar and Semai as the two main languages for use in Orang Asli radio programs, undoubtedly play a part in this, but the underlying difference is one of societal consciousness and cultural allegiance: the Semang and Senoi peoples are orientated more towards their own immediate situations than to the outside world, while the southern Orang Asli and the Melayu are orientated more towards the outside world than to their fellows in other villages. The southern peoples, with their more sociocentric, outward-orientated mode of consciousness and peasantlike socioeconomic circumstances, proved highly susceptible to the spread of the exogenous but centralized and assimilatory culture of the premodern Melayu states; the northerners have retained a more egocentric, inward-orientated mode of consciousness and tribal socioeconomic circumstances up to the present day. We shall see later that the kinship patterns typical of the three main Peninsular traditions directly mirror the differences in their modes of orientation.

The Archaeographic and Ethnographic Evidence

An overview of Malayan Paleo-ecology

Before proceeding to a discussion of the processes through which the Semang-Senoi-Malay pattern of cultural differentiation emerged, let us first look more closely at the relations between environment, demography, and social organization that are likely to have persisted throughout most of the Malay Peninsula until those processes of cultural differentiation took full effect. Except for the severe depredation caused by greedy and ecologically ignorant logging practices in the last few years, the Peninsula is basically an area of evergreen tropical rain forest; it becomes more seasonal in the region of the Kra Isthmus, however (Whitmore 1975). The most striking feature of this environment from the viewpoint of the technologically premodern societies is its generality, that is, its capacity to support several different subsistence modes side by side with no necessary complementarity between them. A reasonably conservative reading of the prehistoric and archaeological evidence (see, for example, Hutterer 1977a; Gorman 1971) suggests that by 6,000 years ago, but probably earlier, techniques were available to the Peninsular forest-dwelling peoples that allowed them to subsist wholly or in part by any of at least the following modes: (1) nomadic or semisedentary hunting and gathering; (2) semisedentary gathering and fishing, mainly along the ecotones provided by the banks of larger rivers; and (3) semisedentary cultivation or tending of such vegetatively reproduced crops such as yams, bananas, and sugarcane. In addition, the coastal areas and the nearby waters would have afforded several other possibilities not available to the inland forest-dwellers, such as the collecting of molluscs and crabs along the strand, sea fishing, and the raising of coconuts and fruits (Miksic 1977). The label "Hoabinhian" has been used for some time now by archaeologists to identify those pre-Neolithic material assemblages of Southeast Asia that appear to have sustained these ways of life: the Hoabinhian was, it seems, quite variable, despite the fact that it predated the emergence of developed grain agriculture.

In the Peninsula there is no archaeological evidence as yet of the early grain cultivation that some workers have claimed was present elsewhere in Southeast Asia by about 5,000 years ago (although other workers have more recently offered evidence to suggest that this is too early a date for agriculture even in Thailand). But the possibility should be borne in mind that as early as 5,000 years ago there may have been available a fourth, grain-based, mode of subsistence in the Peninsular interior, consisting of the production of such crops as *Coix*, millet, and rice, though this is not likely to have become either intensive or widespread until much later. We may reasonably assume, then, that the typical subsistence modes in the interior of the Peninsula some 6,000 or more years ago were mostly the first three listed above: nomadic foraging, semisedentary gathering, and semisedentary vegetative horticulture.

An important feature of the social and cultural ecology of these three modes of subsistence is that (unlike the case with "Neolithic" societies that developed elsewhere on the basis of grain agriculture) they require no technology of food storage. Vegetatively propagated food crops such as tubers, bananas, and sugarcane are "harvested" on a day-to-day basis, and would not in any case keep for long if stored. From an ecological point of view, therefore, no great premium is likely to have been placed on permanency of residence, nor upon the development of a forward-planning mentality. Furthermore, the pattern of animal husbandry that was later to develop in Neolithic Malaya would have been consistent with this, for it was restricted to such nonherd species as duck, fowl, pig, and dog. These are animals that typically look after themselves, foraging off human leftovers or off the surroundings; they need no special attention, nor is their population likely to have risen significantly in the absence of deliberate pasturing—which in any case was not an environmental option in the Peninsular forests.

This brings us to a consideration of the "protein problem." The typical early Peninsular crops, being neither grains nor legumes, were protein-poor, and since the only other protein sources—fish and game—were simply "gathered," not produced, there would have been little increase in the absolute amount of protein available in the early Peninsular diet even after the Hoabinhian had become the Neolithic. Since human bodies are constructed primarily of proteins, there is not likely to have been much increase in the human population of the Peninsula for as long as these subsistence modes remained the only widespread ones. Population growth had to wait until the much later expansion in the production of grain crops richer in proteins and of herd animals raised for their meat.[13]

Technologically, the Malay Peninsula displayed little elaboration in its prehistoric stone tool-kit, especially as compared to other parts of the Old World, or even Australia. The reasonable assumption is that it was wood, cane, and bamboo—all archaeologically unrecoverable—that provided much of the material basis for technological elaboration in early Malaya: certainly, the ethnographically available evidence of that elaboration into recent times would support this view. The significance of this is that all the materials necessary for the maintenance of the early Peninsular tool-kit would have been available within a short walk of every household, unlike what must have been the case in other parts of the world more critically dependent on a stone-based technology (cf. Hutterer 1977b, 22–23). "Internal" trade is therefore not likely to have been an important feature of the Peninsula until metalworking had been developed in one or two areas, several millenia after the basic sociocultural characteristics of the Peninsula had been set. In this respect the Peninsula would also have contrasted with some other parts of Southeast Asia, such as central Thailand (Kennedy 1977), where archaeological evidence of early "internal" trade is now coming to light. (It is possible, of course, that there was some early internal Peninsular trade in exotic rather than mainstay items; the trade in special blowpipe bamboo described by Noone (1954), for example could well be ancient, but it would have left no material evidence.)

The picture that emerges of early Malayan society is of a series of demographically relatively stable populations having available to them a variety of subsistence modes, each of which could be followed integrally or partially without in any way disturbing the subsistence activities of the neighboring populations. Moreover, with no pressure to develop either trade-based complementarity between populations or occupation-based complementarity (such as between herdsman and farmer) within populations, the degree of social complexity—the division of labor—would have remained virtually unchanged from what it had been Paleolithic times, when the only subsistence mode was hunting and gathering. In brief, the Malayan Neolithic probably did not lead autonomously to the direct development of more complex societies. Neolithic innovations in several other parts of the world led fairly rapidly to the development of complex societies containing rulers and peasants alongside the incorporation-resisting "tribal" societies. But the indigenous Malayan Neolithic societies must have remained wholly uncentralized or "tribal" until contact with complex civilizations from outside the Peninsula acted on some of them to bring about complexification, and eventually centralization, from around 2,000 years ago (cf. Benjamin 1985).

How and why, then, did the three societal traditions—Semang, Senoi, and Malay—come to crystallize out of this sociocultural matrix? At least three, overlapping, factors seem to have been involved in the process: (1) the arrival in Peninsula of new influences, such as the external trade in forest products or the emergence of complex civilizations on the Malayan periphery; (2) the intensification of people's attachment to a particular mode of subsistence, and their consequent lessening of interest in the alternative modes; and (3) mutual dissimilation in culture by neighboring peoples, in order to reduce the attraction of other subsistence modes and to maximize the advantages of following complementary environmental orientations. These factors should be understood as having been related to each other in a thoroughly dialectical manner, especially in the case of the mutual dissimilation of the Semang and Senoi patterns, which I shall accordingly treat together.

Semang and Senoi Cultural Regimes

The first thing to note is that the Semang have usually been more involved than the Senoi with the outside world; indeed, as late as the first few decades of the nineteenth century, there were Semang groups living on the northwest coast of the Peninsula. In general, the Semang have lived at lower altitudes than the Senoi, which means that for at least a millenium they have been sandwiched between the plains-dwellers downstream and the mountain-dwelling Senoi upstream. From an ecological point of view, then, the Semang have long been in a position to maximize the advantages of ecotonal situations, both biotic and cultural. The opened-up areas created by Senoi and Melayu farmers around their swiddens or fields are richer in plant and animal resources than the forest depths, and the presence of traders and trade routes affords great opportunities for selling forest products or working for wages. To avail themselves of these resources the Semang have on the whole chosen to remain opportunistic foragers, retaining contact with the outside world while rejecting the outsiders' values. Even today, they hold to a quite different evaluation of tree-cleared land than their neighbors; while the Melayu and the Temiars both regard the forest as too cold and disease-ridden to dwell in, the Semang positively seek out the forest because it is "cool" and therefore "healthy." This evaluation of their environment has been a major factor for the continued resistance of the Semang to settling down permanently in one place on other people's terms (cf. Kirk Endicott 1979a, 184–85). That they have managed to maintain their independence for so long is a tribute to their readiness to pull out at a moment's notice and move on. The literature on the Semang is full of accounts of the way in which they melt into the forest if they have reason to be suspicious of the visiting reporter—who wakes up in an empty encampment that had been full of people the night before. This is tactics, not shyness: by eschewing the amassing of material possessions, by dwelling in lean-to windscreens that can be built from readily available materials in half an hour, and by not letting any of their various subsistence activities (not even their occasional but long-established swidden-planting) keep them waiting in one place for more than a few days, the

Semang have managed to confront the world until very recently on their own terms. But those terms include a non-shy readiness to have dealings with outsiders whenever advantage may be gained from doing so.[14]

The Senoi, on the other hand, have not been able to play so fast and loose with their physical circumstances. Their commitment to village- and field-based life means that they would have much to lose if they were to run away from outside intrusion. But, on the whole, the Senoi have been better placed than other Peninsular populations to keep to themselves. The territory they occupy includes the most mountainous of the Peninsula, and outsiders have rarely attempted to travel through it (cf. Clifford 1904, Chap. 12; Wilkinson 1926, 16–17). The Senoi have therefore lived until recently as the sole, and homogeneous, populations of the northern interior (except, that is, for the ring of Semang populations at their periphery). They have accordingly come to conceive of their cultural identity as attached to the here-and-now of their own settlements rather than to any external focus of attention in the lowlands. In this respect, the Senoi and Semang have been similar, but there are differences between them too: whereas the Semang have had many dealings with outsiders without feeling it necessary to come to special terms with them, the Senoi, who have had much less contact with outsiders, have nevertheless felt it necessary to institutionalize those contacts.

Until recently, the only thing that forced the Senoi into contact with the plains-people was their need for metal knives and axes (some authorities would add salt to this list). In fact, the non-Senoi plains-dwellers, in their quest to maximize political and cultural advantage as well as their supply of forest products, have usually been much more interested in the Senoi than the latter were in them (cf. Couillard 1984). Senoi responses to this situation have been many, some very subtle indeed. I shall mention only two here: the institution of headmanship as an interpolity boundary mechanism between the Melayu and the downstream Senoi (see Benjamin 1968 for a detailed analysis of the history of Temiar-Melayu political interaction); and the care with which Senoi individuals have physically covered their tracks when returning upstream from foreign territory by, for example, walking along riverbeds wherever possible instead of on footprint-revealing ground (Rambo 1982, 272). These later institutions, of course, grew up largely as a response to the sorry history of slave-raiding that accompanied the more recent phases of Melayu political expansion in the Peninsula (see Kirk Endicott 1983 for a detailed discussion of this issue).

So far, I have discussed only the separate relations of the Semang and Senoi to outsiders. But the relations between the Semang and Senoi themselves have been at least as important in generating the cultural array we now find in the north of the Peninsula. The Semang and Senoi had to come to terms with their own mutual differences, especially throughout the long, arc-shaped marchland where they abut on each other (see fig 10.1). They appear to have achieved this by each intensifying their own way of life so as to reduce the attraction of the alternative style, for there is a fundamental incompatibility between the kinds of mental commitment required for nomadic foraging on the one hand and sedentary farming on the other. Present-day Temiar and Semai farmers love to go hunting, but they must avoid letting it become more than a part-time activity or their crops will suffer. The Semang, on the other hand, love to eat the kinds of food grown by the Temiars, and they much admire the traps which the Temiars construct for catching fish and mammals, but they must hold these desires in check or they will lose the very freedom and flexibility that they value most. Both groups appear to have handled this dilemma by interposing a decided cultural boundary between themselves and the other group, so that the Semang and Senoi ways of life have remained quite distinct, even when they share the same settlements. The boundary may be quite ancient, to judge by some recent archaeological findings.

In contrast to what is found in archaeological sites further south in the Peninsula, excavation of the major cave site at Gua Cha in the southwestern part of Kelantan state has revealed no interpenetration there of the Hoabinhian and Neolithic assemblages: the former was simply replaced by the latter some time before 3,000 years ago (Adi 1981). Peacock (1979, 202) implies that the same is true of two other southwest Kelantan cave sites, Gua Chawan and Gua Tampaq. My own glottochronological calculation (Benjamin 1976) of 4210 B.P. as the nominal date by which the mesh connecting the Northern (Semang) and Central (Senoi) Aslian languages finally broke down may well represent the same cultural break, for it occurred in the area which includes Gua Cha. Gua Cha is situated at the meeting point of the furthest known geographical extents of the Semang, Senoi, and Malay cultures in historical times in that part of Kelantan; it is tempting to see these archaeological and lexicostatistical data as evidence that some kind of boundary had already been set up between proto-Semang foragers (the Hoabinhian?) and proto-Senoi horticulturalists (the Neolithic?) some 3,000 to 4,000 years ago. These findings may, of course, be due to coincidence or sampling error, and hence be spurious: but the possibility that they may be significant is too tempting to leave them unrecorded. There is little reason to doubt that the Semang and Senoi are the modern descendents of the populations that produced Gua Cha; Solheim (1980) would appear to agree, in general terms, with this analysis. But, whatever the worth of these prehistoric musings, ethnographic investigation has provided solid evidence that the boundary between the Semang and Senoi later came to be encapsulated in a set of deliberately created differences in their kinship systems. Certainly, the differences—which concern the avoidance and teasing relationships mentioned earlier—are greater between the Jehais and the Temiars, who live side by side, than between those groups who live further apart.

As a consequence of these mutually calculated responses, the Semang and Senoi became more intensively attached to their respective modes of livelihood than they might otherwise have been, so that by the time of the earliest historical and ethnographic reports in the nineteenth century, and presumably long before then, nomadic or semisedentary foraging had become by far the dominant mode for the Semang, and sedentary swidden-horticulture for the Senoi. This parallels Peterson's findings (1977) that the Philippine Negritos became more intensive hunter-gatherers to the extent that their country was occupied by swidden-farming groups; it fits with the claim made in several papers presented to the 1978 Paris conference on hunter-gatherer societies that foraging societies are so intensively nomadic and minimally organized precisely because they have had to maintain a constantly shifting complementarity with their settled neighbors (see Leacock and Lee 1982).

A further factor in the decision of the Semang to intensify their foraging way of life must have been the fact that they sit astride the environmental ecotone formed by the transition in the north of the Peninsula between the equatorial, relatively nonseasonal, evergreen forest and the more seasonal semievergreen forest of Thailand. This "Kra ecotone," which runs southwest-northeast from Kangar on the west coast to Pattani on the east (Whitmore 1975, 163), is doubly rich in natural resources. But, in addition, it corresponds closely to the course of the ancient trade route connecting the Kedah coast with the early center of urbanization at Satingphra on the Isthmian east coast (Stargardt 1983). The rise of civilization in that area, commencing around 2,000 years ago, must have provided a strong push to those Semang populations who wished to avail themselves of the further opportunities this allowed. The known historical distribution of the Semang—to the immediate north and south of the Kedah-Pattani line and in a crescent around the farming Senoi—bears witness to this, as it does also to the Semang habit of seeking out ecotonal areas, whether natural or cultural in origin. There is in fact very little trace in recent times of the sedentary gathering that was probably practiced by the ancestors of both the Senoi and Semang, which has left its trace in the archaeologists' "Hoabinhian" (and which was, I suspect, independently rediscovered for themselves by the so-called Tasaday of Mindanao; see Fernandez and Lynch 1972). The degree to which the Semang and their ancestors have followed sedentary or semisedentary ways of life is a complicated question. Some Semang populations, such as the Lanohs and Mendriqs, seem to have been fairly sedentary for some time, while others, such as some Jahai and Batek subgroups, appear to have become more nomadic in recent years than they were formerly. I return to this issue below.

It is the very dominance of foraging or of swiddening that concerns us here, for if a cultural regime is so organized as to provide a mode of consciousness appropriate to a people's chosen way of life, as I am arguing, then the degree of implicational coherence displayed by that regime will vary with the degree to which the people have had to reject alternative ways of life in favor of one dominant mode. Most Semang and Senoi kinship patterns display such a degree of coherence in the way they organize environmental and interactional meanings that it has proved possible (as I have already suggested) to set up ideal-type "Semang" and "Senoi" systems as tools for the analysis of those cultures that are dominated by one mode of livelihood as well as those that are not.

The "Malay"-type Cultural Regimes

Let us now turn to the cultural effects of collecting-for-trade, the third major, locally dominant mode of livelihood in the Malay Peninsula. Collecting, in this sense of the term, appears to have been taken up primarily in the lower-lying and more traversable southern interior and around the coastal plains. Protohistorical and ethnoarchaeological research has show that the Malayan south (along with the matching areas of Sumatra across the Straits) has attracted traders from overseas for millenia. Relatively flat and with few natural barriers, this region has provided riverine portage routes from east to west, obviating long waits for the monsoons to shift; it is inhabited by populations prepared to fetch such desirable (to outsiders, that is) forest products as resin, cane, camphor, wood-oil, and spices; and there were sites where the foreigners themselves could mine for tin and gold (Dunn 1975; Wolters 1967; Wheatley 1961, 1975). But most of the indigenous people of the area were also swidden-horticulturalists, which, along with some foraging, had probably been their dominant way of life until they decided to incorporate external trade into their cultural repertoire some four or five millenia ago. The archaeological site of Gua Kecil in Pahang, which is within the area occupied today by the Temuans, shows evidence that the Neolithic commenced there some 5,000 years ago, but with Hoabinhian artefacts persisting to form a mixed assemblage (Dunn 1964). This suggests that, unlike the situation further north, boundaries were not set up between populations following different ways of life: indeed, it seems likely that in the south the people retained a much more mixed array of appropriative techniques. Dunn's dating of the innovations at Gua Kecil fits closely with my own glottochronology-based date of 4,900 B.P. for the complete breakdown of the mesh between the Central and Southern Aslian languages. I interpret this—insofar as such relatively shaky calculations are worth interpreting—as resulting from the southerners' move away from the mountains towards the larger rivers to engage in trade with outsiders. Dunn (1975, 136), on the basis of quite different, archaeological, evidence suggests that "a true overseas trade was probably also growing by 4000 B.P.: a trade involving coastal aboriginal peoples and—as exports—some of the products of Malaya's forest and shores." The agreement between these different pieces of evidence is encouraging.

Now horticulture differs from both foraging and collecting in its emphasis on long-term planning, sedentary residence, and the organizing of relatively large work-groups. One reason for claiming that collecting was dominant to horticulture in the south is the southerners' preference for conjugal-family-based units of enterprise,

whereas the horticulture-dominated Senoi have organized themselves into corporate cognatic descent groups. It is true, nevertheless, that the southern Orang Asli (such as the Orang Hulu) and the
Melayu, whose social organization is somewhat more complex (i.e.,
more differentiated) than that of the Semang and Senoi, have been
influenced by horticulture also. As a consequence of this two-way
cultural adaptation, a degree of sexual division of labor has
developed in the south: the men get drawn out of the village to go
collecting in the forest or towards the coast to trade their products,
and the women do most of the farming and food gathering. It seems,
moreover, that among the Orang Hulu horticulture is (for ecological
reasons?) not sufficiently self-supporting anywhere to be the sole
mode of livelihood, so that collecting-for-trade and food production
are carried on as complementary activities in every village (Carey
1976, 22–7). This sexual division of labor has gone hand in hand
with a markedly asymmetric evaluation of the two filiative modes:
domestic and village organization has come to display a matrifiliative
bias, while the organization of external relations and of
nonhorticultural work-groups displays a patrifiliative bias. Although
these unifiliative biases have, with very few exceptions, not
destroyed the formally cognatic structure of the kinship
terminologies, some of the societies (such as the Temoqs) have
developed something resembling a double unilineal descent
ideology—except that it is not strictly lineality that is at issue, so
much as unifiliative or unifocal principles.[15]

   In general, these southern Orang Asli evince a preference for
restricting intrasocietal relations in favor of relations with outsiders.
Carey, who has visited all the Orang Asli groups, remarks (1976,
222) that "Jakun" (i.e., Orang Hulu) settlements are "dispersed over
a wide area, and little or no contact is maintained with one another.
The frequent visits to relatives in other villages, which are so
common among many other Orang Asli groups, are more or less
absent among the Jakun." This exclusivity expresses itself in a
preference for marriage within one's own community and/or between
cousins (who are usually classed with siblings)—with one's own
people, so to speak. In addition, relative age is a major
organizational principle in all Malay-type societal patterns, the
underlying idea being that the husband should be older than the wife.
Since they should also be consanguines if possible, the husband-wife
relation thus tends to be assimilated to the relation between older

and younger "sibling/cousin." The manner of reckoning the relative "age" of one's sibling/cousin does, however, vary—from the use of simple birth-order alone, through considerations of the relative birth-order of the connecting kin in the next ascending generation, to "relative-age" differences ascribed to the overall matrifiliative categories to which each of the spouses belongs. The underlying image nevertheless seems to be the same: a conjugal family formed of socially very close individuals, where the husband should marry later than the wife because, while she has only one domain to learn before they can set up an independent household, he has two—collecting-for-trade and farming. It is the picture of what *should be* that matters here, rather than empirical concern over the degree of consanguinity or relative age of the spouses.

As I argue below, societies with a sexual division of labor, a concern for relative age as a component of affinal relations, and a preference for living in closed consanguineal communities linked through their male members with rather wealthy outsiders tend to become ranked. The southern Orang Asli societies do in fact display a degree of formal political hierarchy that is in marked contrast to the egalitarian political and interpersonal relations of the Semang and Senoi. The Melayu too are well known ethnographically for their concern with descent-linked rank (*pangkat, derajat,* and *keturunan*), which, if the southern Orang Asli are anything to judge by, probably long predates the development of the premodern, centralized Melayu states; Melayu social organization thus seems to share all its major features except the centralized state with the other, "Aboriginal," Malay groups. This implies two things: that the official term "Aboriginal Malay" (*Orang Melayu Asli*) is, after all, an appropriate term for the southern Orang Asli; and that the essential feature of Melayu-ness from the viewpoint of social history is the centralized state. (This latter feature is discussed at length in Wee 1985.)

This "Malay" system, then, constitutes the third of my ideal societal types. As we have seen, it is not only the kinship system of the Malays proper (the Melayu) that belongs here, for so also do the kinship patterns of most of the Southern Aslian speakers as well as the various "Aboriginal Malay" groups. As the untidy fit between language-affiliation and cultural pattern suggests, it is important to remember that we are dealing here with implicational clines—general tendencies that are more marked in some areas than in others.[16]

Ecological Values in Malayan Kinship

There are several ways of pursuing the connections between the patterns of social-organizational and kinship variations mentioned earlier and the historical development of the indigenous Malayan cultures as I have just presented it. One approach is to examine the component variables that can be examined separately, each in relation to the particular ecological features characteristic of the ways of life in which it occurs. This method was utilized extensively in my earlier study (Benjamin 1980), now under revision for publication, so I shall pursue it no further here. Alternatively, the different thematic collocations of variables can be viewed as symbolic representations of the spatial, temporal, and interpersonal orientations appropriate to each of the different ways of life, making them appear irreversible, and thereby "locking" the people into one particular way so that they come to reject the alternative ones. This too was mentioned in my 1980 study, but relatively little attention was paid there to the specifically ecological or on-the-ground social-organizational issues that are the concern of this volume. Here, I shall try to indicate some of the benefits of such an analysis through a discussion of kinship in relation to (1) marriage rules, subsistence mode, and relations of production; (2) local-group organization; (3) intercultural boundary mechanisms; and (4) Malayan archaeology.

The ideographic use of kinship is necessarily part of political process: it involves, first, the generating of values and, second, the mystifying of these values so that they remain embedded in the hard-to-question, taken-for-granted presuppositions of daily life. Kinship rules are paramount here: they serve as boundary policing mechanisms, between alternative ways of life at any one time and between what a group of people are thereby invited to see as their past and their possible future. By its very nature, kinship does a better job of disguising the loci of power than any overtly political or ecological statement could.

From the point of view of their overt "surface" sociology, the following features show the most variation in the Malay Peninsula.

1. *Inclusivity/exclusivity of kinship reckoning.* Some of the cultures (namely, most of the Orang Asli groups) possess *inclusive* kinship systems in which all members of the society (and sometimes beyond) are regarded as kin; the kinship

system then frequently provides a means of talking about the whole field of social relations, and the kin terms are largely "classificatory" in application. The other cultures (the Melayu and a few southern Orang Asli groups) possess *exclusive* kinship systems, according to which only some of one's consociates are regarded as kin; the kinship system then usually provides a means of organizing only voluntary and relatively intimate association and of talking only about a subdomain (usually familial) within the total field of social relations.

2. *Descent group/conjugal family as the primary unit of productive enterprise.* Most Malayan cultures organize their productive activities on a conjugal-family basis; a few, however (mainly the Temiars and the hill Semais), do so on the basis of corporate cognatic descent groups.

3. *"Cousin marriage."* About half of the cultures permit marriage or sexual relations with traceable consanguineal kin (though not with primary or secondary kin); the other cultures (mainly the northern Orang Asli groups) forbid such relationships.

4. *"Mother-in-law avoidance."* Several of the cultures (mainly the northern Orang Asli groups) require the avoidance by a man of his wife's mother and by a woman of her husband's father, and reciprocally; the other cultures do not require such an avoidance.

5. *"Sister-in-law avoidance."* Some of the cultures (the northern Semang) require the avoidance by a man of his wife's sisters and by a woman of her husband's brothers, and reciprocally; a few of the cultures (the Temiars and the hill Semais) require joking behavior between them. Most of the cultures require neither avoidance nor joking, and the relationship is not subject to special attention.

6. *"Sister avoidance."* A few of the cultures (the northern Semang) require avoidance between adult brothers and sisters; most of the cultures have no special rules (other than incest prohibition) for this relationship.

Table 10.1 summarizes these variations for most of the Peninsular cultures, including all of the Semang groups. Table 10.2 summarizes the rather complicated pattern revealed by the "sister-in-law" relationships, not all of which can be discussed here.

Marriage Rules, Subsistence Mode, and Relations of Production

The significant issues here are (1) whether it is the conjugal family or some larger group that is paramount as the social unit of subsistence enterprise, and (2) the patterns of social relations implied or imaged by the different marriage preferences.

Under tropical forest conditions, opportunisitic foraging (i.e., a mix of hunting-and-gathering with trading and occasional wage labor) such as the Semang undertake makes the conjugal family paramount. The ethnographic literature shows that this applies very widely in other parts of the world too. Field observation in the Malay Peninsula (Kirk Endicott 1974; Benjamin, in preparation; Gomes 1982) has shown that among all Semang subdivisions the conjugal family is the only social unit that shows any structural persistance, whether in economic activities or residential patterns. More extended groupings have only a fleeting existence, and they result from the efforts of anyone who can hold the others' attention long enough for any joint action to take place. This image of society finds its kinship expression in the array of cross-sex avoidances (SpPa/ChSp; Sb/Sb; SpSb/SbSp) found in almost all Semang subdivisions, which singles out the conjugal family by tabooing those dyadic relationships that link it conceptually to the wider kinship web.[17]

The various Malay groups, however, have no such tabooed relationships, although they too organize their productive activities on a conjugal-family basis. The Melayu are typical peasants operating within a market economy, and the attainment of the conjugal-family ideal is not problematical: their kinship reckoning is exclusive (so that only a few individuals are considered to be kin anyway), and the conjugal family is in little danger of being merged into a wider network. The same is true of the Malay-type Orang Asli groups, because of their long-established adaptation to trading in forest products.

In contrast, the Temiars and the hill Semais, whose swiddening practices favor, and are best served by, extended-family residence groups, observe no avoidances between siblings or siblings-in-law. This allows them to direct their minds to wider ranges of day-to-day cooperation; and, indeed, the Temiars and hill Semais do aggregate into larger residential and work groupings than the Semang, while maintaining quite widespread relations through marriage.[18]

Table 10.1 Peninsular Societal Patterns

People/ethnic group	Population[1] (1980)	Language	Mode of societal integration	Dominant appropriative mode	Kinship reckoning	Social unit of productive enterprise	Cousin marriage	Ranking	Filiative bias	Type of societal tradition
Kensiu[2]	100	Northern Aslian	band	foraging	inclusive	conjugal family	forbidden	egalitarian	patri	Semang
Kentaq[2]	100	Northern Aslian	band	foraging	inclusive	conjugal family	forbidden	egalitarian	patri	Semang
Jehai[2]	950	Northern Aslian	band	foraging	inclusive	conjugal family	forbidden	egalitarian	patri	Semang
Menriq	125	Northern Aslian	band	foraging	inclusive	conjugal family	forbidden	egalitarian	patri	Semang
Lanoh	224	Central Aslian	band	none dominant	inclusive	conjugal family	forbidden	egalitarian	cognatic	Senoi
Temiar	11,593	Central Aslian	tribe	horticulture	inclusive	descent group	forbidden	egalitarian	cognatic	Senoi
Hill Semai	18,327	Central Aslian	tribe	horticulture	inclusive	descent group	forbidden	egalitarian	cognatic	Senoi
Lowland Semai		Central Aslian	peasantry	horticulture	exclusive?	conjugal family	permitted?	?	cognatic	Malay
Temoq	350	Southern Aslian	tribe	collecting	exclusive	conjugal family	permitted	ranked	matri	Malay
Semelai	2,582	Southern Aslian	tribe	collecting	?	conjugal family	permitted	ranked	matri	Malay
Mah Meri	1,356	Southern Aslian	tribe	collecting	exclusive	conjugal family	permitted	ranked	matri	Malay
Orang Hulu	9,799	Austronesian	tribe	collecting	exclusive	conjugal family	permitted	ranked	matri	Malay
Orang Kanaq	34	Austronesian	tribe	collecting	exclusive	conjugal family	permitted	ranked	matri	Malay
Melayu	6 1/2 million	Austronesian	peasantry	agriculture	exclusive	conjugal family	permitted	ranked	matri	Malay
Minangkabau		Austronesian	peasantry	agriculture	exclusive	descent group	permitted	ranked	matri	Malay
Temuan	9,312	Austronesian	tribe	collecting	exclusive	variable	permitted	ranked	variable	Malay
Che' Wong	200	Northern Aslian	tribe	none dominant	inclusive	conjugal family	forbidden	?	patri	'mixed'
Jah Hět	2,442	Central Aslian	tribe	none dominant	inclusive	variable	forbidden	?ranked	matri	'mixed'
Semaq Beri	2,078	Southern Aslian	band	none dominant	inclusive	conjugal family	permitted	?egalitarian	?	'mixed'
Batek	800	Northern Aslian	band	none dominant	becoming exclusive	conjugal family	permitted	egalitarian	becoming matri	'mixed'

1. Figures based on Malaysian Census 1980, which is not yet complete for some Orang Asli groups in some states; figures ending in zero are estimates.

2. In Thailand there are additionally some 100 Kensiu, 100 Kentaq, 200 Jehai, and, further north, a Semang group not found in Malaysia, the Tonga' or Mos, numbering some 300 people.

Table 10.2 Peninsular Patterns of Cross-sex Relations

People/ Ethnic Group	"MOTHER-IN-LAW" WiMo/DaHu; HuFa/SoWi	"SISTER-IN-LAW" Wi$_e$Si/$_y$SiHu; Hu$_e$Br/$_y$BrWi	"SISTER-IN-LAW" Wi$_y$Si/$_e$SiHu; Hu$_y$Br/$_e$BrWi	"SISTER" Br/Si	
Kensiu	avoidance	avoidance	neutral	avoidance	+avoidance, −joking
Kentaq	avoidance	avoidance	neutral	avoidance	
Jehai	avoidance	avoidance	avoidance	avoidance	
Menriq	avoidance	avoidance	avoidance	avoidance	
Lanoh	avoidance	restraint	joking	neutral	+avoidance, +joking
Temiar	avoidance	joking	joking	neutral	
Hill Semai	weak avoidance	restraint	joking	neutral	
Jah Hět	avoidance	restraint	restraint	neutral	
Lowland Semai	neutral	?neutral	restraint	neutral	−avoidance, −joking
Semaq Beri	restraint	restraint	restraint	restraint	
Batek	restraint	restraint	restraint	?neutral	
Temoq	restraint	?	?	neutral	
Orang Hulu	restraint	restraint	?	neutral	
Che' Wong	?	?	?	?	
All others (including Melayu)	neutral	neutral	neutral	neutral	

Local-group Organization

The main issue here is the extent and direction of deviation in practice away from the strictly cognatic mode of organization implied by almost all Malayan kinship terminologies. While the Senoi are thoroughly cognatic in both practice and (native) theory, the Semang and the Malay-type societies demonstrate a measure of unilateral emphasis, the former patrilaterally and the latter matrilaterally. I have already discussed the historical origins of Malay matrifocality as rooted in the disjuncture between male competitive trading activities and female, agriculture-based, coresidentiality; by implication, this also helps to explain the lack of unilateral bias among the Senoi—they, quite simply, exhibit no gender-based disjunctures of the Malay kind. The patrifocal tendencies of Semang local-group organization have been mentioned, but not yet explained.

It would be quite wrong, however, to assume, just because the terminologies and most of the publicly stated "rules" in the Peninsular varieties of kinship are so cognatic in character, that the unilateral tendencies of the Semang and Malay traditions are merely secondary responses to the exigencies of daily life, devoid of any ideological content: matters are by no means so simple. Nor is it safe to assume that the formally cognatic "structure" is somehow logically or temporally prior just because it is so widespread; this too may well turn out to be the result of some secondarily (and politically) imposed pattern, aimed at holding the feared excesses of the various unilateral tendencies in check. (In any case, there is more than a hint, even in the terminologies, that the cognaticism has not always been unalloyed; see, for example, Blust 1980.)

Let us return to the Semang-Malay cline of kinship features presented earlier. I asserted that at each pole the features held together in a relation of meaningful mutual implication: to demonstrate how this could be so, I shall start with some ideas suggested by Martin (1974, 14–16). In surveying the range of social-organizational patterns displayed by the foraging societies of the world, Martin showed that several commonly held ideas are wrong. First, "bilateral descent" (i.e., cognatic kinship) greatly preponderates over the patrilineal or matrilineal varieties of descent among foragers; among aquatic foragers, matrilineal organization is surprisingly more common than the patrilineal option. Despite this general preference for cognatic organization, postmarital residence is preponderantly patrilocal in all cases, but with matrilocal residence as the highest-ranking alternative. Second, Martin found that these unilateral skews in residence relate not to the relative contribution of men and women to subsistence production, as is often claimed, but to the extent of male cooperation in obtaining the most valued food resources: "meat and fish [and, for the Semang, one might add 'honey'] typically make subsidiary contributions to the diet, but are always exploited to the maximum extent by the largest number of males a community can effectively maintain" (Martin 1974, 20). This imperative does not in itself, however, lead to the generation or imposition of any specific residential mode or rule; other elements are involved as well, such as the degree of environmental concentration and mobility of the most favored resources, and the tendency of males to "monopolize authority roles and the peaceful or hostile relationships with surrounding territorial groups" (Martin 1974, 20).

Clearly, there is nothing in Martin's ideas that would restrict their application solely to foraging societies; the same issues are at play in subsistence-based farming and trading societies too. Just what the issues are will become clearer if we first spell out the different, ideal-typical societal implications of pursuing, on the one hand, the fully patrilineal/patrilocal option (hereafter called "patri" for short) or, on the other hand, the matrilineal/matrilocal option ("matri" for short)—even though in their extreme forms these are rare or unknown in indigenous Peninsular cultures.

The main thing to note is that from a political point of view the "patri" pole is inherently more unstable than the "matri" pole, since it keeps the men of the local group together, making them able, and likely, to cooperate with each other in squabbling with other local groups. "Patri"-groups tend, therefore, to be more thinly distributed over the land. "Matri"-groups are more stable since it is harder for the men, who are now dispersed, to cooperate in squabbling with other groups. For this reason, and because some male cooperation in subsistence activities is always needed, "matri"-groups tend to be more closely distributed over the ground.[19]

These formulations refer expressly to ideal-typic situations. But ideal types are in essence mental pictures of what might come to be the case if real-world circumstances were not held in check. By the same token, however, such pictures could as well serve to indicate what *should be* the case if one's own people are to be kept to a particular way of life in the face of beckoning alternatives—which is precisely what I have been claiming must have motivated a great deal of the cultural differentiation in the Malay Peninsula. Thus, if dispersal and egalitarian segmentation are the chosen mode of organization, the imposition of a degree of "patri" ideology will help to "lock" such a choice, making it relatively more binding on one's fellows. This is the situation the Semang have historically found themselves in, and they do exhibit "patrifocal" tendencies in, for example, the patrilaterally skewed shapes of their personal genealogical memories or the statistical predominance of patri-virilocal residence in their local-group organization. It is quite possible that the "sister-avoidance" rule has further reinforced this patrifocal tendency, for it puts any extended residential sibling-group that might emerge under pressure to be formed of all-brothers or all-sisters, as opposed to a mixed brother-sister group. The all-sister option is unlikely to have been taken up by the Semang, for in addition to the reasons just suggested, the need for cooperation in defense and occasional trading activities would have pushed them towards forming brother- rather than sister-linkages.

Turning now to the "matri" pole, we find not only a reversal of the characteristics typical of the "patri" pole, but also the completely new organizational elements of politics and ranking. In a "matri" regime, the locally coresidential males have no consanguineal bonds of kinship to mediate their relations with each other: their links are supra-kinship in character, and this tends to lead to the emergence of an administrative structure. As Martin puts it (1974, 19),

> typically . . . matrilateral [sic] descent groups are drawn into the same political community by the union of their respective heads on a common council and by the election of a primary headman. Since this involves the consolidation of several territorial groups, matrilocal societies tend to approach *tribal* (Service, 1962) and higher levels of sociocultural integration.

The fully matrilineal/matrilocal form of organization is exhibited in the Peninsula only by the Melayu of Negri Sembilan and Melaka, who claim descent from the similarly (but not identically) organized Minangkabau of the West Sumatran highlands, but partial approximations to it are found among all the other Melayu and those Orang Asli groups that I have characterized as "Malay" in societal type. What all these groups do have in common, however, is an efflorescence of ranked political offices in each local community, which seems to be quite otiose when compared to the otherwise rather slight degree of actual social differentiation exhibited within these communities: the ranking appears to be more symbolic in character than concerned with any exercise of real power.

There is, then, a shadowy but perfectly graspable chain of implication linking the various features at each pole of the Semang-Malay contrast I drew earlier. In particular, there is an obvious connection between (1) patri- versus matri-focality, (2) residential aggregation versus residential dispersal, and (3) segmentary egalitarian societal organization versus ranking. In my earlier study (1980) I argued that patrifiliation in Malayan cultures represents social disjunction, and that matrifiliation conversely represents social conjunction. These ideas now seem to be borne out at a higher level of analysis too, providing the people who live by those cultures with easily picturable models that link kinship ideas with notions of social space. Whereas in the earlier study I was concerned with the ways in which the different kinship patterns modelled different modes of lending coherence to experience, the present analysis is concerned with the cultural imaging of spatial alignments between real people: these are images of "objective" on-the-ground issues, rather than "subjective" in-the-head ones.

The evidence suggests that this way of imaging different ways of life represents something of the means by which ecological values have been propagated, talked about, and put into practice by the people themselves. It is quite probable that Malayan peoples really have been consciously aware of the different social implications of patrifocality and matrifocality—that the former separates, the latter incorporates. Much of the seemingly arbitrary ethnographic variability in the Peninsula will, I feel sure, yield to this approach.[20] The implicational tendencies of the Semang and Malay kinship and social patterns now become clearer, and a further explanation of the various kinship rules discussed earlier becomes possible.

For the Semang, there is a conflict between two perceived or imagined tendencies: (1) the protein-finding imperative, which threatens to lead to mutually distrustful and warring groups, each exhibiting patrifocal local affiliation; and (2) the wish to remain foragers and avoid sedentism, which requires good relations with neighboring Semang groups so as to use their territory and retain a sufficiently large demographic pool. The Semang cross-sex avoidances, coupled with their inclusive mode of kinship reckoning, were therefore probably instituted to keep in the people's minds the proper picture of what is needed—namely, complete fluidity of individually organized relations that do not generate any persistent pattern. In other words, what the Semang cross-sex avoidances preach is randomness of consociation.

In the Malay case, the factors that fall into threatened or imagined conflict are: (1) subsistence farming or fishing by communities closely settled along rivers, which tends towards a sedentary matrifocal mode of local affiliation with dispersed intrasocietal male links; and (2) trading in forest products, which tends towards the generation of locally cooperating male groups. If the matrifocal tendencies were to proceed all the way, the outcome would probably be the dispersal of related males coupled with the organizing of the unrelated coresidential males into a political hierarchy; this would be more conducive to competition than to cooperation. But these conflicting demands appear to have been solved at one stroke by instituting preferential endogamous "cousin"-marriage: this has the effect of declaring the coresidential males to be "related" to each other after all, allowing them at the same time to downplay the more diluted relations between themselves and males in other communities. Whole communities could now easily compete with each other in trading without straining too many kinship ties, but the males within each community could nevertheless cooperate with each other as virtual "kinsmen." This leaves the matrifocal relations untouched, but it produces a degree of overtly-labelled normative, though somewhat unreal, ranking between the males in each community, who are organized into hierarchical set of titled headmanships.

The Senoi pattern—to marry affines but to avoid marrying consanguines (the two categories are clearly distinguished in Senoi kinship)—thus seems to be a fudge between the Semang refusal to marry either affines or traceable consanguines and the Malay preference for marrying consanguines (who, through preferential cousin-marriage, are covert affines anyway). There is a fascinating linguistic clue to the processes by which this Senoi pattern may have come about: the words employed by the Senoi as consanguinal kin-terms are etymologically mostly Austroasiatic in affiliation, but the affinal terms derive from some long-since disappeared (non-Malay) Austronesian source (Benjamin 1980, 49–50). This suggests that the affinal terms were at some stage deliberately modified or added on to the older consanguineal terms, which further implies that the dialectically organized Senoi pattern of kinship and consociation is a development that had to wait, as it were, for the first signs of the emergence of the Semang and Malay patterns.

It is my hypothesis that these relatively simple additions or switches to the kinship rules—Semang cross-sex avoidances, Malay preferential cousin-marriage, and Senoi foreign-language affinal terms—were imposed as consequences of decisions as to which of the various available ways of life people should follow. If this is indeed what happened, there should be some recoverable correlation between the kinship patterns and the archaeology of the Malay Peninsula. If it is true, as is often asserted, that values cannot be read directly from archaeological data, it is also true that certain kinds of archaeological data are crying out for value-regarding interpretation. I shall return to this problem shortly; first let me pursue one other implication of what I have been saying.

Inter-cultural Boundary Mechanisms

In most of the areas where Semang populations abut on Temiar ones there is a definite cultural boundary, despite the fact that in many cultural domains (religion especially) there has been much cultural exchange between them. The problem here is that, though the Semang people generally admire the Temiars' material superiority and apparent ability to attract an undue degree of governmental aid, the Semangs' own success depends on their ability to hold to an ecological niche—foraging—left unoccupied by the Temiars. This ambivalent state of affairs is made easier to bear through the immorality that the Semang are enabled to ascribe to the Temiars, whom they see (rightly!) as only too ready to enter into "sister-in-law" sexual dalliances wherever they go. Both Schebesta (1973, 197–98) and I have recorded Semang statements of moral reprobation at the way their Temiar neighbors carry on. The Semang rule of "sister-in-law" avoidance, therefore, encodes a morality of ecological restraint just as much as it encodes a sexual one. It serves, on a smaller scale, the same functions as some have ascribed to the Great Wall of China: to put a distance between one's own group and other ways of life that may seem attractive in some respects but which would be destructive to the way of life that one has invested so much effort in developing.

Such is just one contemporary situation. But investigation suggests that it is representative of something that has gone on for a very long time in the Malay Peninsula. Not only have people used kinship to give symbolic expression to their preferred way of life (thereby helping to *make* it the preferred way), but they have used it to build barriers against other ways that at least some individuals among them have wished to avoid. A detailed examination of, for example, the seemingly arbitrary differences in the relative-age and relative-sex components of the "sister-in-law" avoidances outlined in table 10.2 shows that each group chooses a rule that contrasts in some way with the equivalent rule of their immediate neighbors. When all such variations are plotted on a map they turn out to correspond closely to what, on the basis of the linguistic evidence, must have been the earlier pattern of contacts before modern inroads broke down the indigenous population pattern. In other cases, however, it is not today's contemporary threat that motivates the imposition of these kinship-rule barriers, but the memory of what one is trying to put behind one in a temporal sense: this is especially true of the Malay-type societies, which are historically "hot" (to use Lévi-Strauss's term) in that they are consciously concerned always to move forward, in an ecological or cultural sense—in other words, to transform themselves while remaining transitional. For the proto-Malays, the decision to encourage marriage between consanguineal kin must have been the rubicon, for nothing is so calculated as this to stir up fears of incest in the other Peninsular peoples.

Archaeological Perspectives
on Malayan Cultural Differentiation

Can anything of this situation be recovered from the prehistory and archaeology of the Peninsula? I mentioned earlier that most ecological decisions are concerned with holding on to what one has rather than with moving towards some new aim. As Smith puts it (1972, 32),

> It is not likely that we can see any long-term intent
> by man as an explanation for his transformations
> through the long reaches of prehistory. Much of what
> might seem, with the benefit of hindsight, to be

purposeful behaviour was more likely the side-effects
of short-run adaptations that in the long-run proved
selectively advantageous for survival. Food production in its initial stages was probably just this: a
series of innovations adopted to maintain the status
quo rather than to abolish it. The original Neolithic
revolutionaries were probably conservative oppor-
tunists, and so were their successors.

To this I would add a caveat: once a particular way of life has been
thoroughly explored by neighboring groups, it possible for people to
work out what the consequences might be and, accordingly, for them
to accept or reject that way of life, as they choose. This has certainly
been the situation in the Malay Peninsula: for example, those
Semang who still stick to nomadic hunting and gathering do so not in
ignorance of agriculture but in the full knowledge of what it entails
(cf. Benjamin 1973, xi–x; Kirk Endicott 1979a).

   Thus, for each of the Peninsular ways of life we need to ask two
quite different questions: (1) what were people trying to hold onto
whenever they undertook an apparently new course of action, and
(2) what were the long-term consequences of doing things in new
ways? Some answers to these questions have already been
suggested in this paper. The Semang were seeking to hold onto their
nomadism—they have on the whole been tacticians. The Senoi were
seeking to maintain some continuing linkage to places—they have
been both tacticians and strategists. The Malays (both "Aboriginal"
and Melayu) were seeking to link themselves with the outside,
eschewing both fixed attachment to place and free wandering over
the land—they needed to ensure that outsiders could find them,
whereas the Semang needed to ensure that they could themselves
find outsiders whenever necessary.

   Let me indulge in a little speculative history at the expense of the
available data on Malayan archaeology. The original shift from
Hoabinhian broad-spectrum gathering to the Neolithic pattern of
food-production was generally very gradual in Southeast Asia, being
little more than an accretion of extra food-getting methods. In the
early stage, therefore, food-gathering would not have had any
sociocultural consequences immediately obvious to the people
themselves, and it is doubtful that any political intervention or a
positive shift in consciousness were necessary to bring it about.

Ecologically directed political action would have entered the scene with regard not to food-producing but to a long-term commitment to sedentization (or trading), and this is likely to have occurred in the Malay Peninsula (as I argued earlier) only with the arrival of seed-sown grain crops. As yet we know very little about early millet and rice in the Peninsula. Rice (*Cryza sativa L.*, probably *Indica* type) has been found plentifully at the Gua Cha cave site, and a sample has been C^{14}-dated at 930 ± 100 B.P. (Adi 1981, 107); on ethnological grounds, it is likely that millet (*Setaria*) was quite widely grown even earlier. But the general picture seems to be that, though the Hoabinhian-Neolithic boundary is datable to around 3,000 B.P. at Gua Cha (which is situated in the present-day Temiar and eastern-Semang area) and to around 5,000 B.P. at Gua Kechil (in the present-day Temuan area), positive intensification of the various ways of life occurred somewhat later. It should be noted that the Malayan cave-sites were used largely for burial rather than for habitation, and the rice grains found at Gua Cha, being associated with Chinese pottery, could well have been obtained in trade rather than grown by the people themselves.

Thus, either grain-farming or trading (or both), but not necessarily the basic Neolithic way of life, would have involved important shifts in values, backed up presumably by diffuse, kinship-based, political sanctions of the sort I have been discussing. Though it is difficult to suggest dates for these shifts, some such argument might apply to the earliest "lockings-in" of both Senoi and Malay patterns.

The Semang story, however, would have been different. At the point where matters were becoming "political" for their proto-Senoi neighbors, the proto-Semang had to decide whether to become sedentary too, or to pull themselves away completely from any such tendency so as to take advantage of the two special opportunities offered in the northern part of the Peninsula. These were the Kra ecotone between nonseasonal and seasonal forest, and the artificial ecotone opened up by the Satingphra trade-and-civilization complex (currently under archaeological study by Janice Stargardt). It is my hypothesis that only after these options opened up did the Semang decide that in order to avail themselves of the resultant opportunities, they now had to impose kinship-based "locks" on their mobile way of life. As the Temiars seem to have taken the opposite course, imposing the "opposite" kinship rules (joking instead of avoidance, for example), it seems highly probable that the Semang and Senoi patterns crystallized out by abreaction from each other and in response to much the same stimuli—the connecting up of the northern parts of the Peninsula with the trade routes emanating from the centers of Mon civilization that were beginning to emerge in the Isthmus around 2,000 years ago.

Sedentism versus Nomadism

There is evidence both for the secondary reinforcement of Semang nomadism in the north and for the secondary "locking-in" of the Senoi further south into a way of life which places positive value on farming-based sedentism, while allowing for the continuation of a Hoabhinian-style broad spectrum foraging. The recent Semang populations, for example, have been much more nomadic in the Malaysian border areas and in southern Thailand (i.e., near the ecotonal region just mentioned) than they have been further south, where such Semang subdivisions as the Lanoh and Batek Nong have usually been reported as living in semipermanent villages. Given that among the world's pedestrian foragers the "overwhelming majority, some 62 percent" (Martin 1974, 11) follow a seminomadic type of settlement pattern, while only 20 percent wander in nomadic bands,[21] these strongly nomadic northern Semang begin to appear somewhat unusual, while their southerly kin take on a more "typical" appearance. This secondarily intensified nomadism of the northern Semang—if such it be—is very likely to have developed as an abreaction from the Neolithic sedentization of their proto-Senoi neighbors; they could now all the more efficiently exploit the ecotonal resources, both natural and artificial, that were to be found further north but not in the south. It may well be, then, that it is some of the semisettled Semang groups (some of whom I earlier characterized as "mixed") who have most closely retained the Hoabinhian attitude—its semisedentary character included—even though the physical nature of their resources has changed somewhat over the years, and even though the prototypical image of the Semang in the wider literature is that of thorough-going nomadism.

It is possible, however, that Semang sedentism, when it does occur, is more complicated than this. The question is whether the few settled Semang villages are local adaptations to links with settled outsiders, or whether they are endogenous adaptations. Probably, both patterns are found with (1) Semang groups settled near Temiar, Malay, or Thai villages, and (2) Semang groups aggregated for mutual defense against outsiders, such as slave-raiders (cf. Kirk Endicott 1983). The evidence suggests that Semang values generally favor nomadism but allow the people to put up with sedentism when it is to their advantage or when they are forced to do so. This latter point raises the historically important fact of the deep seated Melayu intolerance for nomads, in which two factors are involved: a cognitive-cum-cultural fear or distaste for contact with nomads (who are usually categorized, along with former Melayu habitation sites as *kotor* "dirty"), and a need to have control over a mobile but contactable (hence village-centered) subservient population.[22]

But, if the Semang are mostly nomads who often seem, as if incidentally, to spend a lot of time in one place, the Senoi are (to adapt a phrase from Lévi-Strauss) cultivators with a nostalgia for the nomadic way of life. Certainly, they appear to have avoided travelling, at least until recently, too far along the path of sedentism and grain-farming. For example, Temiars maintain an interesting set of work taboos based on the phases of the moon, according to which no house building or farming may be done for some ten days each month. But during the ban on these "Neolithic" activities, the people prefer to go hunting, fishing, and gathering, sometimes in whole family groups—in a word, to be Hoabinhian![23]

Environmental Consequences of Semang, Senoi, and Malay Values

I have said nothing yet as to the ecological effects of the various ways of life I have been discussing. A few pointers may therefore be appropriate before concluding this paper.

A long-term consequence of food production is the destruction of living species. Smith (1972, 12–13) estimates that perhaps 80 percent of the 10 million or so species known to have ever existed will have been made to disappear as human food-producing activities irrevocably alter the landscape. How do the three Peninsular societal traditions measure up against this yardstick?

The Semang and Senoi patterns of exploitation are unlikely to have had any preponderant environmental effect. The Semang have just kept moving about in very small numbers, while the Senoi have depended on recycling their habitation sites through long fallowing periods. The Semang treatment of wild tubers and fruit trees has probably led to a degree of local concentration of these resources, and occasional mistakes in Senoi farming practices have led not to forest regeneration but to infestation by *Imperata* grass and invasion by wild cattle, but these effects hardly amount to the destruction of natural species or of much of the environment. Indeed, as Rambo has suggested (1982, 280–81), Senoi swiddening practices have probably played a part in increasing the species-diversity of the Malayan forest.

The Malay pattern, with its emphasis on the obliteration of former conditions, is much less environmentally innocent. Locally, the effects of Malay-type, especially Melayu, habitation are very marked: loss of trees, the sweeping away of topsoil around the village space because it is *kotor* "dirty," and, more recently, irrigation works. Melayu villagers usually insist on a shining-white, swept-sand area around their settlements, just as they insist on removing fallen logs from their swiddens, which they plant with single crops in neat rows. This, coupled with a keenness to raise goats in the village area, often leads to the complete loss of vegetative cover, except perhaps for coconut palms. In turn, this often causes the drying up of the wells on which (especially in the Riau Islands) they rely for drinking water, and they find themselves having to move on. And so the one-way progression proceeds, congruent with Malay-type values, but unkind to the environment and hardly a viable way of life in conditions of increased population density. Anyone who has experienced the treeless, arid refulgence of government-run rural resettlement schemes or new urban areas in Malaysia will know that the old Malay-type environmental values have been given a new lease on life in contemporary conditions—for what is modernity but "progress" and the sweeping away of the old to make way for the new!

A Conclusion

We have seen that the attempt to relate social forms with ecological values is a far from straightforward task: no single rule of residence or of descent, for example, is generally associated uniformly with particular modes of subsistence or production. On a worldwide scale, the forms of society are in a constant fluctuation that depends on such circumstances as the accidents of local history.

Variation therefore, not regularity, is the norm, and this variation will continue unless and until a situation arises where the powers-that-be manage to coopt a group of people by "locking" them more intensively into one way of life or adaptational mode rather than another. This is likely to occur whenever historical events add further choices to the available sociocultural repertoire: political action may then be undertaken so as to lock people into the now-preferred mode, while restricting (or even blocking) their access to the alternatively available but less-desired modes. Political action of this sort does not usually depend on the use of force, but on the propagation of values; these are actively so diffused throughout the domains of kinship, religious practice, and language, as to make the politically preferred way of life appear more meaningful, matter-of-fact, coherent, or cognitively integrated, than the alternative ones. Typically, this is done by reinforcing and highlighting certain implicational chains of meaning, thereby making the targetted way of life appear much more systemic or logical than it otherwise would or than it really is.

The "themes" of this essay are precisely the three main systematizing patterns of environmentally linked values that appear to have been developed and propagated historically (and perhaps prehistorically) among the indigenous populations of the Malay Peninsula. But complete systematization is never possible, for each new addition to the chain of implications carries with it other possible implications that may tend in directions quite contrary to the intended thematic thrust. The ethnographically observable exponents of each of the different themes have therefore come to display a patchwork of varying approximations around the theme. This means that any attempt at ethnological interpretation and explanation must give equal emphasis to both typological and variational approaches (though, for lack of space, I have not said as much about the latter in this essay as I would wish). It must also be borne in mind that there may also be people in the same geographical area that have not been subjected to such regularizing influences, and who therefore fluctuate *as individuals* between various different modes of organization. Such aggregates of people are truly culturally "mixed"—not only because they appear so in the light of the typologizing approach I have used here, but also because they themselves have been quite happy to mix together deliberately whatever bits and pieces of culture have been made available to them through the accidents of their own local history.

Acknowledgments

This chapter would have been even longer if it did not overlap so much with my 1980 paper, "Semang, Senoi, Malay." Regrettably, that study has still to appear in its final published form, so I must beg the reader's indulgence and hope that my many references to the unpublished manuscript version will not be too annoying. My intention is that this present essay should serve as a study complementary to the monographic version of the 1980 paper when the latter eventually appears.

Time has been short, so I have not been able to circulate drafts of this chapter for critical comment by other readers. I have, however, tried to incorporate some of my reactions to the comments and criticisms of participants at the parent conference in Honolulu, June 1983. While I thank them cordially for their help, I must absolve them from any blame for what has resulted.

I have been less than open in the text of this paper about the ethnographic sources on which my generalizations are based. The list of references, therefore, contains certain items not referred to in the text but which were nevertheless used as basic ethnographic resources. In particular, I would like to acknowledge the following, not otherwise mentioned, sources: for Batek, Karen Endicott (1979, 1981) and Kirk Endicott (1979b); for Jah Hĕt, Couillard (1980); for Semelai, Hoe (1964); for Temoq, Laird (1979); for Semaq Beri, Jensen (1977); for Orang Hulu, Hill (1973, n.d.) and Maeda (1967a, 1967b, 1969, 1971); for Temuan, Baharon (1973, 1983); for Mah Meri/Besisi, Ayampillay (1976) and Wazir (1980); for Melayu, Maeda (1978) and Vivienne Wee (personal communication). Other unsourced references to Semang, Senoi, and Melayu cultures derive from my own field-data, gathered at different times since 1964 on expeditions financed variously and in part by the the Horniman and Esperanza funds of the Royal Anthropological Institute, the Wenner-Gren Foundation, and the Australian National University.

I would like thank Tan Yee Yee for her accurate and uncomplaining typing, thrice, of my heavily scribbled-over and almost illegible handwritten drafts. And thanks also to Joan Goodrum (Australian National University) for performing the equivalent task on my roughly sketched Orang Asli distribution map.

Notes

1. By "political" I mean that cultural values are concerned primarily with power and control, that is, with aligning some persons' actions to what some other person or persons want them to do, or, as Leach puts it (1982, 134), with the "art of rallying support to a cause in which you are interested."

2. Much the same considerations apply to the distinction between "history" and sociocultural "evolution," which seem to me to be fundamentally different (even if evolution results from history). Sociocultural "evolution" properly relates only to the pristine emergence of new (and, at first, unrecognized) forms of organization. Mere change or the nonpristine diffusion of new forms from elsewhere are in my terms "history," not "evolution."

3. Some scholars might object to seeing Malays and Aborigines treated together as a single ethnological complex; Professor de Josselin de Jong (1981, 486) has in fact already done so with reference to my 1979 paper. But to claim that we must treat cultures separately because they pertain to different language families (like the Austronesian-speaking Malays and the Austroasiatic-speaking Semang and Senoi groups) or because the people bear different somatic or "racial" features (like the "Mongoloid" Malays and Senoi as compared to the "Negr[it]oid" Semang) would be to claim that we must exclude Finns, Hungarians, and Basques from consideration of European culture because they speak non-Indo-European languages. To persist in treating these Malayan populations separately would be to fall back into some pre-Boasian morass or to take far too seriously some of the ethno-political ideologies current in contemporary Malaysia.

4. Some of my ideas on this have been presented elsewhere (Benjamin 1974, 1976, 1979, and 1985). In preparations is a monograph on the indigenous societal traditions (cf. Benjamin 1980). This last study overlaps to some extent with the present essay, and reference should be made to it for further ethnological and ethnographic details.

5. For a more detailed map, based on the language and dialect differences, see Benjamin (1983).

6. For an excellent alternative formulation of the characteristic

features of the Semang, Senoi, and (Aboriginal) Malay
societal and ecological patterns, see Rambo (1982). My
approach differs from Rambo's mainly in ascribing more
importance to collecting-for-trade as a factor in the
(Aboriginal) Malay pattern; otherwise, the degree of
agreement between us is high.

7.   See also McKinley (1979) for a parallel treatment of these
features as exhibited in Melayu culture.

8.   See Dunn (1975, 117–19) for a summary model of the
history of indigenous Malayan collecting-for-trade.

9.   In addition to these major modes of appropriation, there have
also been ethnologically important, though less well known,
populations living in lake-fringe, coastal, or estuarine areas
engaged in fishing and strand-foraging, sometimes combined
with desultory farming. At the present time, lake-fringe
fishing is practiced by some of the Semelai, an Orang Asli
group living on Lake Bera, Central Pahang. The Orang
Kuala—also known as Duano or Desin Dolaq—of the west
coast of Johor and the east coast of Sumatra are fishing and
trading seafarers; the Orang Seletar of the mangrove inlets
of Johor and Singapore are strand-foragers. The Semelai and
Orang Kuala, however, may be considered as falling within
the "collectors" category, and the Orang Seletar can be
classed as "foragers": their kinship and social organization
appear to fit the generalizations I draw later, but there is no
space for further discussion here. They will be discussed in
the published version of Benjamin (1980).

10. Archaeological research is also beginning to show firm
evidence for the even earlier establishment of complex,
state-based civilizations to the immediate north, southwest,
and southeast of the Peninsula; I shall discuss some of the
implications of these discoveries later. To complete this
picture mention must also be made of the close links that
emerged between the coastal Melayu kingdoms and various
European powers from around A.D. 1400, and/or the
progressive establishment of modern centralized administration and industrial methods of production during the last 100
years or so.

11. There are, as always, exceptions that prove the rule. For
example, one of the Northern Aslian languages, Che' Wong,

is quite distinct from the others, forming no part of the Semang mesh; but the Che' Wong people are in fact not nomads, and they lead a fairly sedentary life. Among the Central Aslian languages, the various dialects of Lanoh do form a close mesh; but their speakers were formerly more nomadic than their Temiar neighbors (though probably secondarily so), and their culture is usually classed as Semang. The fact that the Lanoh are physically mostly Negritos, like the other Semang populations, and the Che' Wong are not, is a further indication of the directions taken by their different patterns of intermarriage.

12.   The evidence for this claim will be presented elsewhere. Briefly, there are still present or recently extinct a few languages that are generically Malayic but not Malay; most Aslian languages display morphological and lexical borrowings found in Western Austronesian generally but not in Malay. The Chamic languages may have left traces on their way through from Sumatra to Cambodia and Vietnam; a Tai element is discernible in at least some of the Peninsular languages. The ancestors of some of the northern Melayu populations were probably speakers of the Mon language until eight or nine centuries ago.

13.   This argument assumes, of course, that the populations had already expanded to some sort of equilibrium density befitting the particular subsistence regime—swidden vegeculture, expecially—that they had chosen; any further increase would have necessitated a change of commitment to growing grain-crops and raising animals for meat. Even nowadays, when most settlements do possess herd animals in the form of goats, sheep, or cows, the Senoi usually adhere to their presumably long-established taboo on eating the flesh of animals they have reared themselves.

14.   That this situation has been reversed in a few places during the last decade or so is the result of a massive input of administrative and policing energy on the part of Malaysian governmental agencies.

15.   The connection between cognaticism and unifiliative bias will be discussed in detail in the published version of Benjamin (1980).

16.   As is indicated in Table 10.1, there are several "mixed"

cultural patterns present in the Peninsula, demonstrating that there is a significant amount of opportunistic changing between (or combining of) one mode and another, especially in those areas where foragers, farmers, and collectors meet up. These important cases do not contradict what I have to say in this chapter, however, and I leave them for detailed discussion elsewhere.

17.   Note that only the cross-sex relations are used in this way: the productively somewhat more important same-sex affinal relationships are unrestricted. A further correlate of the Semang kin-avoidances is that the people's constant or frequent movement required them constantly to cross areas occupied (though not "owned") by other bands. The Semang "sister-in-law avoidance" rule, which includes "cousins"-in-law too, would if observed prevent marriage between traceable affines. The rule is interpretable therefore as expressing the ideal of maintaining a wide dispersal of afffinal relationships, thereby blocking any less adaptive tendency to form closed marriage alliances. I do not have sufficient data to ascertain the real-world demographic effects, if any, of these rules on the Semang populations, but the lexico-statistical evidence discussed earlier does indicate that there has been an ancient and continuing dispersal of social relations throughout a very wide area, which is exactly what would result if the people had in fact followed the rules. Gomes (1982) reports a variety of institutions, such as long post-partum abstinence and various culturally derived limitations on the frequency of coitus, that have the effect of reducing the number of children born. It is a reasonable supposition that the Semang marriage rules, which effectively mean that a young man must usually look far and wide for a wife, are an additional factor leading to an overall reduction in the frequency of coitus, and hence a factor in keeping the Semang population at the virtually zero-growth level it apparently maintained for centuries.

18.   It is surely also significant that their population density is at least ten times as great as that of the Semang, and is increasing. Not only do they find spouses relatively more easily and at an earlier age, but they have few sanctions against extramarital coitus, which is if anything somewhat

encouraged by the sibling-in-law joking relationships. For further discussion of topics related to this section (such as the reasons for the inverse distribution of cousin-marriage and the mother-in-law taboo, and the biosociological or demographic consequences of the different kinship regimes, see Benjamin (1980, 1985).

19. Compare these remarks with those of Murphy (1957), which would seem to be the source of Martin's ideas.

20. As a more general point, people everywhere must be aware of the fundamental contrast between the experientially primary nature of the mother-child link and the secondary nature of the father-child link (cf. Freeman 1973).

21. The remaining 18 percent maintain permanent villages!

22. It was interesting to note in my own fieldwork the parallelism between the attitudes of Melayu individuals in Kedah towards their Semang neighbors and those of the Riau-islander Melayu towards the nomadic Orang Laut of the latter area.

23. The original version of this chapter contained a further discussion along these lines of certain problematical features in the archaeology of material culture in the Malay Peninsula, including the apparent disappearance of pottery manufacture; this has reappeared in my 1985 study.

References

Adi bin Haji Taha
   1981   The Re-excavation of the Rockshelter of Gua Cha, Ulu Kelantan, West Malaysia. Master's thesis, Department of Prehistory and Anthropology, Australian National University.

Ayampillay, Satkuna Devi
   1976   Kampung Tanjung Sepat: A Besisi (Mah Meri) Community of Coastal Selangor. Provisional Research Project, School of Comparative Social Sciences, Universiti Sains Malaysia.

Baharon Azhar bin Raffie'i
   1973   Parit Gong: An Orang Asli Community in Transition. Ph.D. dissertation, University of Cambridge.

1983   The Temuans of Melaka. In *Melaka: The Transformation of a Malay Capital ca. 1400–1980*, vol. 2, edited by Kernial Singh Sandhu and Paul Wheatley. Kuala Lumpur: Oxford University Press. Pp. 3–29.

Benjamin, Geoffrey

1966   Temiar social groupings. *Federation Museums Journal*, n.s., 11:1–25.

1968   Headmanship and leadership in Temiar society. *Federation Museums Journal*, n.s., 13:1–43.

1973   Introduction. In *Among the Forest Dwarfs of Malaya*, by Paul Schebesta. Kuala Lumpur: Oxford University Press. Pp. v–xi.

1974   Prehistory and Ethnology in Southeast Asia: some new ideas. Working Paper No. 25, Sociology Department, University of Singapore.

1976   Austroasiatic subgroupings and prehistory in the Malay Peninsula. In *Austroasiatic Studies*, vol. 1, edited by P. Jenner et al. Honolulu: University Press of Hawaii. Pp. 37–128.

1979   Indigenous religious systems of the Malay Peninsula. In *The Imagination of Reality: Essays in Southeast Asian Coherence Systems*, edited by A. Becker and A. Yengoyan. Norwood, NJ: Ablex. Pp. 9–27.

1980   Semang, Senoi, Malay: Culture-history, Kinship and Consciousness in the Malay Peninsula. Unpublished. [To be published, revised, under the title *Semang, Senoi, Malay: The Societal Traditions of the Malay Peninsula*. Singapore: Institute of Southeast Asia Studies.]

1983   Peninsular Malaysia. Map no. 37 in *Language Atlas of the Pacific Area*, edited by Stephen A. Wurm and Shiro Hattori. Canberra and Tokyo: Australian Academy of the Humanities and the Japan Academy.

1985   Between Isthmus and Islands: Notes on Malayan Palaeo-sociology. Paper presented at the Twelfth Congress of the Indo-Pacific Prehistory Association. Peñablanca, Philippines, 1985.

n.d.   In preparation [Paper on Semang ethnography.] To appear in *Contributions to Southeast Asian Ethnography*.

Blagden, C. O.
1906   Language. In *Pagan Races of the Malay Peninsula,* vol. 2, by W. W. Skeat and C. O. Blagden. London: MacMillan. Pp. 379–775.

Blust, Robert
1980   Early Austronesian social organization: the evidence of language. *Current Anthropology* 21:205–26.

Carey, Iskandar
1976   *Orang Asli: the Aboriginal Tribes of Peninsular Malaysia.* Kuala Lumpur: Oxford University Press.

Clifford, Hugh
1904   *Further India: Being the Story of Exploration from the Earliest Times in Burma, Malaya, Siam, and Indo-China.* London: Lawrence and Bullen.

Couillard, Marie-Andrée
1980   *Tradition in Tension: Carving in a Jah Hut Community.* Penang: Penerbit Universiti Sains Malaysia.
1984   The Malays and the "Sakai": some comments on their social relations in the Malay Peninsula. Kajian, Malaysia: *Journal of Malaysian Studies* 2:81–108.

De Josselin de Jong, P. E.
1981   Review of Becker and Yengoyan [see Benjamin 1979]. *Bijdragen tot de Taal-, Land- en Volkenkunde* 137:486–91.

Dentan, Robert K.
1979   *The Semai: A Nonviolent People of Malaya.* Fieldwork edition. New York: Holt, Rinehart, Winston.

Diffloth, Gérard
1975   Les langues mon-khmer de Malaisie: classification historique et innovations. *Asie du sud-est et monde insulinde* 6(4):1–19.

Djamour, Judith
1965   *Malay Kinship and Marriage in Singapore.* London: Athlone.

Dunn, Frederick L. K.
1964   Excavations at Gua Kechil, Pahang. *Journal of the Malaysian Branch, Royal Asiatic Society* 37:87–124.
1975   *Rain-forest Collectors and Traders: A Study of Resource Utilization in Modern and Ancient Malaya.* Kuala Lumpur: Malaysian Branch, Royal Asiatic Society.

Endicott, Karen Lampell
  1979   Batek Negrito Sex Roles. Master's thesis, Australian National University.
  1981   The conditions of egalitarian male-female relationships in foraging societies. *Canberra Anthropology* 4(2):1–10.
Endicott, Kirk
  1974   Batek Negrito Economy and Social Organization. Ph.D. dissertation, Harvard University.
  1979a   The impact of economic modernisation on the Orang Asli (Aborigines) of northern Peninsular Malaysia. In *Issues in Malaysian Development*, edited by J. C. Jackson and M. Rudner. Singapore: Heinemann. Pp. 167–204.
  1979b   *Batek Negrito Religion*. Oxford: Clarendon Press.
  1983   Slavery and the Orang Asli. In *Slavery, Bondage, and Dependence in Southeast Asia*, edited by A. Reid and J. Brewster. Brisbane: Queensland University Press. Pp. 216–45.
Fernandez, C. A., and Frank Lynch
  1972   The Tasaday: cave-dwelling food-gatherers of South Cotabato, Mindanao. *Philippine Sociological Review* 20:279–330.
Freeman, Derek
  1973   Kinship, attachment behaviour and the primary bond. In *The Character of Kinship*, edited by Jack Goody. Cambridge: Cambridge University Press. Pp. 109–19.
Gomes, Alberto G.
  1982   *Ecological Adaptation and Population Change: Semang Foragers and Temuan Horticulturalists in West Malaysia.* East-West Environment and Policy Research Institute, Research Report No. 12. Honolulu: East-West Center.
Gorman, Chester F.
  1971   The Hoabinhian and after: subsistence patterns in Southeast Asia during the late Pleistocene and early Recent periods. *World Archaeology* 2:300–320.
Harris, Marvin
  1977   *Cannibals and Kings: The Origins of Cultures.* London: Fontana.

1978   Cows, Pigs, Wars and Witches: The Riddles of Cultures. London: Fontana.
Hill, Andrew
1973   Letter to the author, 3 October.
n.d.   Research report on Orang Hulu. Unpublished.
Hoe Ban Seng
1964   Aboriginal Communities at Tasek Bera. Research thesis, Social Work Department, University of Singapore.
Hutterer, Karl L.
1977a   Prehistoric trade and the evolution of Philippine societies: a reconsideration. In Economic Exchange and Social Interaction in Southeast Asia, edited by Karl L. Hutterer. Ann Arbor: University of Michigan, Center for South and Southeast Asian Studies. pp.117–96.
1977b   Reinterpreting the Southeast Asian Paleolithic. In Cultural-ecological Perspectives on Southeast Asia, edited by William Wood. Athens, OH: University Center for International Studies. Pp. 9–28.
1977c   Economic Exchange and Social Interaction in Southeast Asia: Perspectives from Prehistory, History, and Ethnography. Ann Arbor: University of Michigan, Center for South and Southeast Asian Studies.
Jensen, Knud-Erik
1977   Relative age and category: the Semaq Beri case. Folk 19 & 20:171–81.
Kähler, Hans
1960   Ethnographische und linguistische Studien über die Orang darat, Orang akit, Orang laut und Orang Utan im Riau-Archipel und auf den Inseln an der Ostküste von Sumatra. Berlin: Dietrich Rimmer.
Kennedy, Jean
1977   From stage to development in prehistoric Thailand: an exploration of the origins of growth, exchange, and variability in Southeast Asia. In Economic Exchange and Social Interaction in Southeast Asia, edited by Karl L. Hutterer. Ann Arbor: University of Michigan, Center for South and Southeast Asian Studies. Pp. 23–38.

Laird, Peter
  1979   Ritual, territory and region: the Temoq of Pahang, West
         Malaysia. *Social Analysis* 1:54–80.
Leach, Edmund R.
  1982   *Social Anthropology.* London: Fontana.
Leacock, Eleanor, and Richard Lee (editors)
  1982   *Politics and History in Band Societies.* Cambridge:
         Cambridge University Press.
Maeda, Narifumi
  1967a A Jakun kinship terminology. *Southeast Asian Studies*
         4:834–53 [in Japanese].
  1967b Familial forms of the Jakun (Orang Hulu) in Malaya.
         Southeast Asian Studies 5:463–83 [in Japanese].
  1969   Marriage and divorce among the Jakun (Orang Hulu) of
         Malaya. *Southeast Asian Studies* 6:740–57 [in Japanese].
  1971   *Authority and Leadership Among the Orang Hulu.*
         Discussion Paper No. 24. Kyoto: Kyoto University,
         Center for Southeast Asian Studies.
  1978   The Malay family as a social circle. In *Proceedings of the
         Seminar on the Problems of Rice-growing Villages in
         Malaysia,* edited by L. J. Fredericks. *Southeast Asian
         Studies* 16:216–45.
Martin, M. Kay
  1974   *The Foraging Adaptation—Uniformity or Diversity?*
         Reading, MA: Addison-Wesley.
McKinley, Robert
  1979   Zaman dan masa, eras and periods: religious evolution
         and the permanence of epistemological ages in Malay
         culture. The *Imagination of Reality: Essays in Southeast
         Asian Coherence Systems,* edited by A. Becker and A.
         Yengoyan. Norwood, NJ: Ablex. Pp. 303–24.
Miksic, John N.
  1977   Archaeology and palaeography in the Straits of Malacca.
         In *Economic Exchange and Social Organization in
         Southeast Asia,* edited by Karl L. Hutterer. Ann Arbor:
         University of Michigan, Center for South and Southeast
         Asian Studies. Pp. 155–75.

Murphy, Robert F.
1957 Intergroup hostility and social cohesion. *American Anthropologist* 59:1018–35.
Noone, R. O. D.
1954 Notes on the trade in blowpipes and blowpipe bamboo in north Malaya. *Federation Museums Journal*, n.s., 1 & 2:1–18.
Peacock, B. A. V.
1979 The later prehistory of the Malay Peninsula. In *Early South East Asia*, edited by R. B. Smith and W. Watson. Oxford: Oxford University Press. Pp. 199–214.
Peterson, Jean
1977 The merits of margins. In *Cultural-ecological Perspectives on Southeast Asia*, edited by William Wood. Athens, OH: University Center for International Studies. Pp. 63–73.
Rambo, Terry
1982 Orang Asli adaptive strategies: implications for Malaysian natural resource development planning. In *Too Rapid Rural Development: Perceptions and Perspectives from Southeast Asia*, edited by Colin MacAndrews and Chia Lin Sien. Athens, OH: Ohio University Press. Pp. 251–99.
Schebesta, Paul
1973 *Among the Forest Dwarfs of Malaya*. Kuala Lumpur: Oxford University Press. First published 1928, London: Hutchison.
Service, Elman R.
1962 *Primitive Social Organization: An Evolutionary Perspective*. New York: Random House.
Skeat, W. W., and C. O. Blagden
1906 *Pagan Races of the Malay Peninsula*, 2 vol. London: MacMillan.
Smith, Philip E. L.
1972 *The Consequences of Food Production*. Reading, MA: Addison-Wesley.
Solheim, Wilhelm
1980 Searching for the origin of the Orang Asli. *Federation Museums Journal*, n.s., 25:61–75.

Sorokin, Ptirim
1957   *Social and Cultural Dynamics*. Boston: Porter Sargent.
Stargardt, Janice
1983   *Satingphra I*. Oxford: British Archaeological Reports.
Wallerstein, Immanuel
1978   *The Capitalist World Economy*. New York: Cambridge University Press.
Wang Gungwu
1958   The Nanhai trade. *Journal of the Malayan Branch, Royal Asiatic Society* 31:1–135.
Wazir Jahan Karim
1980   The nature of kinship in Ma' Betisek villages *and* The affinal bond: a review of Ma' Betisek marriages on Carey Island. *Federation Museums Journal*, n.s., 25:121–50.
Wee, Vivienne
1985   Melayu: Hierarchies of being in Riau. Ph.D. dissertation, Australian National University.
Wheatley, Paul
1961   *The Golden Khersonese*. Kuala Lumpur: University of Malaya Press.
1964   *Impressions of the Malay Peninsula in Ancient Times*. Singapore: Donald Moore.
1975   Satyanr̥ta in suvarṇadvīpa: from reciprocity to redistribution in ancient Southeast Asia. In *Ancient Civilization and Trade*, edited by J. A. Sobloff and C. C. Lambert-Karlovsky. Albuquerque: University of New Mexico Press. Pp. 227–84.
Whitmore, T. C.
1975   *Tropical Rain Forests of the Far East*. London: Oxford University Press.
Wilkinson, R. J.
1926   *The Aboriginal Tribes*. Papers on Malay Subjects, Supplement. Kuala Lumpur: Federated Malay States Government Press.
Wolters, Oliver W.
1967   *Early Indonesian Commerce*. Ithaca: Cornell University Press.

CHAPTER 11

CHANGING VALUES IN MARKET TRADING: A THAI MUSLIM CASE STUDY

Chavivun Prachuabmoh

While doing fieldwork in Pavillion Village, southern Thailand, in 1978 and 1979, I was struck by the fact that the majority of villagers in this community were professional traders. Further, I found that women valued trading more highly and engaged in it to a greater extent than men. There was a growing receptiveness among men toward trade as an occupation, however, and they were increasingly found to be engaged in this enterprise.

This situation seemed to merit further attention. I have argued previously (Prachuabmoh 1984), in response to Kirsch (1975), that religious values cannot be taken as a determining factor in explaining the specialization of women in trading. This can be demonstrated in the case of Thai Muslims. In this paper, I approach the problem by considering changing values with regard to occupation. I employ an ecosystemic model for conceptualizing economic interactions among the villagers in the belief that such a model is useful for understanding situations involving interactions among individuals within a community, and between communities and a set of interrelated constraining factors. On this basis, I suggest that we cannot specify particular institutions—such as cultural or religious values, economic activities, social structure, or other practices—as determining the form and structure of others. Rather, viewed over time, there is a process of systemic interaction and adaptation between these institutions. This is illustrated in the case of Thai Muslims.

Thai Muslims in Southern Thailand

Thai Muslims in the four southern provinces of Thailand (Pattani, Narathiwat, Yala, and Satun) are ethnically Malay. Though their communities have been politically integrated into the Thai nation for approximately 100 years,[1] the people maintain their ethnic identity through cultural symbols such as language[2] and clothing. Further, they minimize interactions with Thai Buddhists in nearby communities. There is, however, variation among Thai Muslims in their degree of integration into Thai institutions (Prachuabmoh 1980). This integration appears to be stronger in urban communities than in isolated rural communities. This is reflected in the style of clothing and in the ability and fluency with which Thai is spoken.

The extent of Thai education is a primary factor differentiating Thai Muslims. The ethnic identity of Thai-educated Muslims has become rather complex, and their ethnic image is becoming more closely associated with other Thais (White and Prachuabmoh 1983). Both the life-style and certain values of Thai-educated Muslims have changed considerably; they place greater value on secular education and government employment than more traditional Thai Muslims do.

Pavillion Village

Pavillion Village is located about 8 km from the town of Pattani and has a population of 1,328 people living in 236 households. The most common form of household is the nuclear family (represented by 206 households in the village). Only 30 households can be classified as extended families. The settlement is administratively divided into two separate villages that form a single religious community sharing a common mosque for Friday prayers.

The village has existed as a settlement for at least 100 years. During this time it has continued to grow steadily, most residents having come originally from surrounding villages. The village today is divided both sociologically and spatially into three neighborhoods. Each neighborhood has its small *surau* (prayer house) where residents go for daily prayers.[3]

Pavillion Village still has a rural appearance (see fig. 11.1). Behind the houses, which are arranged in single rows along the roads, one sees rice fields and clumps of coconut and other fruit trees. There are two basic house types: those located along the main road

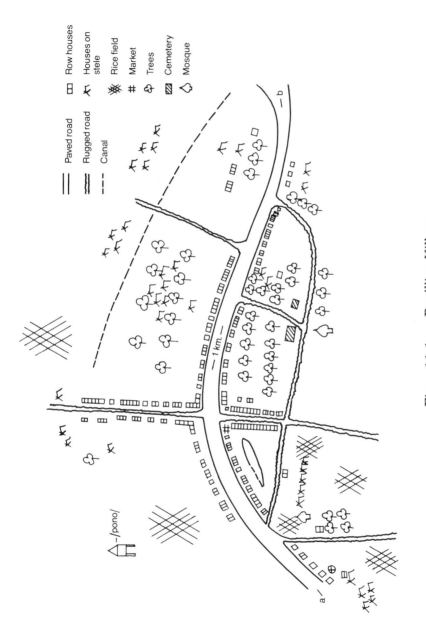

Fig. 11.1. Pavillion Village.

are generally built directly on the ground, whereas those on side streets are usually constructed on stilts. Villagers generally build the first type in anticipation of eventually converting their residences into shops.

There are five grocery stores in Pavillion, five tea shops, and numerous small food stands. Villagers do their daily food shopping in the village market, now located in the new market neighborhood. Most of the vendors of fresh food are from more rural villages.

Pavillion is characterized both by its own citizens and by Muslims elsewhere as very progressive. This characterization seems generally accurate by local standards. The settlement is close to a larger town and conveniently accessible by highway. Approximately 30 percent of the villagers speak Thai fluently, while 63 percent of the population over ten years of age are able to speak at least basic Thai (Prachuabmoh 1980, 39).

Economic Organization of Pavillion Village

In the past, the economy of Pavillion Village was less commercially oriented than at present. Some sixty years ago, according to informants, the village economy was primarily agricultural. Villagers exploited the natural environment directly, growing rice and tending fruit gardens. There was also some small-scale manufacturing: men made baskets and women wove cloth. It was not, however, a self-sufficient community. Surplus products were sold or exchanged for necessities such as salt, fish, and clothing. Though villagers remained primarily agriculturalists, some—particularly women—engaged in small-scale, part-time trading, selling fruits, sarongs, and clothes. There were also a few men involved in trading, offering primarily hand-made containers and certain kinds of leaves. Pavillion was then located along a dirt road.[4] It had a relatively large market, with some twenty traders purveying fish, clothes, vegetables, and fruits produced in Pavillion and surrounding villages. The great majority of traders were women. The importance of the market derived from the fact that it was located at a road junction.

The traditional economic activities of rice farming and fruit cultivation remain important in Pavillion Village today. According to one report by local community-development fieldworkers, 61 percent of household heads (567 households; 3,778 persons) in this tambon[5] engage primarily in rice farming, cultivating an area of 5,812 *rai* (2,325 acres, or 363 hectares).[6] Twenty-six percent consider rambutan growing as their primary occupation. Over the past few decades, however, the relative reliance on farming as an economic base has declined, while trade has become increasingly important for many villagers as an economic activity. In order to understand this shift, it is necessary first to consider the organization of the agricultural sector of the village economy.

Rice land holdings range from .25 *rai* to 24 *rai*, with an average holding of 2 to 3 *rai* for each family. Most people grow rice exclusively for home consumption. Only a few, usually families owning more than 20 *rai*, sell their rice crop. Generally these are well-to-do families that have their rice land cultivated by tenants. In general, farmers who work the land themselves do not plow it themselves but hire someone to do it. However, they usually plant and harvest the crop themselves. The chief complaint of farmers is not the small size of landholdings but the lack of an adequate irrigation system. At present, they get only about 30 *tang* (600 liters)[7] of paddy per *rai* and can harvest only one crop per year. There is an organization, the "Farmers Group," that is supposed to deal with the problem of insufficient water for irrigation. This organization is encouraged by Thai authorities, who have sent specialists to supervise the construction of irrigation facilities. Most farmers in the community (about 130) are members of this group, paying an annual membership fee of 70 baht (US $3.50). This organization was inactive at the time of my fieldwork, because the members had been cheated by an organization leader, and this had discouraged them from continuing support of the program.

Farming in Pavillion is generally a family enterprise, involving little cooperative labor-exchange among villagers, as is typical in more traditional communities. Villagers stated that in the old days there was more cooperation (*tulong menelong*, "helping each other") at harvest time. During my study, such cooperation was witnessed only once, under somewhat unusual circumstances. The relative absence of cooperative labor in rice farming in Pavillion today is probably due to the small size of fields and to the existence of wage labor.

Farmers today use new technologies such as chemical fertilizer. With small holdings, and lacking sufficient irrigation, however, the increase in output is minimal. Thus, realizing a substantial surplus is impossible for most families in Pavillion Village. In addition, a family's rice land is often scattered throughout several neighboring villages, making production still more difficult and costly. Many lease these plots to relatives in those neighboring villages. There are several reasons why plots are scattered in other villages. Some individuals previously lived in those villages and moved to Pavillion when they married a resident there. Others came to Pavillion to become professional traders because of insufficient rice land in their own villages. There are also cases in which rice plots are included in a neighboring village because of the way in which administrative divisions are drawn up.

In addition to rice land, most residents of Pavillion also have at least .5 *rai* but not more than 4 *rai* of fruit gardens in which they mainly grow durian and rambutan. Usually a male member of the family cares for these trees. They need to be watered and weeded only at certain periods of the year. Those who have more than 2 *rai* may be able to earn 1,000 to 2,000 baht (US $50–100) a year from selling the crop. Thus, fruit gardening cannot be considered a primary source of income for villagers.

Since Pavillion is situated on a low-lying river plain, it is not a suitable environment for rubber planting. However, a number of villagers own rubber plots in Yala Province some miles away. Their holdings are adjacent to each other and range in size from 40 to 200 *rai*. The owners visit their rubber plots occasionally, particularly when the trees are ready for tapping. They usually hire professional tappers to do the work, generally young men from Pavillion Village with whom they share the output in equal halves (*pawoh*). These young people are attempting to establish themselves economically. They return occasionally to Pavillion to visit their families and may eventually settle there permanently, turning the task of tapping over to younger men.

The economy of Pavillion Village has changed a great deal in the past few decades due to several interacting factors. Pavillion is even less economically self-sufficient today than it was in the past, and most economic "resources" lie outside the community. Many of the villagers' plots of rice land are located in adjacent villages, while their rubber plots are generally in neighboring provinces. With the exception of rice, their products, such as fruits and rubber, are exported from the community, the sale being dependent on markets outside the village. Conversely, citizens of Pavillion consume manufactured goods produced externally. In particular, traders, government officials, and individuals such as taxi drivers rely to a great extent on commodities imported from urban centers.

The occupational pattern in Pavillion also has changed considerably. Two factors are of particular interest. First, a single individual may engage in more than one occupation, and most villagers today do in fact combine several occupations for their livelihood. One person may simultaneously work at rice farming, fruit gardening, and trading. It is common to find four occupations being pursued in a single family. For example, the husband may take care of the rubber plot while his wife operates a small stand of ready-made foods and sweets. Both cooperate in rice farming and fruit gardening. Second, an individual may change his occupation very easily, as there is a high degree of occupational mobility, particularly among males. For example, one villager had been in succession a rubber tapper, a farmer, a taxi driver, and, finally, at age thirty-five, a trader selling fish from his small truck. Such frequent occupational change characterizes at least 25 percent of the male population of Pavillion.[8] Villagers are thus very flexible economically, and when they find something that appears to be more suitable and profitable than their current occupation they generally opt for the more promising opportunity. My classification of occupations, as seen in Table 11.1, relies on stated primary occupation (see Prachuabmoh 1980, 45).

In recent years, the village economy has become diversified. There are few individuals or families today who identify their primary occupation as rice farming and fruit gardening. Those who do must find some other occupation to supplement their traditional agricultural activities. One reason for this is the scarcity of land. In the last sixty years, the population has increased fourfold, from about 300 villagers to more than 1,300, while available agricultural land remains the same. The lowland area along Pattani River is classified by the Thai government in the 1982–1986 Five Year National Development Plan as a "poor area" since it is estimated that the average family of five persons has only 10 *rai* of land for rice cultivation. A farmer can produce 30 *tang* (600 liters) per *rai* (Jantanin 1983).

At present, the market in Pavillion Village has declined in importance because with improved roads and means of transportation larger towns are now more accessible to villagers.

Table 11.1. Primary Occupations as Represented by Males and Females

Classification	Categories	Males	Females	Total
Nonworking	Students and children			615
	Old people	6	17	23
	Housewives or persons doing housework		81	81
	No information about occupation	50	24	74
	Insane	4	1	5
				798
Working	Rice farming	55	41	96
	Fruit gardening	34	24	58
	Rubber tapping	4	2	6
	Trading	67	123	190
	Working for the government (i.e., teachers, clerks, nurses, health officers)	52	14	66
	Driving minibus, taxi, and trishaw	55	0	55
	Others: sewing religious hats, teaching, religion, dressmaking, barbering, and goldsmithing	44	15	59
				530
TOTAL				1,328

Nevertheless, trading remains a very important occupation. Unlike traders in more rural villages, who sell primarily their own agricultural products, most of the traders in Pavillion are professionals. That is, they act solely as middlemen and do not produce the goods in which they deal, nor do they produce these goods for their own consumption. The majority are females (123 females, 67 males).[9] Many of the male traders engage in trading along with their wives. There are about fifty traders whose business is primarily within the village, while the rest are engaged in trading outside the village at rotating markets and in larger town markets. Interestingly, traders selling in the village are those dealing in groceries, prepared food, sweets, and students' supplies. Traders in fruits, vegetables, beef, clothing, and cooking utensils sell their goods at rotating markets within a 50-km radius of the village. I lack a conclusive explanation for this pattern as yet, but it may be of significance that the prices of the former items are generally fixed. The traders selling food in the village market are from neighboring rural villages. Many of the traders selling food outside the village explain that they make a very small profit on the goods they sell and can distribute only a small volume in the village since villagers prefer to shop in the larger towns of Pattani and Yala. It is thus more profitable to sell outside of one's own home village. Villagers also mentioned on occasion the difficulty of selling to neighbors and relatives in the village, because they are expected to give discounts and grant credit.

Traders organize themselves for economic activities. There are about twenty traders (five males and approximately fifteen females) who have formed an organization and travel together to rotating markets (*kedanna*)[10] in the same minibus. Most of these traders sell batik clothes imported from Indonesia and Malaysia. Others who join the group sell food items such as bean sprouts and red onions.

A number of individuals and families independently engage in trading outside the village. For example, about ten men, occasionally accompanied by their wives, sell men's clothing at various rotating markets. Most have small trucks worth about 100,000 baht (US $5,000) that enable them to cover markets further away than those attended by the women traders. These men sell mostly western clothes such as jeans and shirts. Each week they purchase merchandise costing some 10,000 baht (US $500) or more depending upon their expected sales volume.

As noted, more men are becoming involved in trading activities, usually selling western clothes. When I revisited Pavillion Village in April 1983, I found that about fifteen young men under the age of twenty-five, most having little Thai education, had started to sell used clothes from the United States and Japan in neighboring provinces, and even as far away as Bangkok or the northern and northeastern provinces. This had not been the case at the time of my original fieldwork in 1978 and 1979. At that time, and among the older generation in general, trading was considered to be a shameful occupation. This had changed. When interviewed, these men gave as reasons for engaging in their business unemployment, profit, and travel experience. Money and profit have become a major motivation. Thus, a critical question arises: Why have so many people recently come to regard trading as a desirable occupation? A few old men, when they were asked this question, reasoned that money had not been so important in the old days, since agriculture provided adequately. Trading had provided only supplemental income, involved locally made or grown products, and was the province of women.

There are several factors to account for these developments. First, land holdings are too small to make investment in agricultural technology worthwhile. Techniques such as irrigation and fertilization do increase outputs, but with small holdings it is inadequate for commercialization. Second, the rice lands of most villagers are located at some distance in the more rural villages. Their cultivation involves, therefore, a large amount of travel and transport time. As a result, villagers rent their land to others. Third, and most important, the development of an asphalt road about thirty years ago and the introduction of means of rapid transportation about twenty-five years ago have allowed easy access to other towns in Pattani and Yala provinces. Though the village market has lost its role as a trading center for the neighboring villages, villagers have been able to compensate for this by travelling to other communities to trade. Thus, other technology has opened new opportunities and resources for villagers to exploit. By shifting to a professional trading role, villagers have changed their means of livelihood. They no longer rely on the direct exploitation of the natural environment.

In general, villagers are economically successful, though it is difficult to assess income accurately. From indirect methods, however, it was found that more than 10 percent of the families in

Pavillion possess properties worth more than one million baht (US $50,000). Only 15 to 20 percent of families earn 60 baht (US $3.00) a day or less. One man, retailing ice from a small cart, earns 200 baht (US $10.00) net profit almost every day. He and his wife also have rice plots to lease. The evidence of economic success is seen in things such as clothing and housing.

Cultural Values and Changing Economic Activities

The preceding description of the socioeconomic pattern of Pavillion Village indicates that trading is today a dominant economic activity. The traditional economy relying on agriculture—rice farming and fruit gardening—has declined in importance, while the degree of economic dependency on larger towns and urban centers has increased. A considerable number of villagers, formerly selling only their own agricultural products to supplement their income, have moved into the role of professional traders. This is in large part a reflection of the farming situation in Pavillion. However, it also has to be seen in relation to cultural and religious values and changes in them. It is useful to explore the question of Thai Muslim values and their relationship to changing economic activities. Nearly 100 percent of the villagers are Muslim. There are only two Thai Buddhists living in the village. Villagers perceive themselves to be religious in the sense of being conscientious in religious education and in adhering to Islamic principles. It is notable that, while villagers strongly maintain their religious tradition, they also value Thai education and a more modern, materialistic way of life. There are, for example, a number of houses in the village furnished in a Western style with sofas, ovens, and washing machines. At least one-third of the families in the village have refrigerators and television sets. Many villagers believe that their tou'imam (Islamic teacher as well as registrar of marriage and divorce) and other leaders contribute to both the progress and the religious success of the village. A number of highly educated villagers stated that their village is a successful example of the compatibility between modernization and Islam.

Thai Muslims are generally perceived by non-Muslims to be strongly opposed to Thai education. When Thai education was introduced to the people in this area about eighty years ago, they refused to send their children to Thai schools. Until approximately twenty years ago, most villagers thought that it was sinful to read or keep Thai texts, and children in particular were not allowed to keep Thai books inside the house. This attitude is gradually changing and the majority now see Thai education as valuable. They realize that a Thai education facilitates better opportunities for employment. More than 170 persons have attended Thai schools beyond grade seven (Prachuabmoh 1980, 43). Most of those who have never attended Thai schools (268 individuals) are more than forty years old (see table 11.2).

It was apparent that most villagers, both Thai educated and non-Thai educated, are devout practitioners of their faith. Much time was devoted to religious pursuits. Prayers are said five times a day with each period of prayer requiring about half an hour. Religion is also a common topic of conversation. In a year there are several religious occasions such as *Maulud*[11] (Prophet Mohammed's birthday), *Hari Rayo*[12] (end of Ramadan, Islamic New Year), and *Asuro*.[13] During Ramadan, economic activities are decreased significantly. It is clear that villagers place great value on religious observances.

The success of villagers in Pavillion derives from their adaptability in the diversified economy, their readiness to change occupation to fit new circumstances. It is interesting that the residents of Pavillion have adapted to changing circumstances in a way that differs from the normal pattern in Thailand. According to Hanks (1972, 151–56), the people of Bang Chan have utilized three modes of cultivating rice through time: first shifting cultivation, then broadcasting, and finally transplanting. This change of technology occurred in response to increasing population and resulting land shortages. Residents of Pavillion Village have adapted to the changing economic environment by developing their roles as traders rather than through agricultural intensification. They have developed skills in commercial trading, an occupation that is based on the specialization of production characteristic of contemporary society. The villagers value this new occupation and increasingly enunciate values associated with such economic concepts such as investment, profit, and time (Belshaw 1965). Still, traditional values may counter or delay the adoption of new economic roles.

As noted before, there are more female than male traders in the village. This is a common phenomenon in Thailand (Prachuabmoh 1983). It was found that men and women value trading differently.

Table 11.2. Educational Attainment in Pavillion Village

Educational Level	Total	Male		Female	
		Num-ber	Percen-tage	Num-ber	Percen-tage
Under school age (7)	161	71	44.0	90	56.0
No Thai education	268	125	46.6	143	53.4
P1-P4 (Grades 1-4)	510	245	48.0	265	52.0
P5-P7 or M1-M3 (Grades 5-7)	202	115	56.9	87	43.1
MS1-3 or M1-M6 (Grades 8-10)	87	35	40.2	52	59.8
Higher	83	45	54.2	38	45.8

Fifty villagers were interviewed, most of whom were females, and only seventeen gave definite answers about the occupation considered to be best. It is noteworthy that most individuals (eleven persons) emphasized trading as their preferred occupation, because it allowed them to be independent and provided a relatively good income. Only four preferred government employment as an occupation, explaining that it is light work with a stable income.[14] Significantly, perhaps, agriculture was never mentioned as a desired occupation.

It was found that the attitude of women toward trading was particularly positive. Women stated that:

1. Men do not trade because they are lazy.
2. Women are better at taking care of the money involved in trading.
3. Men devote most of their time to religious activities rather than earning money.
4. Women are good at talking, which is necessary for trading.
5. Trading is an independent job that can be stopped at any time.
6. Trading is an occupation that can be done in front of one's own house.
7. Trading is a very profitable occupation.
8. Islam encourages trading.

Contrary to this, male responses reveal ambivalence toward trading. Some men value trading, but most consider it an exclusively female occupation. Most think that trading is a petty and tedious occupation and that more important occupations such as farming or office work should take precedence. Men stated that:

1. Trading is boring. Men like occupations that give them more physical movement, such as driving a taxi.
2. Women are good at trading.
3. Men have other jobs to do, such as farming.
4. Trading is a woman's occupation. Men are ashamed of doing women's work.

It is interesting to note that women take up trading only after marriage. Previously (Prachuabmoh 1984) I have suggested that contributing to the household economy is part of a wife's culturally defined role. Trading allows women to fullfil this role and at the same time agrees well with their status in the family. The question of the evolution of this pattern of women's specialization in trade is intriguing. I lack a conclusive answer as yet. It is suggested that it may be related to the traditional division of labor in which men do heavy work such as plowing in agricultural production (see table 11.3). Women act more as distributors since in communities practicing matrilocal residence women may have important roles in the internal exchange of food products. For example, in Malay fishing communities in southern Thailand, women do not fish, but they sell the catch. In agricultural communities such as Pavillion Village, men and women work together in farming. But it is men who do more gardening.

Today, both men and women in Pavillion, as well as Muslims and Chinese in the nearby town of Pattani, view female villagers as generally harder working than males. This observation is not without some basis; some 90 percent of the Muslim traders in the town market are females. As noted, villagers explain that men devote more time to religious concerns. Women, as well as several religious leaders, feel that this is contrary to Islamic values, which presuppose men to be the breadwinners, while women are supposed to stay home to take care of the children and should not engage in public activities.

It seems then that the observed pattern of activities concurs with a set of values which predisposes women more positively toward trading than men, although there are indications that both values and activities are changing. These values diverge from official Islamic norms, and I have argued elsewhere (Prachuabmoh 1984) that religion cannot be the determining factor for the observed pattern. Islam originated and developed in a particular kind of environment—trading centers and oases established in connection with a pastoral economy in a desert area. Mecca was a center of redistribution (Aswad 1970, 6). Medina held a crucial ecological position in northern Arabia, since it contained great water resources and produced wheat as the predominant crop. Aswad (1970, 6) writes that, "on the eve of the development of Islam, and accompanying the migrations of the tribes toward the north, there was an increase in trade along the inland routes passing through the oases of Mecca and Medina." This "brought about a modification of the social organization in each area, but in different ways. The occupation of Mecca by desert groups resulted in the beginnings of

Table 11.3. Categories of Work of Husbands and Wives

Categories of work	Husband	Wife
Housework		
Heavy work (carrying, etc.)	frequently	sometimes
Cooking	rarely	frequently
Cleaning house	rarely	frequently
Laundry	rarely	frequently
Taking care of baby	sometimes	frequently
Managing household finances	sometimes	frequently
Economic work (depending on the combination)		
Rice farming		
plowing	always	rarely
planting	always	always
reaping	always	always
Trading	sometimes	frequently
Orchard work	frequently	frequently
Taxi driving	always	never

stratification around the control of trade. In Medina, it caused a transfer of power from the cultivators to the encroaching semi-herders" (Aswad 1970, 58). Goitein (1966, 223) cites Shaybani as referring to the fact that all the early champions of Islam were businessmen. Thus, the importance of trading in early Islamic communities in Arabia resulted from the natural and strategic environment of Mecca. Individual status was associated with economic and political activity, and traders became leaders in the community. Trading was valued because it was the source of high status in society. The values associated with this occupation were subsequently incorporated into Islam. Goitein (1966, 219) writes "No wonder, then, that the full-fledged religion of Islam, as it appears to us through the writing of the third and fourth centuries of the Muslim era, is pervaded by the spirit and ideas of the rising merchant class." Goitein (1966, 331) also points out that the positive value toward trading is expressed in religious law, which occupied itself with the Muslim merchant and protected his interests. Many members of the business class occupied prominent positions in the early Islamic states.

In the fourteenth and fifteenth centuries, when Islam came to the Malay Peninsula, then under Hindu and Buddhist influence, these specific values relating to trade were probably not adopted by Malay communities. In describing the ecology of mainland Southeast Asian societies, Winzeler (1976, 625) notes that

> As tropical forest agricultural cultures, the Khmer and the Maya have in common ecological homogeneity and hence weakly developed interregional economics. Dependent everywhere upon the same crop and supplementary foods which were everywhere similarly harvested or gathered at the same time, and lacking adequate transportion systems, the economics of these two peoples were not based upon extensive trade, either internal or external.

It is suggested that Malay society at the time of Islamic expansion was with certain exceptions in a similar environment. Fraser (1960, 19-20), citing Cortesao, notes that a century later Pattani was an important commercial center. However, the available evidence indicates that the exchanges concerned external or foreign trade between state rulers and foreign traders rather than local or internal trade between villagers. This external trading did not seem to affect the lives of villagers directly.

Though the province of Pattani has in the recent past had a dual economy corresponding to its two major environments (fishing in the coastal area and rice farming in the lowlands near the rivers) it may be assumed that communities in ancient Pattani were self-sufficient like other Malay communities described by Winstedt (1950, 120–23). In fact, as observed by Fraser (1960), Rusembilan villagers also grow their own rice. This is also true in another fishing community (Prachuabmoh 1980). Since there was little specialization in production, trading between villages would have been limited, particularly since money had not yet become the effective medium of exchange.

According to Winstedt (1950, 50–51), Malay society at that time was without social differentiation between the sexes but not without class distinctions. The main distinction was between the ruling class and commoners, but there was no trading class. Political offices were hereditary within the ruling class. Their sources of power were descent, militaristic capabilities, and religious beliefs. Economic power to impose tax, demand corvee, or engage in trade resulted from their political power. Thus, it appears that class distinctions were more associated with political status than with economic status. Among commoners, mostly agricultural villagers, rice land was probably important and valued since it was the critical resource for their survival. Services associated with the ruling class, being in political office, might also be valued. In the context of this ecological niche (involving aspects of the natural environment, sociopolitical structure, and economic organization), it is logical that no values favoring trading emerged for men whose position did not derive from trading (as it was in the incipient Islamic state) but was primarily based on their role in agricultural production. Their social and economic roles were circumscribed with the traditional values of kinship, harmony, reciprocity, and religious activities as they still prevail in various Malay communities (Prachuabmoh 1980; Firth 1966). Trading, then, was not a viable alternative for villagers.

Today, as noted above, the situation is changing. Trading is becoming an economic option due to the increased availability of rapid transportation and the importance of the monetary system. At the same time, alternatives are limited for villagers. Commercialization of rice production as it exists in many communities of central Thailand is not economically viable due to land scarcity and the small size of landholdings. Nor is the area suitable for rubber growing or for commercial fishing. The villagers must take advantage of the economic options open to them or seek their fortunes elsewhere. Without a Thai education it is hard to find government employment. Men with little Thai education are therefore more positive about trading than others, although the types of commodities handled and the scale of men's trading activities differs from that of females. Since female traders must allocate time for domestic work, they cannot fully engage in trading full time. Though they may control large sums of money, only infrequently do they use this as capital. Men, on the other hand, maximize their monetary resources for capital—buying trucks or otherwise extending their trade. The data show a division along sex lines in the nature of the goods both in the village and the town (see table 11.4). It is notable that female traders are engaged in the trade of foods and items associated with women, whereas men's goods are likely to be commodities foreign to the traditional exchange system such as Western clothing and noodles (adopted from the Chinese).

Success in large-scale trading is constrained both by social organization and by values. Thai Muslims recognize that they are not as successful in large-scale trading in town as the Chinese. A number of reasons are given for this. First, unlike the Chinese, Thai Muslims lack such organizations as clans or dialect associations, and thus cannot draw upon the large capital resources these organizations can generate. Second, if they compete with Chinese traders in town, they are discriminated against in their attempts to get credit. There are only a few successful, large-scale, Muslim traders in town. They belong to closely related families whose ancestors have been involved in commercial activities in town for generations. In fact, some maternal ancestors of these traders were Chinese. In general, Thai Muslims believe that they are not as effective as the Chinese when it comes to trade. They attribute this shortcoming to character traits such as envy and the lack of economic cooperation among Muslims themselves.

Villagers also still value agricultural land. When they have money, they prefer to buy a rubber plot or land in another area rather than investing it in trade. Land is valued with respect to

Table 11.4. Types of Goods Traded by Sex of Trader

Items traded	Male		Female	
In Village				
Jewelry	1		5	
Grocery	3		15	
Tea	2		2	
Bananas	0		12	
Kitchen-ware	1		1	(husband and wife)
Batik cloth	0		13	
Men's clothing	7		2	(with husbands)
Food or sweets	4		20	
Gasoline	2		0	
Beef	0		0	
TOTAL	20		74	
In Town				
Jewelry	0		0	
Grocery	0		5	
Tea	0		0	
Fruits	0		70	(varies seasonally)
Fish	0		35	
Vegetables	0		40	
Food	5	(selling noodles)	10	(rice and curry)

inheritance; it is felt that some land should be left to one's descendants. In addition to this, villagers have not reconciled the trading relationship with other, more general, social relationships such as kinship and friendship. Traders frequently complain about the difficulty of trading with relatives. Relatives generally expect to buy merchandise on credit or at a discount. Relatives also frequently promise this special treatment to friends, and it is expected that the trader will concur with it. This may be one of the reasons why local traders, with the exception of some grocers and vendors of ready-made food, prefer to sell their goods elsewhere, while traders from other villages come to trade in Pavillion Village. The other reason may be that there are 190 traders in the village, making it difficult to compete within the community.

Beyond this, the general economic and religious values of villagers do not facilitate economic improvement. The religious world view is that everything in the world—the whole natural environment including land, water, and rainfall—is given by God. To a certain extent Muslims believe they are allowed to change these conditions. The ability to change the environment for one's livelihood is also given by God. Success in doing this depends on *rezeki* (luck) (Prachuabmoh 1980, 139) or economic destiny (Wilson 1967, 106), which is also determined by God.

Though Islam does not prohibit materialistic accumulation (Goitein 1966, 236), villagers emphasize happiness in the hereafter, *akirat*, as their primary goal in life. Religious activities in this world thus are seen as very important, since they are an investment in the next world. Villagers believe they should be ever conscious of religious goals. *Dawa*, who are agents of the Islamic revitalization movement that has become increasingly visible in recent years, emphasize the proper way of living, as exemplified in the Prophet Mohammed's life, such as living moderately and adhering to religious principles. Such behavior leads to a good life in the hereafter. In 1983, when I revisited the village, this movement was having a strong impact on the values and behavior of villagers. Young girls used the *kaeinglepah* (shawl) to cover their hair when they went outside. More men prayed in the mosque. Some were worried that their economic activities may interfere with religious duties. Others were concerned that they might be doing wrong because they must lie about price every day in the course of trading. Some villagers concluded that it is desirable to achieve a balance between the goals of this world and that of the hereafter, since Prophet Mohammed said, "Work as if we were not going to die, and make merit as if we are going to die." The point of balance, then, is left to individual interpretation.

Conclusion

Both the data and the analysis presented here are somewhat incomplete. However, I present this as an exploratory study which may help to generate some ideas for further investigation. A number of things must be kept in mind in evaluating the observations presented here.

First, in the case considered it is difficult to segregate cultural from environmental aspects. The natural environment has been highly transformed through technological and social means, so that the present environment cannot be understood adequately unless it is seen as incorporating social and cultural elements. The population under discussion here cannot satisfy its needs by exploiting resources provided by the local natural environment alone, but it reaches out to more distant environments through social interactions in the form of trade. The wider social setting is therefore part of the effective environment, and social interactions must be considered as part of ecological interactions in the broader sense.

Second, Thai villagers may take different routes in adapting to changing environments. Individuals make choices in the framework of their cultural values. These are a reflection of previous socioeconomic structures and technologies that interacted with the natural environment in specific ways. For example, in most societies the most basic goals of life are probably similar: physical survival and social recognition. The means to achieve these goals, however, may differ. Social status, for instance, may at one time be associated with land, while at some other time it may become associated with money or other material goods. Accordingly, people may change their activities to achieve their goals by different means. Thus, general values may not change at all, while attitudes toward specific things or activities may change considerably.

Cultural values are complex constructs that never occur in isolation but are related to each other. For example, values associated with trading may be closely related to values pertaining to profit, interpersonal relationships, and time. Change in one value will therefore usually affect the other values in the set. On the other hand, there also may be sets of values such as religious achievement, reciprocity, kinship, and friendship bonds, which are deeply embedded in a society as cultural ideals and act to inhibit change.

Although the change of values observed in Pavillion Village occurs in relation to changes in cultural factors such as technology, socioeconomic organizations, and other values, this does not mean that the natural environment has no impact on the shaping of cultural values. The formation of early Islamic society in Arabia illustrates the impact of the physical environment on the formation of social classes and occupational values. In the case of Malay Muslims, values pertaining to trade were not adopted, although other Islamic values have had great impact on their behavior. An ecological interpretation of this has been suggested. In early Malay society, various communities were economically self-sufficient. Trade was not essential and traders had no place in the social structure. Thus, trading was not valued. Such values, once originated or adopted, persist and influence individuals in the way they use their changing environment.

Both agricultural products and agricultural labor have become commercialized in Pavillion Village in recent years. There is little *tulong menelong* today. Labor exchanges occur only in limited social and religious contexts. Villagers have entered the market system in which economic transactions are on a purely impersonal basis (Hardesty 1977, 87). Profit principles have replaced reciprocity. Money is valued as the most important basis for living. This is contrary to the traditional emphasis on personal relationships based on kinship and neighborhood. Villagers have therefore adopted a strategy of trading outside the community. The development of the villagers' role in trade is made possible through the changing economic environment, in which the crucial resources are no longer natural, but social and technological.

Firth (1964, 208) points out that "value" is a term often used in a vague way. In fact, the use of the term "value" is a way of talking about behavior. Belshaw (1965) distinguishes two categories of values: (1) those that govern actual specific action, and (2) those that are ultimate objectives.

Cultural values pertaining to trading belong to the first category. I have focused on these values to illustrate "the nature of the relationship between the cultural values espoused by a people and

the way that these people interact with their environment"
(Jamieson and Lovelace, this volume) for the following reasons:

1. The gradual change of occupational values, particularly with
    regard to trading, is a highly noticeable feature of Pavillion
    Village and probably other villages under similar ecological
    conditions as well.
2. If we are interested in the relationship between values and
    environment, occupation is a good area for investigation
    since it provides the basic livelihood while it is based on a
    choice guided by cultural values.
3. Professional trading in a modernizing market situation
    constitutes a very specialized and complex kind of
    relationship to the environment. It is unlike hunting-
    gathering, shifting cultivation, or even intensive agriculture,
    where humans depend on natural resources exploited for
    direct consumption. By contrast, trade involves social
    interactions that are based on specialized production and the
    need to transport goods to strategic locations for exchange
    between specialized producers. Such interactions involve
    values relating to capital, profit, money, competition, and
    risk (Belshaw 1965).

On the basis of the case discussed here, it might appear that
material conditions such as agriculture or traditional means of
production in tropical environments determine occupational values.
Although Islam and Islamic ideology have prevailed among Malays
for centuries, the Islamic esteem for trading was originally not
adopted, particularly by men, since the material conditions were not
appropriate and other traditional values such as the importance of
kinship and social harmony did not encourage it. Such values persist
and guide villagers in their use of the economic environment even
though it has changed. This is the case in Pavillion Village, where
many villagers still maintain their investment in land. There is
significant variation among villagers, however, in the way they react
to the changing environment, and there is evidence that values are
gradually changing. Without taking an extreme materialist position,
I would suggest that this process of change is of a systemic,
interactive nature, rather than a deterministic one. In the case
presented here, material conditions seem to have taken precedence in
promoting change. Villagers have developed successful strategies in

using their new environment. However, it is also evident that values, once established, do persist and may act counter to change.

Acknowledgment

I would like to thank Dr. Paritta Chalermpao for reading the original draft of this paper and providing comments.

Notes

1.   Though the state of Pattani was annexed by Thailand nearly 200 years ago, the impact began to be felt only about 80 years ago with the introduction of the Thai political and educational systems.

2.   The local language is Patani, similar to the Malay dialect spoken in Kelantan, Malaysia.

3.   All adult males are required to attend communal prayers. They frequently hold discussions and make decisions about community affairs after prayers. Women never attend these meetings.

4.   King Rama V visited the market in 1915 (Sakhi 1915, 76). According to historical records, the village was reached by the royal car on a sandy road.

5.   The *tambon* consists of seven villages. These two villages are the most dense. Residents of other villages are primarily rice farmers or fruit gardeners.

6.   One *rai* equals 0.4 acre or 0.16 hectare.

7.   1 *tang* equals approximately 20 liters. The equivalent measure used by Thai Muslims is *kantang*.

8.   Traditionally, Thai Buddhists, particularly males, have valued government employment. Today there is evidence of change, as Thai men have started to value trade. This is probably due to economic pressures such as the increasing importance of money, unemployment, increasing population, and low pay in government posts.

9.   These numbers include part-time, small-scale entrepenuers. Approximately 75 percent are professional traders.

10. *Keda* means market; *nna* derives from *nad*, meaning appointment. The term *kedanna* refers to a kind of "swap meet." The Ministry of Agriculture makes arrangements for

this kind of market. Traders pay only a small amount (less than US $.25) for a plot in the market, which is usually held in an open space near a town or district market. There is at least one swap meet in every district or town market place. The one in the town of Pattani is open twice a week. Most of the traders living in villages go to swap meets in different towns and districts since by doing this they are able to trade every day without having to pay the high rent of the market place. Traders in Pavillion thus are very mobile.

11. Many households invite religious leaders, relatives, and neighbors for feasting to celebrate the occasion.

12. This day marks the end of fasting, and is the most important holiday in the life of villagers. There are religious activities and a large feast, and relatives exchange visits.

13. *Asuro* is celebrated in memory of the Prophet Mohammed's sufferings during his missionary wars. It used to be a very important day in village life. Today, however, few households engage in these activities, the giving away of sugar, coconut, rice, and other commodities used for the making of special foods.

14. More men than women work for the government, and only men engage in transportation as an occupation.

References

Aswad, B.
  1970   Social and ecological aspects in the formation of Islam. In *Peoples and Culture of the Middle East,* edited by L. Sweet. Garden City, NY: The Natural History Press. Pp. 53–73.
Belshaw, C.
  1965   *Traditional Exchange and Modern Markets.* Englewood Cliffs, NJ: Prentice-Hall.
Firth, R.
  1964   *Essays on Social Organization and Values.* London: The Athlone Press.
  1966   *Malay Fishermen: Their Peasant Economy.* London: Archon Books.
Fraser, T. M.
  1960   *Rusembilan: A Malay Fishing Village in Southern*

Thailand. Ithaca, NY: Cornell University Press.
Goitein, S. D.
1966   Studies in Islamic History and Institutions. Leiden: E. J. Brill.
Hanks, L.
1972   Rice and Man. Chicago: Aldine.
Hardesty, D.
1977   Ecological Anthropology. New York: John Wiley.
Jantanin, Sanong
1983   A Long-Term Plan for Southern Development. Paper presented in a seminar on Fertility, Planning and Development. Haatyai, March 16.
Kirsch, A. T.
1975   Economy, polity, and religion in Thailand. In Change and Persistence in Thai Society, edited by G. W. Skinner and A. T. Kirsch. Ithaca, NY: Cornell University Press. Pp. 172–96.
Prachuabmoh, Chavivun
1980   The Role of Women in Maintaining Ethnic Identity and Boundaries: A Case of Thai-Muslims (The Malay Speaking Group) in Southern Thailand. Ph.D. dissertation, Department of Anthropology, University of Hawaii.
1984   Women's Economic Role and Religion. In preparation.
Sakhi
1915   Diary on the King's Passage to Southern Provinces (June 4 –August 5, 1915). Bangkok: PimThai (in Thai).
Wilson, P.
1967   A Malay Village and Malaysia: Social Values and Rural Development. New Haven: HRAF Press.
Windtedt, R.
1950   The Malays: A Cultural History. London: Routledge and Kegan Paul.
Winzeler, R.
1976   Ecology, culture, social organization and state formation in Southeast Asia. Current Anthropology 17:623–40.
White, G. M. and Chavivun Prachuabmoh
1983   The cognitive organization of ethnic images. Ethos II:2–32.

CHAPTER 12

IDEOLOGY, CULTURE, AND THE HUMAN ENVIRONMENT

Ho Ton Trinh

Environment, biosphere, ecosystem, ecology—the words are recent and the science modern. But nature has remained the same and humans have the same aunt as their foster-mother. The problem of the human relationship to the environment has existed since humans began to create their own environment while creating themselves. Humans are part of their environment.

From an ecological point of view, Vietnam is a country rich in natural resources with a much varied geographical environment, but the climate can be very harsh and unpredictable. For millenia our forefathers stubbornly worked and struggled to "tame" and mold this environment to meet the needs of both individual subsistence and the development of an entire nation. They also had to fight foreign invaders to safeguard national independence, and therefore natural and cultural patrimony.

"Dong Son civilization," "Red River civilization," "aquatic rice civilization": these are names coined by scholars in the context of scientific research on history and civilization, but they are not irrelevant to the ongoing and steadfast struggle carried out by our fellow countrymen. In the spirit of close national unity these people adapt themselves to their environment and transform and defend it by their labor, creativity, and patriotic struggle.

Among our many legends there is one that is typically Vietnamese and both literally and metaphorically related to the theme I would like to discuss in this chapter. It is the legend of the Mountain Genie and the Water Genie. There was once an emperor who had a very beautiful daughter whose hand was simultaneously asked for by two youths who were in fact genies: the Mountain Genie

and the Water Genie. The emperor promised her hand to the suitor
who first brought a betrothal present. The Mountain Genie won and
became the emperor's son-in-law. Out of revenge, the Water Genie
rose and rose, mobilizing the flood to swamp his rival. The latter,
however, rose still higher, minute by minute, at long last winning the
fight—and the woman.

This is a myth about the struggle of humans against "heaven,"
against natural calamities, against inundation. It is a myth that
reveals one of the most typical aspects of the Vietnamese people's
history of building their country, an essentially agricultural nation
with its own "Asiatic mode of production," its own conception of the
"ecosystem," as a forerunner of modern ecology. As "man's power
over nature was very restricted, he was protected by the cushion of
dreams" (Lévi-Strauss 1955, 452). These dreams are often
expressed in legends that fill the folklore of all nations. For example,
there is an abundance of proverbs that speak volumes regarding the
ecological perception of the Vietnamese in times past:

> Firstly water, secondly manure, thirdly labor, and
> fourthly seeds.
>
> A virtuous woman as wife, a house facing the South
> as dwelling.

In the development of agriculture, or the building of villages, the
Vietnamese have always taken into account the relations between
people and nature, the harmony between the human environment
and natural ecological conditions. This is evident in the Vietnamese
countryside in general; in the orientation chosen in the construction
of a home; in the landscape surrounding each village, each hamlet,
and each communal house; and in the Vietnamese garden. All bear
the mark of a traditional perception of the environment, an issue of
great interest to contemporary ecology. The philosophy of nature
and the love for nature with which our popular and national
literature are imbued also reflect this consciousness of the
interdependence of humans and nature of which our poets are the
most articulate exponents.

In our cultural patrimony, inherited from our early scientists—
for instance, the *Geography of the Country*, by Nguyễn Trãi (fifteenth
century), the *Miscellaneous Texts*, by Lê Qúy Dôn (eighteenth

century) and the *Chronology of Dynasties*, by Phan Huy Chú (ninteenth century)—are contained the rudiments of a Vietnamese ecology still of value today.

Human Ecology in Vietnam

Scientific ecology is for us, and in Vietnam, a new discipline. But the problems of the relationship between humans and the environment, and of the quality of life, have been addressed in Vietnam by our past generations. They have been dealt with in our own way, at different historical stages, according to the level of knowledge and the scientific practice of our people.

Given the stagnation of feudal society, the system of small-scale production, and our deficiency in the field of science and technology, there occurred in our history no sudden or major improvements in the relations between humans and the environment. Further, for more than a century the colonialists and later the neocolonialists invaded our country, ravaged our environment, and destroyed the equilibrium between nature and society, all in their own interests and motivated by their desire for domination. Consequently, the improvement I will discuss has taken place only under our new regime, with national independence and socialism, and the recovery of our position as the real masters of our country and of our natural and cultural patrimony.

To give you a few examples, let me note some major achievements in the economic-social field that have marked a new stage in the consolidation and development of socialism in Vietnam, and which directly concern my theme:

- Spectacular development of agriculture with intensive farming, increased productivity, breeding, and use of new seed varieties; improvement of the means of production (manual semimechanized and mechanized labor); improvement of irrigation and drainage facilities; establishment of "new economic zones" in coordination with demographic redistribution on a national scale; and industrial exploitation of such trees as rubber and coffee.

- Creation of new industrial centers, including such large ones as the socialist undertakings of the River Da hydroelectric project, the Pha Lai thermal power plant, the Bim Son and

Hoàng Thach cement plants, and the Vũng Tau gas and oil prospecting area.

- Tapping of underground resources and deposits, geological research and survey, and study of the mineral potential of the country.

- Afforestation and multiplication of trees, reforestation, and the struggle against deforestation.

- Intensification of animal husbandry, crossbreeding of Vietnamese pigs with foreign ones to obtain more productive breeds, industrialization of aviculture, restoration and development of sericulture and agriculture.

- Creation of "green belts" around towns, and development of truck farming, particularly near cities like Hanoi and Ho Chi Minh City.

- Development of gardens for medicinal plants, to aid in the development of traditional medicine as a supplement to modern medicine.

- Restoration of historic monuments, conservation of renowned and exceptional landscapes, the establishment of ecological reserves in areas of remaining primeval forest (for instance the Cuc Phuong Forest in Ha Nam Ninh Province), and protection of scenic areas.

Achievements such as these clearly reveal the fact that they are based on an understanding of ecological relationships, even though scientific ecology is only at its infancy in our country. Of course, much remains to be done and there are enormous difficulties to be overcome. But these achievements constitute a springboard, an encouragement of great historic, political, and social significance.

It is not ideological values alone that create these achievements or transform everything like the magic wand in fairy tales. When we speak of the environment, we speak of nature: nature as it is extolled by artists and poets, but also in terms of the biosphere, the earth's crust, subsoil, water, flora, and fauna. Nature has its laws and dialectic. Any "revolt," any struggle by humans against nature has to take these laws into account and grasp them through knowledge and learning. The transformation of productive forces, the exploitation of natural resources, and the maintenance of an equilibrium between humans and the environment cannot be achieved simply through sheer human will.

When we speak of the environment we also speak of people: we speak of people as objects as well as subjecta; humans are part of nature and subordinate to the "dictates" of its laws. At the same time, humans consider nature as an object, and themselves the subject capable of "reproducing wholly nature" (Marx), of creating their habitat, their environment. Thus, there is a dialectical relation between nature and humans. Nature and society "form one body," indissoluble. Within this relationship nature acts upon humans, who in turn try to react and to transform nature to further the ends of their social existence. Therefore, economically and socially, the environment has its infrastructure, superstructure, means of production, and forms of social consciousness. And it is here that dialectical materialism has a voice in the matter. It helps us grasp the important role of ideology in the solution of environmental problems, and particularly in the transformation of human ecological space.

For us, and in our country, ideology means Marxism-Leninism. In our country, the evolving socialist is perceived as a new type of person, *master of nature, master of society, and master of himself.* In the socialist system, leadership belongs to the Working Class Party, administrative management to the State, and the right of collective management to the people. This system assigns to itself the great task of raising human consciousness, dignity, abilities, and performances. It wants to enable humans to be masters of nature, of society, and of themselves for and in the building of a new life and a new society. It is a formula, a basic principle, that reflects a dialectical and scientific perception of the interaction between humans and nature, between the environment and those who people it. Viewed in terms of achieving ecological transformation for socialist aims, ideological values play here the role of compass and guide. They enable humans to be conscious of themselves, of the laws of nature and society, and of what people have to learn and do, so that they can truly become the masters of their country. This principle of building humans and the new society—established as philosophy and policy; incorporated into organization, management, and action; and promoted as a way of life—has clearly shown that ideology helps humans grasp the natural and social aspects of environment, to "create new conditions of existence" (Engels 1972,

368) and new human environments. President Ho Chi Minh (1961) said that "to build socialism, there must be socialist men." The people whose motto of life is "one for all and all for one" are sensitive to the problematic in ecology which calls for "an environmental ethic," an ideological foundation for human action concerning the environment.

In Vietnam, all achievement depends upon patriotism and socialist consciousness, upon the collective participation of the people, upon their struggle, labor, and creative activity. Scientific and technological organizations, as well as individual scientists and experts, have an important role in the scientific and technological revolution directed toward the mastery and modification of the environment. One of the major tasks incumbent upon them is to propagate science and technology among the masses, to initiate them step by step and according to varying background, requirements, and specific conditions, into science and various techniques, so that they may apply this knowledge on every level, from the most fundamental context to ever more complex situations. The political, economic, and social mobilization of the working classes, and their "intellectualization," has the aim and effect of making them masters of the country and at the same time "shepherds of the future" (Saint John Perse). The political organizations and governmental institutions of education, culture, propaganda and information, literature, and the arts have their share of responsibility in this important and long-range mission, each according to its specific function.

The task of transforming ecological conditions to the advantage of socialism must be a mass effort, enlisting the efforts of political and social bodies and organizations as well as scientific and technological centers. Thus, ideological values have their "freedom of the city"; they penetrate the entire range of issues under discussion and play an important role throughout.

This new kind of socialist emerges gradually in connection with the three revolutions we are carrying out simultaneously and gradually: (1) the revolution in the relations of production; (2) the ideological and cultural revolution; and (3) the scientific and technological revolution, which constitutes the keystone. The socialist community is the driving force behind all three of these revolutions.

If we view the development of socialism as a transformation of the human environment, we see that these revolutions evince a judicious, dialectical conception of the systemic character of the environment: relations betweeen humans and the environment are predicated upon the relationships between people, including relations of production. In the transformation of the means of production, an intrinsic bond exists between the internal development of the way of life of a society and its approach to the exploitation of the natural ecosystem. The few examples I have adduced bear witness to this. Thus, I wish to discuss now the role of ideological and cultural revolution in the work of transforming the human environment.

The ideological revolution, as I have briefly indicated, is aimed at raising and consolidating the ideological and political levels of socialist society. This involves in the final analysis an increase in labor productivity, which ensures the victory of socialism over capitalism.

Culture and the Environment

The cultural plan developed in this connection includes measures concerning the problem of eradication of illiteracy (which began immediately after the August Revolution), of education, instruction, propaganda and information, and of creativeness in arts and literature, in the context of the socialist way of life. The problem of protecting and successfully exploiting the environment in the long run is closely linked to the problem of the endogenous development of a country in various basic disciplines: economic, social, demographic, cultural, and scientific. For successful development deep knowledge of the country, of its geography and history, is required. A culture constantly enriched and renewed is an indispensable condition of development for any country. We can mold or exploit the environment, destroying it in the process either willfully or inadvertently. Ecological conditions are not unalterable. The environment can produce, give, and reproduce. "Artificial" ecosystems, characterized by human manipulation, can sometimes be created and multiplied with lightning rapidity. However, many such ecosystems have proved to be deficient: the genetic patrimony has become impoverished, the soil degraded, and sites and historical monuments destroyed or forsaken. In many cases, "man has done nothing but blithely dissociate billions of structures to reduce them to a state in which they allow of no more integration" (Lévi-Strauss 1955, 478).

Another mark of this deficiency is human alienation and the social-cultural marginalization of certain social strata for political, social, and psychological reasons (UNESCO 1977, 276).

Initiation of the people into proper ecologial thinking and action is still in its first stage in our country. Great strides have been made toward the acquisition of ecological knowledge, however, through raising the level of culture and of general instruction; through information and propaganda concerning the various problems of agricultural, industrial, and handicraft production, housing, and the way of life; and through the popularization and dissemination of new achievements in such sciences as biology and biochemistry, and in the fields of rice cultivation, silviculture, and oceanography. General education, the popularization of general as well as scientific and technical knowledge and experience through mass media and the intensification of work experience and scientific research (sciences and technology including social sciences, particularly economics, sociology, history, philosophy, ethnology, archaeology, and law) have contributed greatly to this process that conjoins production and the transformation of the environment.

In their own ways, and according to their specific characteristics, literature and the arts maintain links with living reality and the people's struggle and labor. The public has discovered and recognized in many poems, novels, films, plays, songs, paintings, and statues the images of socialism in the process of construction and the problems encountered during this process. The struggle between the old and the new, the backward and the advanced; the efforts displayed in productive labor; the application of new initiatives and research in the practice and development of agriculture, industry, land reclamation, exploitation of forests and reforestation, and conquest of the sea all have their echoes in art. Of course, these are not mere object lessons, that is, lessons in science or ecology. Art is art; it does not reproduce reality as such, but is a metaphoric representation of the essence of reality, in "*abime*" to use a word by André Gide.

However, in making us experience feelings and be moved, in making us enjoy and think, art becomes a source of knowledge, of action and re-creation, of *savoir vivre* and *savoir faire*. It is precisely on this epistemological basis that literature and the arts have fairly fulfilled their mission of "*gestus*" (B. Brecht), of guide and ferment. Humans, society, and science can benefit from them and appropriate it in a variety of ways. Thus, good works of art have been a tremendous source of attraction for readers and laborers in various fields. In several works we find solutions, theses, and propositions that help us to communicate with life in general as well as to affirm our national identity.

The artistic values of these works mobilize and actualize human potential while communicating to the reader some major problems of concern here, including the relations between humans and society, between humans and the environment—of ecological conditions. For instance, *Land of Village* (Nguyễn Thi Ngoc Tú) is a novel that successfully reflects the transformation of the Vietnamese country-side in the process of reforms in agrarian management. In the novel *The Men Leaving the Forest,* by Nguyễn Minh Châu, and *The Swallows at the Beginning of Spring,* a play by Tất Dat, the conquest of land in the "new economic zones" by army units engaged in agricultural production is portrayed. For "the laborers of the sea," the struggle of man with the ocean is shown in such films as *Starfish,* by Dang Nhât Minh, and *The Ocean Which Lights Up,* by Nguyễn Manh Tuấn. These are only a few examples. They do not represent "environmental literature and art" but nevertheless reveal the outlines of interesting problems in human ecology. But there is still more, since at issue is the building of a society, and a society is not simply a place of "consumption with a supplement of soul" as Delmas put it.

With the appearance of "artificial" ecosystems (e.g., towns, cities, agricultural and industrial centers, townships, villages, tourist centers, and "new economic zones") the problem of human relationships to the environment comes up from another perspective. How is it possible to organize life under these conditions in such a way as to maintain an equilibrium, a harmony, in the relations between humans and the environment? How is it possible to foster a way of life that is based on social progress reflected in the people's standard of living while maintaining an equilibrium between humans and nature?

Not being an expert in all matters, I am unable to deal with so vast a problem. I will confine myself, therefore, to the question of culture and to a still more restricted field: I think that the dynamism and complex nature of all ecosystems, especially human environments, require the creation of a cultural medium within which the new field can prosper.

All ecosystems controlled by humans bear the marks of culture, particularly of science and technology. However, this does not constitute ipso facto a true cultural medium. This is of concern since an inadequate cultural medium can cause deterioration of the ecological relationships in question, including relationships of the human environment. Does this opinion constitute an antiphrasis or a paradox?

By cultural medium I mean a social atmosphere involving a wholesome and vibrant way of life, based on disciplined and conscious labor as well as good relations between human beings. This includes many aspects. Among them are effective administrative and social management; good manners (an important aspect of the quality of life) and respect for the law; and, finally, a political, cultural, and moral life that is rich and varied and helps people express their values and satisfy their needs as masters of the environment, pursued in the interest of all and not to the prejudice of anybody.

Of course, I am not calling for a utopian "city of the chosen" in the manner of Eldorado. This cultural medium is our goal, and by pursuing it with perserverance, we are realizing some aspects of it. In Vietnam, even in the countryside, many villages and cooperatives have grown and taken on new form, but small ecosystems also bear their own cultural imprint. The organization of sociocultural activities in these places sets good models for socialist construction in Vietnam. On the other hand, there are also examples where, for sociocultural reasons, distortions are seen in the human environment (e.g., low yields in agricultural production, survival of social evils, and the destruction of forests).

The ideal cultural medium, then, is constituted by a culture (in the anthropological sense) that is all-encompassing as a way of life and involves integration and convergence of values and behavior to the point that they enable humans to engage in *savoir vivre* and *savoir faire* for the good of the community and for the good of nature.

An ecosystem and a cultural medium naturally depend upon each other. But the quality of the cultural medium is not always and completely a function of the economic and material level of the ecosystem. Even a society that is relatively undeveloped, where the material life has not yet reached a desired level, can promote a wholesome and healthy cultural medium favorable to improvements in the society, in ecological relationships, and in the human environment. The decisive factor is the nature of this society, specifically its ideological and political orientation.

Thus, in concrete historicosocial terms, we have to know how to use cultural values as a dynamic, generative force to raise the quality and way of life. These in turn will help us transform the physical and material base of the human environment according to our goal.

Selected Bibliography

EDITOR'S NOTE: Obtaining complete citations for many of Professor Trinh's references and organizing the latter into the style of the other papers in this volume has been extremely difficult. Much of the difficulty is due to the fact that Trinh's paper was originally written in French, then translated into Vietnamese for review in Vietnam, and finally translated from Vietnamese into English. The Editors thus have decided to include Trinh's references largely as they were originally translated (in Vietnam) into English, and apologize for any inconvenience to the reader.

I. Environment

Ananichev, K. V.
  1981. The activity of international organizations and the international cooperation in the field of protection of environment. Review "Okhrana prirody; vosproizvdstvo prirodnyc Kh resursov," No. 8, 1978; translated into Vietnamese by Nguyen Quynh Nhu. Review "The problems of development of science and technology," Institute of Information of Science and Techniques (State Committee for Sciences and Techniques), Hanoi, No. 10.
Final Report
  1981 Environment and Development Regional Seminar on Alternative Patterns of Development and Life-styles in Asia and the Pacific. Bangkok, August 1979, translated into Vietnamese, same Review, No. 47.
Garkovenko, R. V.
  1979 On the essential particularities of the relation between society and nature. *Nauka*, Izdat, Novosibirsk, 209–66, 1978. Translated into Vietnamese by Nguyen Ba Chinh,

same review, No. 30.

Giljarov, M. S.
1979   On the work of development of ecology in the USSR. *Ekologiga*, No. 5-6, 1977. Translated into Vietnamese by Bui Duy Lich, same review, No. 30.

Jamieson, Neil L., and George W. Lovelace
1983   Cultural values and tropical ecology—some initial considerations. East-West Environment and Policy Institute Working Paper.

Literaturnaja gazeta (USSR)
1974   Interviews made with great men of science from various countries on the problem: The 20th century a science. Translated into Vietnamese by Tran Tien Duc, same review, No. 5.

Nguyen Tien Chau
1973   To protect human environment. In Vietnamese, same review, No. 11.

Oldak, P. G.
1978   The protection of environment, a task in the development of social ecology. *Problemy nazvitija sovremmenoi nauki*, Novosibirsk, 226-44.

Poris Pal, Paris Gyorgy
1973   The urbanization and pollution of environment in 2000. Tajekoztado, Budapest: *Muszaki Gazdasagy* No. 6, 1972, 726-44. Translated into Vietnamese by Nguyen Tien Chau, same review, No. 1, 2.

II. Philosophy-History-Culture-Sociology-Ethnology

Aron, Raymond
1967   *The Stages of Sociological Thinking (Les etapes de la pensee sociologique)* Paris: Gallimard.
1977   *A Pleading for Decadent Europe (Plaidoyer pour l'Europe decadente)*. Paris: Robert Laffont.

Auzelle, R.
1962   *A Pleading for a Conscious Organization of Space (Plaidoyer pour une organisation consciente de l'espace)*. France: Vincent-Freal et Cie.

Collective of Soviet authors
1982   *The Socialist Life-style.* Translated into Vietnamese, Su

That edit., Hanoi.

Dion, Michel
1975 *Sociology and Ideology (Sociology et ideologie).* Paris: Edit. Sociales.

Engels, F.
1972 *The Dialectics of Nature.* Translated into Vietnamese. Su That edit., Hanoi.
n.d. *The Origin of the Family, Property and the State.* Translated into Vietnamese.

Godelier, M.
1977 *Marxist Journeys in Anthropology (Trajets marxistes en anthropologie).* Paris: Francois Maspero.

Harris, M.
1980 *Culture, People, Nature.* USA: Harper-Row.

Ho Chi Minh
1972 *Work Is Duty and Glory (Lao dong la nghia vu va vinh quang).* Same edit.
1961 *The Socialist Man (Con nguoi xa hoi chu nghia).* Same edit.
1977 *On the Work of Culture, Literature and Arts (Ve cong tac van hoa van nghe).* Same edit.

Huard, Pierre, and Maurice Durand
1954 *A Knowledge about Vietnam (Connaissance du Vietnam).* Hanoi: Imprimerie Nationale, E.F.E.O.

Institute of Archaeolgy (Vien khao co hoc)
1972 Act relating to the Second Conference on the period of the country's founding by Hung Kings (Hung vuong dung nuoc), Tome II. edit. Hanoi: Social Sciences (Khoa hoc xa hoi).

Lê Qúy Dôn
n.d. *Miscellaneous Texts (Van dai loai ngu).*

Lenin, V.
1960 *Materialism and Empirio-criticism.* Translated into Vietnamese. Same edit.
1977 *Philosophical Notes.* Translated into Vietnamese. Same edit.
1977 *On Culture and Art: Selected Works.* Translated into Vietnamese. Same edit.

Marx, K.
1962 *Contribution to the Criticism of Political Economy.*

Translated into Vietnamese. Edit., Su That, Hanoi.

Marx, K. and F. Engels
1954   *On Literature and Art: Selected Works*, ed. by Jean
        Freville. Paris: Edit. Sociales.

Nguyen Khanh Toan
1972   *History of Vietnam (Lich su Viet Nam)*, Tome I. Hanoi:
        Sciences Sociale.

Phan Huy Chu
1961–   *A Chronology of Dynasties (Lich trieu hien chuong loai chi)*
1962   Translated from Chinese characters in Vietnamese by the
        Institute of History, Tome I and Tome II, edit., *History
        (Su hoc)*, Hanoi, Tome III and Tome IV, same edit.

III. Other Documents

Cardiol, Bertrand
1983   *Vietnam, Documentation, Development*, Tome I, II, III, IV,
        Collectif M.A.D. Paris.

Conference of Ministers for Culture (Conference des Ministres de la
        Culture Cotonou, 1981 A.C.C.T.)

Encyclopedia of Sociology (Encyclopedie de la sociologie) Paris:
        Larousse, 1975.

UNESCO
1977   From Understanding to Action (*Comprendre pour agir*).
        Presses Universitaires de France.

1979   *The Natural Resources of Humid Tropical Asia (Ressources
        naturelles de L'Asie tropicale humide)*. Lausanne:
        Imprimeries reunies.

1982   *Draft of Medium-range Plan (Projet de plan a moyen terme
        [1984–89])* 4e session extraordinaire, Paris.

CHAPTER 13

SOME EFFECTS OF THE DAI PEOPLE'S CULTURAL BELIEFS AND PRACTICES UPON THE PLANT ENVIRONMENT OF XISHUANGBANNA, YUNNAN PROVINCE, SOUTHWEST CHINA

Pei Sheng-ji

This chapter examines several important effects that certain cultural beliefs and practices of the Dai people of southwestern China have had upon the plant environment of Xishuangbanna in southernmost Yunnan Province. These effects can be briefly summarized as follows: (1) the introduction and current distribution of many locally cultivated plants in Xishuangbanna is historically related to the spread and acceptance of Hinayana Buddhism within the last 1,400 years; (2) the Dai people's conception of "Holy Hills," a belief derived from an earlier and formerly more dominant polytheistic religious tradition, has helped to preserve certain areas of pristine forest vegetation; and (3) the traditional Dai practice of cultivating fuelwood also contributes to the conservation of natural forests and is of economic and ecological significance to human adaptation in humid tropical environments.

The Regional Environment

Xishuangbanna Dai Autonomous Prefecture is one of eight autonomous prefectures in Yunnan that were established by the People's Republic of China in the 1950s to allow for regional autonomy and protection of national minorities. This prefecture is located in the south of Yunnan Province in southwest China (21°10'-22°40' N, 99°55'-101°50' E), where it is bounded on the

Fig. 13.1 The location of Xishuangbanna Dai Autonomous Prefecture, Yunnan Province, southwest China.

south and southeast by Laos and on the southwest by Burma (fig. 13.1).

Approximately 94 percent of Xishuangbanna's total area of 19,220 km^2 consists of mountainous and hilly terrain, with river valleys making up the remaining 6 percent. Although much of the area lies between 500 and 1,000 m above sea level, some locations in the extremely mountainous zone are between 2,300 and 2,400 m above sea level. This mountainous zone is a southern extension of the Hengduan Mountains, which are themselves part of the Ai Lao mountain chain. The gradient of Xishuangbanna generally slopes from this northeastern zone toward the southwest where the upper Lancang Jiang (Mekong River) and its tributaries, the area's major river system, are concentrated.

The Dai People

Much of southwestern China, including Yunnan Province, is inhabited by ethnic groups of a non-Han Chinese ethnic status and derivation. In Yunnan Province itself, twenty-three different ethnic groups are officially recognized as "national minorities." Overall, these groups display considerable cultural and economic diversity, ranging from hill peoples who practice swidden cultivation to lowland-dwelling groups whose primary mode of subsistence is intensive wet-rice agriculture.

One of the more important national minorities of Southwest China is the Dai (T'ai) people who have a total population of well over 750,000 people. Approximately 220,000 of this larger population are concentrated in Xishuangbanna, where the Dai, as the largest single ethnic group, constitute about 35 percent of the autonomous prefecture's population.

The Dai language is believed to be derived from the larger Zhuang-Dong linguistic group, which is a branch of the Han-Tibetan language family. The Dai possess their own dialect and script, both referred to as *Daile*. The word for Xishuangbanna in Daile is *Sip-song Pan-na*, meaning "twelve administrative areas."

An aboriginal people, the Dai were first recorded in Chinese historical texts dating to the early years of the Han dynasty (ca. 200 B.C.) where the ancestors of the Dai people were called "Dian" and "San." According to the *Hou Han Shu* [History of the later Han dynasty], written in the fifth century A.D., after A.D. 79 the chiefs of the Dai sent several missions to Luoyang, the capital of the Eastern Han dynasty. On behalf of the chiefs, these missions accepted titles, ranks, and land rights from the Han emperor. In *A Record of Le,* a Dai history written in the Dai language in A.D. 1180 (the year 542 of the Dai calendar), it is recorded that the Dai chieftain Pa Zheng established a local regime called Mengle, with its capital at Jinghung, and unified the whole of Xishuangbanna. Pa Zheng made his regime a vassal state of the Heaven dynasty, i.e., the imperial government of China, and accepted titles and an official seal from the imperial court.

In 1952, three years after the founding of the People's Republic of China, the Xishuangbanna Autonomous Prefecture was established in line with government policies that guaranteed the equality of all nationalities in China as well as their right to preserve their own languages, traditions, customs, and religious beliefs.

For centuries, the economic existence of the Dai people of Xishuangbanna has been characterized by a local autarky, a self-sufficient economic pattern that combines agriculture with the exploitation of natural products. Although they have long possessed settled agriculture, primarily wet-rice cultivation accompanied by the cultivation of tea, fruits, spices, and herbs, the Dai have also raised cattle, pigs, and poultry in a semidomestic manner. In addition, hunting, fishing, and collecting of wild plants have played important, traditional roles in the local economy. As a consequence, the Dai people still depend greatly upon natural plants and animals for their livelihood.

Traditional Beliefs and Practice of the Dai People

The primary religion of the Dai people in Xishuangbanna is Hinayana Buddhism. In addition, many traces of an earlier and formerly more dominant polytheistic religious tradition remain. Various elements of these patterns of belief affect the nature of Dai interactions with the local environment. Particularly important among these are the religious significance accorded by the Buddhist canons to particular cultivated plants, the concept of the "Holy Hills," and a traditional Dai cultural practice that involves the cultivation of fuelwood.

Hinayana Buddhism and Its Effect on the
Distribution of Cultivated Plants

The Hinayana Buddhism of the Dai has been infused with national color and is still very popular among the villagers. Buddhist temples, called *wa* in Dai, are found in most villages, and historical records suggest that there were once more than 360 temples in the area. Of these, about 220 still exist.

Although its general date of introduction is unclear, Hinayana Buddhism has long been embraced by the Dai people of Xishuangbanna. At the academic symposium on "How Sthavirada Buddhism [an early form of Theravada Buddhism] was Introduced into China," held in Kunming, Yunnan, in October 1982, three different dates, ranging from the middle of the Tang dynasty (A.D. 618–907) to A.D. 1300, were put forward. According to the high-ranking monks and folk stories in Xishuangbanna, however, this form of Buddhism was introduced and practiced much earlier. The routes of introduction into the Xishuangbanna region are also unclear, there being three somewhat different opinions. One opinion is that the religion was introduced from northern Thailand (i.e., from Chiangmai) to Xishuangbanna. The second posits that it came from central or northeastern Burma. The last opinion suggests that the religion came from India via Thailand.

Prior to the founding of the People's Republic of China, the organizational structure within the Buddhist temple hierarchy was closely related to the feudal system. Boys entered the temples at the age of nine to be trained as monks. They studied ancient Dai letters and learned the Buddhist canons. Those who did well in their studies were honored and promoted to higher ranks. The masses worshipped at the Buddhist temples frequently and piously, and regularly offered food to the monks.

The monks were ranked rigidly. Promotions required the approval of feudal lords called *Zhaomeng*, and the highest-ranking positions were usually held by the lords' relatives and friends. Temple expenditures were also controlled by the lords, who often came to the temples during Buddhist festivals to promote the monks in the name of Buddha. In these ways, the feudal lords exercised much control over the temple hierarchy and were able to consolidate their own ruling positions.

The canons of Buddhism specify that four requirements must be met before a Buddhist temple can be established. These requirements are a statue of Sakyamuni (*Pagodama-Zhao* in Dai), the founder of Buddhism; a pagoda in which Sakyamuni's ashes can be preserved; at least five monks; and the presence of some specified "temple yard plants." According to the classic *Records of the Buddhas' Coming into the World for 28 Generations*, of which there is a copy, handwritten in Dai, in the temple in Man Jin in Menghun, the Buddha of each generation designated one species of plant as the adored plant for Buddhists.

Accordingly, the temple yards of the Buddhist temples normally contain dozens of specified tropical plants that are cultivated in line with the creeds of the Buddhist canons. In investigating the yards of more than twenty Buddhist temples in Xishuangbanna, I found that there were more than fifty-eight different species of commonly cultivated plants. Based on discussions with the temples' monks, these species can be grouped into three types: ritual plants, twenty-one species; fruit trees, seventeen species; and ornamental plants, twenty species (see tables 13.1, 13.2, and 13.3). Examining table 13.1, we find that many of these species were introduced to the region. Twenty-nine species are derived from either India or tropical Southeast Asia, nineteen species from China or Southeast Asia, and ten species from either tropical America or Africa. The presence of nonendemic plant species from Southeast Asia and India, as well as from the more eastern portions of China, suggests the possibility that the plants were introduced to Xishuangbanna from both the south-southwest and the east. With the exception of the fruit trees and some of the ornamental plants, these introduced species are grown almost exclusively in the yards of the Buddhist temples. In the cases of several species, i.e., *Corypha umbraculifera, Ficus religiosa,* and *F. altissima,* their local distribution appears to be maintained by Buddhist doctrines that prohibit their cultivation by individuals outside the temple precinct. Such distributions suggest that Hinayana Buddhism played a role in the spread of these plants.

China was one of a number of areas in the Old and New Worlds that possessed fairly sophisticated, agricultural civilizations. These areas were not entirely isolated in historic times and there appear to have been numerous instances in which cultivated plants, agricultural technologies, and inventions were exchanged between China and Southeast Asia, Europe, Arabia, and even the Americas.

Many details of these early exchanges are mentioned in ancient Chinese records, and they have been more recently discussed by

Table 13.1. Cultivated Plants for Buddhist Rituals Grown in Temple Yards in Xishuangbanna

Botanical Name	Dai Name	Use or Significance	Origin
Acacia pennata (L.) Willd.	Sòngbài	Dye for making the sutra more readable	Native and S.E. Asia
Aegle marmelos (L.) Corr.	Mabinghan	Fruits used as offering	India and Bengal
Aleurites moluccana (L.) M.-A.	Maiyao	Seed oil used as lamp oil	S.E. Asia
Areca catechu L.	Gèma	Fruits used as offering	S.E. Asia
Bixa orellana L.	Gèmaxie	Aril used as an offering and as a dye	S.E. Asia
Borassus flabellifer L.	Gedan	Symbol of temple	India and Burma
Caesalpinia sappan (L.)	Gefan	Heartwood used as an offering and as a dye	India
Corypha umbraculifera L.	Gelan	Leaves used as writing paper (sutra)	India and Ceylon
Crinum asiaticum L.	Linuolong	Flowers used as offering	Native and S.E. Asia
Dipterocarpus turbinatus Gaertn.	Mainamanyan	Resin used as lamp oil	S.E. Asia
Ficus altissima Bl.	Maihongnong	Held sacred by Buddhists	Native and S.E. Asia
F. glomerata Roxb.	Gèlei	Bark used for making paper	Native and S.E. Asia
F. religiosa L.	Gèxili	Held sacred by Buddhists	India
Gmelina arborea Roxb.	Maisuo	Wood used for sculptures	Native and S.E. Asia
Impatiens balsamina L.	Lòulei	Flowers used as offering	India
Jatropha curcas L.	Maihongham	Seed oil used as lamp oil	Tropical America
Livistona saribus (Lours.) Merr.	Geguo	Offering	Native and S.E. Asia

Botanical Name	Dai Name	Use or Significance	Origin
Mesua nagassarium (Burm. f.) Kost.	Maibola	Seed oil used as lamp oil	India
Nymphaea spp.	Nuðzhàngwan	Flowers used as offering	China
Streblus asper Lour.	Gehui	Bark used for making paper	Native S.E. Asia
Tectona grandis L.f.	Maisa	Wood used for sculptures	India and S.E. Asia

Table 13.2 Cultivated Fruit Trees in Temple Yards in Xishuangbanna

Botanical Name	Dai Name	Common English Name	Origin
Ananas comosus (L.) Merr.	Makeliang	Pineapple	Tropical America
Annona reticulata L.	Magan	Bullock's apple	Tropical America
A. squamosa L.	Magantùlù	Sugar apple	Tropical America
Artocarpus heterophylla Lam.	Gèmaleng	Jackfruit	China
Carica papaya L.	Guishèbao	Papaya	Mexico
Citrus grandis (L.) Osb.	Mabù	Shaddock	China and S.E. Asia
Cocos nucifera L.	Gèbao	Coconut	S.E. Asia
Flacourtia ramontchii L'Her.	Majing	Governor's plum	India
Litchi chinensis Sonn.	Magài	Litchi	China
Mangifera indica L.	Màmou	Mango	India
Musa spp.	Gùi	Banana	S.E. Asia

Botanical Name	Dai Name	Common English Name	Origin
Oroxylum indicum (L.) Vent.	Gèlièga	India trumpet flower (young fruit edible)	Native and S.E. Asia
Phyllanthus emblica L.	Mahangbàng	Emblic	Native and India
Psidium guajava L.	Maguixiongla	Guava	Tropical America
Spondias pinnata (L.) Kurz.	Gèmaigègù	Andaman mombin	Native and S.E. Asia
Syzygium jambos (L.) Alston.	Gezhongbù	Rose apple	S.E. Asia
Tamarindus indica L.	Mahang	Tamarind	India

Table 13.3. Cultivated Ornamental Plants in Temple Yards in Xishuangbanna

Botanical Name	Dai Name	Origin
Butea monosperma (Lam.) O. Ktze.	Gexiham	Burma and India
Caesalpinia pulcherrima Sm.	Nuðhàosang	India
Cassia agnes (Dewit) Brenan	Maiblongliang	S.E. Asia
C. fistula (L.)	Maiblonglan	S.E. Asia
Cinnamomum porrectum (Roxb.) Kost.	Maizhong	Native and S.E. Asia
Delonix regia (Bojor) Raf.	Gènðumailiang	Tropical Africa
Dendrocalamus giganteus Munro	Maibð	Native and S.E. Asia
Gardenia jasminoides Ellis	Nuoshuilong	S.E. Asia to E. Asia

Botanical Name	Dai Name	Origin
G. sootepense Hutch.	Gèmo	Native and S.E. Asia
Gendarussa venticosa (Wall.) Nees	Maihahào	India and S.E. Asia
Mayodendron igneum Kurz.	Nuobĺlong	Native and S.E. Asia
Michelia alba DC.	Zhanghào	South China and S.E. Asia
Millingtonis hortensis L.f.	Maigasolðung	Native
Nyctanthus arbor-tristis L.	Gèmàhong or ge'màhong	India
Pandanus tectorius Sol.	Nougen	S.E. Asia
Parkia leiophylla Kurz.	Maihuanguang	S.E. Asia
Plumeria acutifolia Ait.	Zhangbadiàn	Tropical America
Samanea saman Merr.	Maisongsà	Tropical America
Sesbania grandiflora (L.) Pers.	Gèluogai	India
Thyrsostachys siamensis Gamble	Maihe	S.E. Asia

the British scholar Joseph Needham in *Science and Civilization in China* (1954, 1956). As historically described, however, these records emphasize the importance of two major trade routes. One important route was the Old Silk Road that extended across central Asia and northwest China from Europe and Arabia in the west to China proper in the east. From the Han through late Yuan dynasties (ca. 200 B.C.–A.D. 1300), merchants, travelers, and adventurers journeyed along this well-travelled route spreading ideas, inventions, commercial goods, and plants. A second well-known route involved maritime trade between China and Southeast Asia. Although sea trade between these regions existed as early as the Jin (221–207 B.C.) and Han (206 B.C.–A.D. 220) dynasties, the trade in cultivated plants was intensified during the Ming dynasty (A.D. 1368–1644) when Chinese merchant fleets travelled throughout the seas of Southeast Asia.

Much less is known of the apparent existence of an early land route between India/Southeast Asia and China. I believe southwest China was an important corridor for the exchange of cultivated plants between these areas and that the spread of many plants, especially those of a ritual significance, has much to do with the introduction and spread of Hinayana Buddhism along this route. Some of these plants, such as *Corypha umbraculifera,* are propagated by seed and probably were brought directly from India by travelling monks. Additional research upon both the botanical and historical aspects of introduced plant species and the historical development of Hinayana Buddhism should provide much information about the nature of this early land route.

"Holy Hills" and Their Effect on Conservation

Prior to the introduction of Hinayana Buddhism, the Dai people appear to have possessed a polytheistic religion that was heavily bound to the natural world. Like many early groups, the Dai associated the forests, the animals and plants that inhabited them, and the forces of nature with the supernatural realm. Proper actions and respect for the gods were believed to result in peace and well-being for the villagers. Improper activities and disrespect, on the other hand, incurred the wrath of the gods who punished the Dai villagers with a variety of misfortunes. Thus, the early Dai were encouraged to live in "harmony" with their surroundings.

Although this earlier pattern of polytheistic beliefs was largely supplanted by Buddhism, certain elements still persist. One important element that continues to affect the ways in which the Dai interact with their environment is their concept of "Holy Hills." In traditional Dai thought, the Holy Hill is a forested hill where the gods reside. In addition, the Dai believe that the spirits of great and revered chieftains go to the Holy Hills to live following their departure from the world of the living.

Holy Hills are an important visual element of the modern Xishuangbanna landscape and can be found wherever one encounters a virginal forested hill in association with a Dai village. The Holy Hill is a major component of the traditional Dai agricultural ecosystem, which combines paddy fields, home gardens, and cultivated fuelwood forests in addition to the naturally forested Holy Hill (fig. 13.2). In Xishuangbanna there are approximately 400 such hills, occupying a total area of roughly 30,000 to 50,000 ha (1.5 to 2.5 percent of the total area of the prefecture).

There appear to be two types of Holy Hills. The first, *Nong Man* (*Man* is the Dai word for "village"), refers to the presence of a naturally forested hill, usually 10 to 100 ha in size, that is worshipped by the inhabitants of one nearby village. Where several adjacent villages form a single larger community (*Meng*), another type, called *Nong Meng*, is frequently found. Forested hills of this second type occupy a much larger area, often hundreds of hectares, and belong to all of the villages in the community.

All of the plants and animals that inhabit the Holy Hills are considered to be either companions of the gods or living things in the gods' gardens. The Holy Hill is a kind of natural conservation area founded with the help of the gods, and all animals, plants, land, and sources of water within it are inviolable. Gathering, hunting, wood-chopping, and cultivating are strictly prohibited activities. Although intimately associated with their beliefs and rituals, the Dai people do not use these hills as "cemetaries" (*bahao*); areas of burial are confined to separate hills. Neighboring non-Dai ethnic groups respect the wishes of the Dai and never enter the forests. The Dai people believe that such violations would make the gods angry and that misfortunes and disasters would be brought down upon them as "punishment." These punishments can take many forms, including diseases, floods, fires, windstorms, earthquakes, plagues of insects, and attacks by beasts. It is, therefore, in the villagers' interests

Fig.   13.2   The traditional Dai agroecosystem in
Xishuangbanna.

not to violate the sanctity of the Holy Hill and to present regular offerings in the hope that the gods will be pleased and protect their health and peace.

The villagers go to the Holy Hill (*Nong Man*) annually to conduct rituals. In a ceremony that is presided over by a local priest, called *Bomo*, the Dai offer such things as pork, beef, cocks, rice, fruits, flowers, wine, spices, and candles to the gods. The ceremony is attended by all of the local villagers and sometimes lasts as long as three days. In cases where the Holy Hill is worshipped by a larger community of multiple villages (*Nong Meng*), the ceremony normally is presided over by a higher, professional priest called *Bomo Meng*.

Protected by this belief, the natural forest vegetation within the Holy Hill has been preserved for a long time. A number of studies by the Yunnan Institute of Tropical Botany, Academia Sinica (1972–82), indicate that the vegetation occurring on these hills closely resembles the patterns of vegetation in larger tracts of pristine regional forests in terms of character, structure, function, and species composition. Near the village of Da-Meng-Long, located in a tropical, seasonal, rain forest area, for example, there is a Holy Hill called Mangyangguang which has an area of 53 ha at an altitude of 670 m above sea level. Our studies indicate that the hill's forest contains 311 different plant species belonging to 108 families and 236 genera. Of the 311 species present, 283 are vascular plants. The structure of the forest community can be divided into three layers of trees, one shrub layer, and one layer of herbs and seedlings of which 20 to 30 percent are deciduous or semideciduous in nature. The total amount of dried materials annually returned to the soil by the forest is 12.83 tons per ha, of which 6 tons are leaves, 1.2 tons are flowers and fruits, and 1.2 tons are branches and barks. The remaining 4.43 tons consist of decayed plant materials such as shrubs, vines, and herbs. A comparison of the annual and seasonal temperatures occurring within the Holy Hill forest to those occurring in surrounding open land shows substantial variation. Although the yearly average temperature within the forest is only 0.6° C lower than that of open lands, the average maximum temperatures inside are 3.4° C lower than those recorded outside. Soil surface temperatures also vary, being 6.6° C lower inside the forest for much of the year. However, during January (the coolest month of the year), the soil surface temperatures in the forest can be 0.2° C higher than in surrounding open lands. Similar kinds of variation are found in air humidity. The yearly mean value of air humidity inside the forest is four points higher than that outside the forest, and, during the dry season from March to early May the humidity in the forest can be as much as ten points higher.

In general, these and other characteristics indicate strong similarities between the patterns and the exchanges of material and energy within the forest of the Holy Hill and those of the tropical seasonal rain forest nearby. These similarities suggest that the careful protection of the Holy Hill forest vegetation by the local Dai people has allowed the forest vegetation and its biotic and abiotic components to be maintained in a stable yet internally dynamic ecological balance.

The Ecological Significance of
Traditional Fuelwood Cultivation

Of the fifty-six different nationalities found within China, only the Dai of Xishuangbanna have traditionally cultivated trees for fuel. Although they dwell in a region of dense tropical forests in which the natural regeneration of vegetation proceeds at a fairly rapid rate, the Dai traditionally have chosen to cultivate a particular species of tree (*Cassia siamea* Lam.), known locally as *gemaixili*, to meet their energy needs. In part, this may be a strategy to protect the vegetation of the Holy Hills.

Gemaixili is an evergreen tree of the Caesalpiniaceae family. It is normally cultivated on gentle slopes below 1,000 m near each Dai village. The proximity of the area of fuelwood cultivation to the village makes transport easy. Originally a native of Thailand, the plant was introduced into the Xishuangbanna area at some distant, undetermined point in the past. Our ethnobotanical and historical studies indicate that *gemaixili* has been grown in this area for at least the past 400 years.

The traditional technique of cultivating *gemaixili* begins with the collection of seeds during the dry season (March to early May). This activity is followed by the sowing of the seeds on burned slopes at the beginning of the rainy season in late May or June. Sometimes the *gemaixili* seeds are sown along with the seeds of other dryland crops. Once a year, a small amount of weeding is done, but herbacious growth is not entirely removed. After four to five years, when the trees reach a height of 15 m and a diameter of 10 to 15 cm, the Dai cut an initial crop of fuelwood from the cultivated trees. This first harvest involves felling the *gemaixili* trees to a height of approximately one meter with the branches and upper trunk taken away as fuelwood. The stump is left to sprout new branches, which will appear within the same year. In three years as many as five new branches will grow to a height of 8 to 10 m and a diameter of 5 to 8 cm. The Dai then cut fuelwood for a second time, once again leaving the stump. Thereafter they cut the new growth every three years. Our studies suggest that a cultivated *gemaixili* forest can be cut for about 100 years and that a forest's annual output can be as high as 60 m^3 of wood per hectare. Each Dai uses roughly 1.0 to 1.5 m^3 of fuelwood every year, and 0.1 ha normally provides more than enough fuel for one person.

From an economic standpoint, the cultivation of *gemaixili* for fuelwood has many advantages. Not only is it relatively easy to cultivate and manage, but it also can be easily cut, transported, and stored. The speed with which the new sprouts naturally appear and grow further serves to replenish fuel sources quickly and reduces the amount of insect and animal damage. In addition, *gemaixili* burns very well and produces considerable heat, factors which make it an extremely good source of energy. Apart from these economic advantages, the cultivation of *gemaixili* as fuelwood has broader implications for the preservation of the natural vegetation of Xishuangbanna in that its cultivation reduces the degree to which the natural forests of the region are used for energy.

Given these advantages and implications, the practice of *gemaixili* fuelwood cultivation has not only developed as a unique technique among the Dai people, who are concerned with protecting the vegetation of the Holy Hills, but it has spread to other ethnic groups as well, including the Han Chinese, who see it as an important solution to a growing shortage of local fuels.

Conclusion

The above discussions provide several specific examples of how the plant environment of Xishuangbanna in southwest China has been affected by the cultural beliefs and practices of the Dai people. These examples range from the enrichment of the local plant community through the introduction of nonendemic species that are associated with Hinayana Buddhism to the direct and indirect conservation of tracts of native tropical forest vegetation in line with the concept of Holy Hills and the cultivation of *gemaixili* for fuelwood. In offering these examples, my intent has been to show how cultural beliefs, values, and the actions that these generate can affect the patterning of the natural world.

It would be incorrect to suggest, however, that the cultural realm either dominates or ultimately determines the natural realm. Instead, it seems clear that the cultural and natural aspects of Dai existence have long been involved in a dynamic pattern of interaction that has led to modifications in the characteristics of each realm by the characteristics of the other. Through time, these mutual modifications have closely interwoven the cultural and natural aspects of the Dai world in a manner that makes it impossible to fully understand either aspect without reference to the other.

References

Ford, Richard I. (editor)
  1978   *The Nature and Status of Ethnobotany.* Museum of Anthropology, Anthropological Papers No. 67. Ann Arbor: University of Michigan.
Hooker, Joseph Dalton
  1875–   *Flora of British India,* Volumes I-VII. London: L. Reeve
  1897   and Co.
Institute of Botany, Academia Sinica
  1972–   *Iconographia Cormophytorum Sinicorum,* Tomus I-X and
  1982   Supplementum I. Beijing, People's Republic of China (in Chinese).
Li Hui-lin
  1970   The origin of cultivated plants in Southeast Asia. *Economic Botany* 24(1):3–19.
Lin Yang-wu
  1983   The sects of Hinayana and religious festivals in Dehong. *Social Sciences in Yunnan,* No. 1. Kunming, Yunnan, People's Republic of China (in Chinese).
Lovelace, George W.
  1982   Cultural Beliefs and the Management of Agroecosystems. Paper presented at the Chinese Ministry for Urban and Rural Construction and Environmental Protection/ East-West Environment and Policy Institute joint workshop on "Ecosystem Models for Development,"

Kunming, Yunnan, and Guangchou, Guangdong, People's Republic of China, Sept. 27–Oct. 11, 1982.

National Research Council, Committee on Selected Biological Problems in the Humid Tropics
 1982   *Ecological Aspects of Development in the Humid Tropics.* Washington, D.C.: National Academy Press.

Needham, Joseph
 1954   *Science and Civilization in China,* Volume I. Cambridge: Cambridge University Press.
 1956   *Science and Civilization in China,* Volume II. Cambridge: Cambridge University Press.

Pei Sheng-ji
 1982   A preliminary study of the ethnobotany in Xishuangbanna. In *Collected Research Papers on Tropical Botany.* Kunming, Yunnan, People's Republic of China (in Chinese).

Rambo, A. Terry
 1982   Human ecology research on tropical agroecosystems in Southeast Asia. *Singapore Journal of Tropical Geography* 3(1):86–99.

Sheng Cheng-yu and Zhang Yu-hu
 1979   *The Acclimatization of Plants.* Shanghai, People's Republic of China (in Chinese).

Vavilov, N.I.
 1949–   The origin, variation, immunity and breeding of cultivated
 1950   plants (translated by K. S. Chester). *Chronica Botanica* 13(1–6):1–336.

Yan Zhi-ming and Dong Khai-zheng
 1980   A contribution on the origin of cultivated plants in ancient China. In *Agricultural Science in Ancient China,* Beijing, People's Republic of China (in Chinese).

Yu Ping-hua, Xu Zai-fu, and Huang Yu-lin
 1982   Research on the ethnical timber utilization in the district of Xishuangbanna. In *Collected Research Papers on Tropical Botany.* Kunming, Yunnan, People's Republic of China (in Chinese).

Yunnan Institute of History
 1980   *Minorities in Yunnan.* Kunming, Yunnan, People's Republic of China (in Chinese).

Zheng-lan
    1980   *Travels   through   Xishuangbanna.*   Beijing,   People's
            Republic of China (in Chinese).

CHAPTER 14

MAN, LAND, AND MIND IN EARLY HISTORIC HONG KONG

George W. Lovelace

In the past two decades there has been an increased interest among cultural and human ecologists in the relationships of traditional patterns of thought and belief to human interactions with the environment. Much interest has focused upon these traditional patterns as bodies or systems of accumulated environmental information and wisdom that allow traditional societies to regulate their impacts upon the environment to maintain overall stability in their adaptation and inhabitation of it.

The emphasis that has been placed upon homeostasis and its maintenance in the traditional world is somewhat unfortunate for a number of reasons. Not only does it provide only a partial, and sometimes overly romantic, view of traditional human interactions with the environment, conjuring up images of the "noble savage" and the "timelessness" of traditional society, but it also tends indirectly to assign greater causal significance to the material conditions of existence than to other factors (see chapter 2). Human populations have demonstrated the ability to modify their surroundings, consciously and unconsciously, to a degree far exceeding that of any other species. Environments inhabited and utilized by humans are always characterized, if not defined, by a combination of natural and cultural elements. To the extent that something approaching homeostasis evolves in particular situations of human-environmental interaction, it is probably the combined result of a human population's interference with and adjustment to environmental conditions over a period of time.

When we begin to consider the degree to which societies actively modify and structure the physical environments they inhabit, the

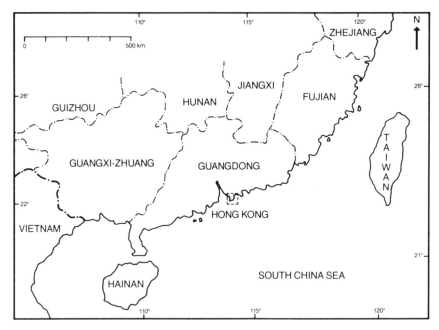

Fig. 14.1. Hong Kong's position on the south coast of China.

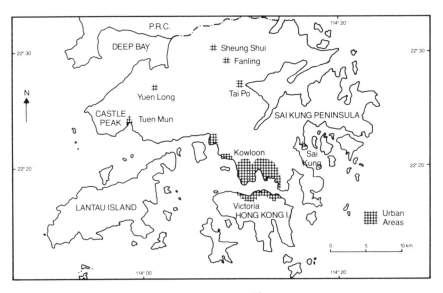

Fig. 14.2. Hong Kong.

cognitive aspects of human adaptation assume considerable significance. Cognitive frameworks, that is, patterns of traditional thought and belief, may play a threefold role in environmental interactions: (1) they provide the tools to conceptualize the environmental context; (2) they act as storehouses of environmental knowledge; and (3) they contain rules for human behavior with respect to the environment.

This chapter examines the role a traditional system of Chinese cultural beliefs and practices, known as *feng shui,* may have played in the inmigration and settlement by wet-rice farmers of the Hong Kong area during the early historic period (ca. A.D. 1000–1700). It shows that this complex of beliefs has affected not only the historic pattern of settlement but also the process of environmental change.

The Modern Environment of Hong Kong

Lying off the southern coast of China, the British colony of Hong Kong is situated at roughly 22° latitude and 114° longitude (fig. 14.1). To the north are the People's Republic of China (Guangdong Province) and the Asian landmass; to the south and east, the South China Sea; and to the immediate west, the Pearl River estuary, the gateway to Guangzhou [Canton].

When most people think of Hong Kong, they envision the bustling metropolitan areas of Hong Kong Island and the Kowloon Peninsula, cities in which some of the world's most dense concentrations of population are found. Geographically speaking, however, the region extends far beyond the modern urban settlement and is considerably more varied. It consists of 236 islands and a portion of the Chinese mainland which together amount to approximately 1,040 km^2 of land area (fig. 14.2). Hong Kong is administratively divided into two parts: the metropolitan areas of Hong Kong Island and Kowloon and the more rural areas, which are referred to as "the New Territories." This administrative division is in large part a reflection of the colony's historical development. Hong Kong Island and the Kowloon Peninsula were ceded in perpetuity to the British following hostilities during the early 1840s and late 1850s, while the New Territories consists of approximately 914 km^2 that was leased for 99 years to the British government by the Qing dynasty in 1898.

Hong Kong's landscape generally exhibits three topographic patterns: a rugged coastline with numerous indentations; a hilly, upland terrain of often steep gradient; and broad, alluvial floodplains that are primarily concentrated in the northwestern corner of the mainland portion of the New Territories.

Much of the region initially was created during periods of igneous geological activity that began some 150 to 200 million years ago (Allen and Stephens 1971, 89–90). This activity resulted in a predominance of volcanic and granitic formations that occur today throughout the uplands of the region (fig. 14.3). Both rock formations are extremely susceptible to weathering and erosion when exposed to the elements, so substantial upland erosion has taken place. Most upland soil profiles currently display shallow and/or truncated "A" horizons and in certains portions of the region a "badlands" landscape has evolved (Grant 1961, fig. VIIb). Through time, the lower portions of the area's landscape have had considerable alluvial and colluvial deposition as a consequence of extensive upland erosion (Allen and Stephens 1971, 81–83).

Hong Kong's position on the southern coast of the Asian landmass and the northern edge of the tropics gives the region a marked tropical/subtropical monsoon climate of alternating wet and dry seasons. The wet season is a consequence of warm and extremely humid winds that blow northward from the South China Sea from May through September. Relatively high temperatures, averaging 29°C in July, and substantial rainfall (approximately 1,840 mm) are recorded during this season. In contrast, the dry season, from October to April, witnesses a shift in the wind pattern that brings much drier and cooler winds southward from the Asian landmass. Temperatures during this season usually range between 10° and 16° C, and the precipitation is reduced to roughly 230 mm per season.

Despite Hong Kong's position on the northern edge of the tropics and the large amount of wet-season precipitation, the present vegetation is sparse. Vast portions of the region's uplands are today covered only with xerophytic grasses and shrubs, a pattern that is partially related to limited upland soil development (Hill et al. 1975, 3–4).

Because of the current pattern of vegetation and the intensity of modern settlements, the region's land fauna is limited. Present species are primarily those specifically adapted to grass and shrubland environments. Pests such as rats that thrive in densely populated areas occur as well (Marshall 1967; Lance 1976).

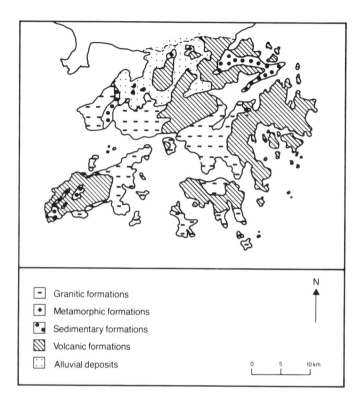

Fig. 14.3. The geology of Hong Kong.

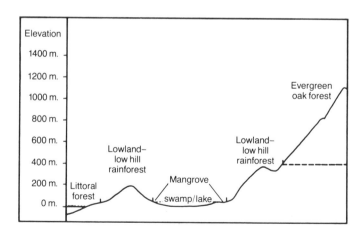

Fig. 14.4. The vegetation profile for the early environment of Hong Kong.

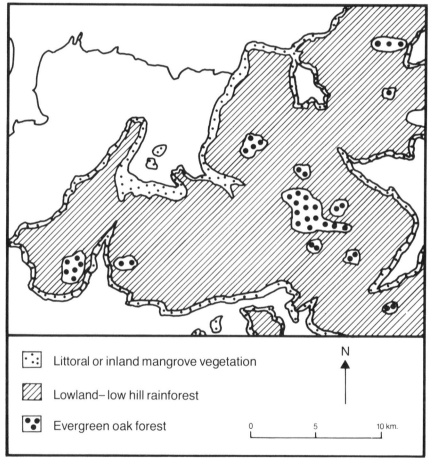

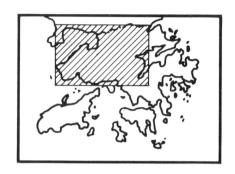

Littoral or inland mangrove vegetation

Lowland–low hill rainforest

Evergreen oak forest

N

0   5   10 km.

Fig.   14.5.   The probable distribution of early forest types in the
western mainland, New Territories.

The Previous Environment

In contrast to the denuded and eroded modern landscape, the region's natural environment appears to have been markedly different in earlier times. Analogies with other areas of similar latitude, solar radiation, precipitation, soil, topography, and other environmental characteristics have led botanists such as L. B. Thrower (1975) to argue that Hong Kong's natural climax vegetation should be closely related to a tropical and subtropical rainforest, grading from a littoral and mangrove vegetation along the coasts to an evergreen oak formation in the areas of highest elevation (figs. 14.4 and 14.5). With certain exceptions, this projected pattern of vegetation is not evident today; major areas of woodland that currently exist are primarily the result of government reforestation programs begun after World War II (Daley 1975). Certain evidence for the previous existence of an earlier tropical rain-forest pattern is provided, however, by remnant patches of tropical forest vegetation that often occur in association with traditional villages (S. L. Thrower 1975; Wang 1961) as well as by the remains of tropical forest species occasionally uncovered during archaeological and construction excavations (e.g., Bard 1976; Kendall 1976).

Although Hong Kong's natural fauna is currently limited, historical records and broad zoogeographic classifications, such as those by Wallace (1876) and Hsieh (1973, 55), suggest that the south coastal region of China once contained a large variety of tropical and subtropical fauna. The sources suggest the previous presence of many types of mammals including the barking deer (*Muntiacus reevesi*), the wild boar (*Sus scrofa*), a civet cat (*Paguma larvata*), and the South Chinese red fox (*Vulpes vulpes hoole*) (Marshall 1967; Lance 1976). Elephants (*Elephas maximus*) are recorded to have been present in the general region as late as the tenth century A.D. (Schafer 1954, 4–5; Balfour 1970).

Geological and pedological studies (Grant 1961, 1964) further indicate that, prior to deforestation and upland erosion, lowlands in the northwestern mainland New Territories were fewer in number and smaller in size. It seems that considerable portions of what are now inland or coastal lowlands were previously inundated by lacustrine, freshwater swamp, or tidal-marsh conditions (fig. 14.6). In coastal lowland areas such as those that border Deep Bay, an earlier tidal-marsh setting is suggested by the presence of saline

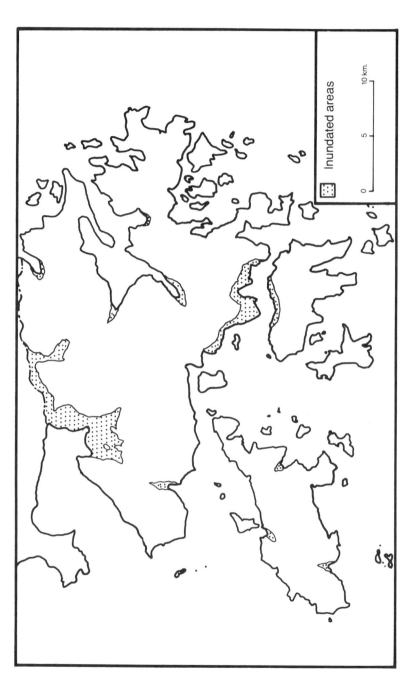

Fig. 14.6.   The probable extent of inundation in early Hong Kong.

characteristics and occasional shell inclusions in the modern alluvial soils. Previous inundation of certain inland, lowland areas, such as the Ng Tung Ho floodplain, is also suggested by the recurrence of dark, highly organic soils and substrata that are believed to have been formed during the drying and alluvial infilling of swamps and shallow lakes (Grant 1961, 54–81).

The probable presence of this earlier environmental pattern of forested uplands and inundated lowlands is reinforced by its basic similarity with early historic Chinese accounts of the south-coastal China region in general. These accounts, recorded at various times just prior to and throughout the first millennium A.D., commonly characterize early historic south-coastal China as an "untamed wilderness" of forested uplands and swampy lowlands inhabited by fierce "barbarian" tribes (Wiens 1967, 335). Although it was perceived as a region rich in natural resources, and Chinese traders ventured into it for trade and resource exploitation on a short-term basis (Li 1979), much of early historic south-coastal China was considered unfit for long-term settlement. Mention is also made of extreme malarial conditions during the wet and humid summer months that occasionally caused government officials to flee to other areas (Schafer 1969; Wiens 1967).

The Historic Pattern of Wet-Rice Agricultural Settlement

The emergence of the modern physical environment from this early pattern appears to have involved a combination of human and natural events that occurred after A.D. 1000. Despite the continued appearance of generally unfavorable comments on the region's environmental conditions, considerable numbers of Chinese wet-rice agriculturalists began to move southward into south-coastal China during and following the Song dynasty (A.D. 960–1126). As these agriculturalists settled the region, they undertook a number of activities that contributed to overall environmental change. Principal among these activities was deforestation. Combined with a geological structure highly susceptible to weathering and erosion, and with a tropical monsoon climate, the removal of vegetation appears to have resulted in extensive upland erosion and in the subsequent alluviation of the inundated lowlands through time.

A number of factors appear to have influenced the movement of Chinese settlers into south-coastal China. One of them involved the increasing pressure and military incursions of Mongol groups in the northern portions of the country. This caused a shift of Chinese political centers as well as populations southward from northern China to the politically more stable and agriculturally rich regions of the Yangzi [Yangtze] Basin in central China, a shift that was to lead to the establishment of a new imperial capital at Hangzhou in 1126 A.D. In combination with internal population growth, this population shift led to increasing regional population pressure in central China itself. The consequent need for additional land seems to have led, in turn, to a second phase of southward migration, this involving central Chinese rice farmers who moved into the southernmost regions of the country where lands, if not well suited to settlement, were at least available. Other contributing factors were agrotechnological in nature. During this general period, the Chinese technology of wet-rice agriculture underwent significant developments and improvements. These developments not only included the introduction of early ripening and drought-resistant strains of rice from Southeast Asia but, perhaps more important for the settlement of south-coastal China, increased capabilities in the hydraulic technology of wet-rice agriculture (Elvin 1973, 118–30). Greater sophistication in the techniques of water control permitted increased manipulation of the hydrology and landform of local environments and allowed the historic Chinese wet-rice agriculturalists of the early second millennium A.D. to expand into, modify, and bring under cultivation areas that previously had been avoided.

Such developments contributed to a reevaluation and a new conception of areas such as south-coastal China. Rather than being seen as environments unsuited to long-term settlement, these regions began to be perceived and depicted as important frontier areas that could be made suitable for lowland settlement. In the case of the historic settlement of the northwestern corner of Hong Kong's mainland New Territories, the pattern of settlement can be divided into three somewhat overlapping stages. The first stage of settlement began early in the second millennium A.D. (roughly A.D. 1000), when individuals or small groups of wet-rice farmers started to enter the region. Seeking conditions appropriate for both settlement and cultivation, and confronting the forested uplands and inundated lowlands, these groups settled in small valleys and along restricted drainages at the edges of the flooded lowlands. In occupying these locations, the settlers appear to have avoided the areas of inundation or extreme seasonal flooding and to have taken advantage of small streams that were ideally suited for the construction of irrigation systems by a small labor force. Although the next few centuries witnessed local population growth as a combined result of continued inmigration and the *in situ* growth of already established families, the individual settlements do not appear to have increased tremendously in size. Instead, a pattern of dispersed settlement is indicated in the genealogical records (see fig. 14.7). As the population of particular settlements at the edges of the inundated lowland increased, kin groups appear to have undergone fission (Baker 1968, 28–29; Sung 1973, 114), with new settlements being established in separate, yet environmentally similar, locations.

Although this continued pattern of settlement at the edges of the floodplains suggests that large portions of the lowlands were still unsuited for settlement, it is likely that initial environmental modification began early in this settlement process. To establish their settlements, the early settlers would have had to clear a certain amount of vegetation. Given their small numbers and a consequent need for non-labor-intensive methods of vegetation clearance, burning was probably employed. It is also possible that an initial phase of swidden cultivation was practiced while the paddy irrigation systems were under initial construction. Other impacts on the surrounding vegetation, such as the cutting of trees for building materials and fuel, should be envisioned as well. As a result of such activities, upland soil erosion and lowland deposition are likely to have increased.

The second stage of settlement had begun to occur by the late fourteenth century A.D., about the time most of the available locations at the edges of the floodplains were occupied or controlled by established groups. New groups entering the region at this time found that the best land had already been settled, and they were forced to occupy more marginal areas (Baker 1966, 27–31). The historical experience of the Man [Mandarin: Wen] lineage of San Tin, described by Watson (1975, 31–34), provides a good example. Due to unavailability of good land, early generations of Man settlers were forced to occupy brackish-water areas adjacent to the mouth of the Sham Chun River on the northeastern shore of Deep Bay. The Mans surmounted these adverse conditions by undertaking land reclamation and by cultivating a strain of rice that was suited to the brackish conditions of the site (fig. 14.8). The degree of population pressure and competition for remaining land resources is further indicated in Man lineage traditions suggesting that this brackish area was settled previously and that the Man settlers had to compete for its control. It is said that only after a bit of trickery associated with the folk belief-system of *feng shui* were the Mans able to gain control and dominate the San Tin area (Watson 1975, 19–20).

Although the Man lineage is well remembered for its land-reclamation activities, which were practiced into the twentieth century, there is no reason to assume that these methods were employed only by the Mans or only in marginal, brackish-water settings. In fact, cooperative land reclamations appear to have been relatively common in both river-valley and coastal settings elsewhere in lowland south-coastal China during much of the period (Elvin 1973, 124–25). The general pattern of increasing population pressure in the middle of the second millennium A.D. makes it likely that other groups living or arriving in the Hong Kong area at this time also would have resorted to this or a related technique as the number of available locations for new settlements and paddy fields declined (Lovelace 1983:157–61).

Although historical records of extensive land-reclamation activities by other groups are not available, circumstantial evidence is provided in the genealogical histories of several groups in which intensive, cooperative efforts in establishing new settlements in lowlying portions of the landscape are recorded. During the latter half of the fifteenth century A.D., for example, kinsmen of the Tang (Mandarin: Deng) lineage settled the lowlying area at the western end of the Kam Tin Valley. The valley's major streams converge in this area, and this convergence, in combination with soil characteristics that reflect a previous inland extension of Deep Bay, strongly suggest that inundation and extreme seasonal flooding presented obstacles to initial settlement.

Despite these obstacles, the Tang genealogies indicate that within approximately one generation (A.D. 1465–1487) six new villages were established by cooperating kinsmen in proximity in the lowest portions of the Kam Tin floodplain (fig. 14.9). The village of Kat Hing Wai, for example, was founded by three kinsmen, while Tai Hong Wai and Wing Lung Wai were founded by cooperating groups of five and eight kinsmen, respectively (Sung 1974, 168). The other three villages shown in figure 14.9 also are to have been cooperatively founded at roughly the same time (Sung 1974,

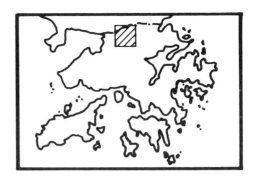

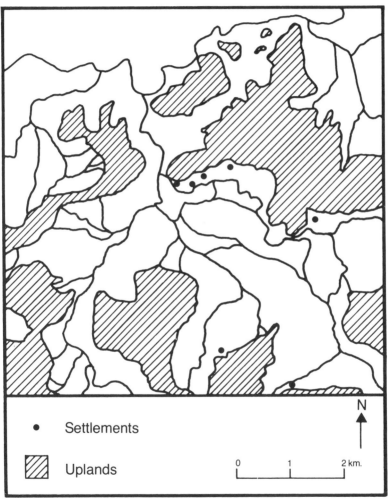

Fig. 14.7. Early Liu lineage settlements in the Ng Tung Ho Basin.

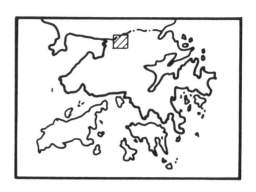

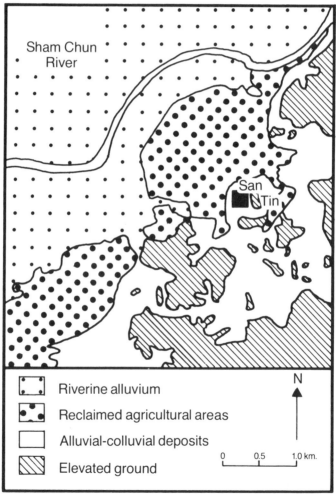

Sham Chun
River

San
Tin

Riverine alluvium

Reclaimed agricultural areas

Alluvial-colluvial deposits

Elevated ground

N

0   0.5   1.0 km.

Fig.   14.8.   The Man lineage village of San Tin.

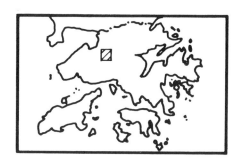

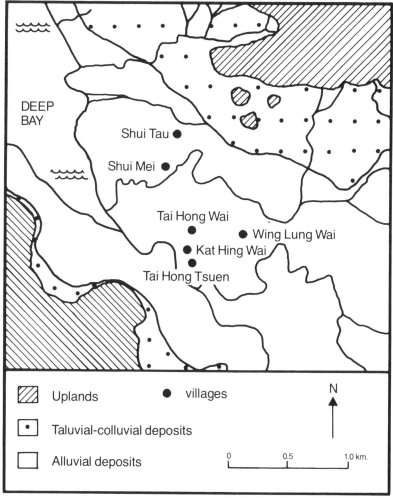

DEEP
BAY

Shui Tau ●

Shui Mei ●

Tai Hong Wai
●   ● Wing Lung Wai
● Kat Hing Wai
●
Tai Hong Tsuen

Uplands   ● villages

Taluvial-colluvial deposits   N

Alluvial deposits   0   0.5   1.0 km.

Fig. 14.9.   Tang lineage villages formed in the lower Kam Tin
              Valley during the late fifteenth century, A.D.

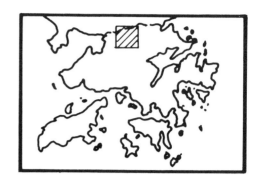

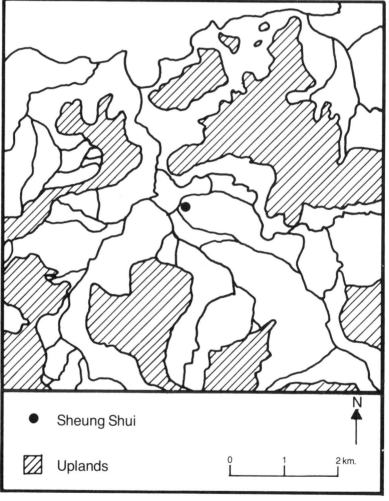

● Sheung Shui

▨ Uplands

N

0   1   2 km.

Fig.   14.10.   The large Liu Village at Sheung Shui.

169–70).

A similar pattern of cooperative settlement of lowlying areas is said to have taken place 10 km to the northeast in the inland Ng Tung Ho Basin where, over the course of seven or eight previous generations, the Liu (Mandarin: Liao) lineage had established a series of settlements at the edges of the floodplain (see fig. 14.7). By approximately A.D. 1600, when the seventh or eighth generations had risen to positions of authority in the lineage hierarchy, the scattered, lineage settlements were reunited as a single, much larger village at Sheung Shui in the middle of the floodplain (Baker 1968, 29; fig. 14.10).

The cooperative settlement of the lowlying portions of the landscape ushered in a third stage of wet-rice agricultural settlement in the Hong Kong region. This stage involved the appearance of a new type of settlement, the single-lineage community. Basically, this type of community consisted of a large group of males, all descended from a common ancestor, who lived together with their wives and unmarried daughters in one locale. The members of the patrilineage shared in the ownership of at least some property, which was usually in the form of ancestral lands held together in a trust. The community was theoretically under the leadership of the man most senior in generation and age (Baker 1979, 49–51).

Single-lineage communities, such as those of the Tang lineage at Kam Tin and the Liu lineage at Sheung Shui, were a rather common and important form of social organization and residence in late historic south-coastal China (Freedman 1958). In the Hong Kong region itself, strong single-lineage communities occupied and dominated vast portions of the agricultural lands in the northwestern portion of the mainland New Territories during late historic times (Baker 1966; Potter 1968).

The cooperative construction of irrigation systems has been put forward as a partial explanation for the appearance of single-lineage communities in south-coastal China (Freedman 1966, 159–61) and it is in relationship to that explanation that the preceding discussion of the changing pattern of historic settlement has particular bearing. These lowland communities do not seem to have occurred very early in the process of wet-rice settlement, as some authors (e.g., Lo 1966, 1968; Barnett 1967a, 1967b) have envisioned, but rather during a much later period in the context of growing population pressure, increasing competition for land, and the subsequent cooperative settlement of lower and previously inundated portions of the landscape.

Although cooperative construction of dams, dikes, and irrigation systems was of great importance to the settlement of these lowlying areas, the preceding stage of settlement at the edges of the floodplains also seems to have played a crucial role, since during this time the pattern of upland erosion and lowland deposition began to intensify. By the time kin groups had begun to reclaim and settle the lowlying floodplains cooperatively, it is likely that these areas already had begun to emerge as a consequence of alluvial deposition, thereby facilitating reclamation activities.

The Belief System of *Feng Shui*

Successful occupation and manipulation of the local environment for the purpose of wet-rice agricultural settlement were in part related to the adherence of the agricultural immigrants to a particular complex of folk beliefs known as *feng shui* (literally, "wind and water"), more commonly known in the Western world as "Chinese geomancy." *Feng shui* is basically concerned with landscape interpretation and the spatial positioning of villages, structures, and graves within the environment so as to encourage good luck and well-being for both inhabitants and descendants. As such, *feng shui* combines a number of ecological, social, and ideational considerations in the context of settlement behavior. The belief system was highly integrated into traditional life in China, particularly in southern China, to the extent that the rise and fall of personal, lineage, and village fortunes often were attributed directly to some aspect of *feng shui* lore.

Although varied interpretations of the belief system have been offered, ranging from "a farrago of absurdities" (de Groot 1897, 938) and "a pseudo-" or "proto-science" (Needham 1956, 346) to a primarily religious phenomenon (Meyer 1978, 151), the position taken here is that during the period of Hong Kong's settlement by Chinese wet-rice agriculturalists during the early second millenium A.D., *feng shui* provided a strategy for selecting, occupying, and modifying locales for wet-rice agricultural settlement. In linking this pattern of beliefs to wet-rice agriculture, I develop an interpretation similar to that of E. N. Anderson (1973). However, unlike Anderson, who argues that this belief system allowed Chinese agriculturalists to maintain a homeostatic relationship with their environment, I consider *feng shui* as an agent of environmental modification.

It should be noted that *feng shui* is a fairly large, complex, and continuously evolving body of cultural beliefs that has been diversely applied in traditional Chinese settlement-patterning since its first appearance more than 2,000 years ago. *Feng shui* beliefs have figured not only in the siting of individual structures and graves, but in the historic planning and construction of a number of major cities such as Beijing in north China. Given its varied applications and long history, *feng shui* has also displayed considerable temporal and regional variation with particular schools of emphasis arising in different portions of the country at different periods of time.

In this chapter I will focus primarily upon one such school of *feng shui*, that known as the "School of Forms" (*Xing*). The focus is appropriate because this school appears to have been dominant in much of south China during the general period in which the south-coastal region was settled by Chinese agriculturalists (March 1968, 261). It is further appropriate because the School of Forms is believed to have arisen initially in Jiangxi Province, a rich, rice-growing region in central China (see fig. 14.1) from which most of the previously discussed lineage groups historically derive.

The School of Forms involves an intricate interpretation and on occasion the manipulation of landform, vegetation, and hydrology to insure a villager's well-being and general success in life. Closer scrutiny suggests that the School of Forms also entailed an approach to environmental management that had a far-reaching effect on the historic agricultural settlement of south-coastal China. By examining some of the school's more important elements in the contexts of wet-rice agriculture and the south Chinese landscape, we are provided with some understanding of how settlers who migrated into the Hong Kong area might have occupied and modified the early environment.

Basic to *feng shui* lore is the general notion that certain locations within the landscape are more beneficial than others to the activities of human beings. Such places are said to accumulate great quantities of *qi* ("life breath") and to be protected from *sha* ("evil" or "noxious influences"). These beneficial locations can be identified by the presence of particular landscape features that are thought to encourage the concentration of *qi* and resist the influence of *sha*.

The flow of these benevolent and malevolent cosmic breaths are primarily controlled by the forces of wind (*feng*) and water (*shui*). It is important to ensure a slow but steady flow of *qi* into a site. If the wind is too strong, the *qi* will disperse or pass too quickly. If the wind is absent, the *qi* either will not come to the site or once there will stagnate. Slow meandering streams are important, as these are thought to encourage the penetration of *qi*. In contrast, evil influences are believed to travel in straight lines, so protective ridges and meandering lines that block or deflect the course of *sha* are advantageous. Locations containing gaps in ridgelines or straight water courses are to be avoided if possible because these landscape characteristics serve to channel and encourage the influence of malevolent forces.

According to *feng shui* belief, the optimal landscape for occupation is one of undulating terrain enclosed by mountains or ridgelines on three sides while remaining open in front (fig. 14.11). Within this landscape, there should be complementary and harmonious amounts of *yin* (manifested in this case by "dips") and *yang* ("raised features"). Enclosure of the terrain on three sides produces what Freedman (1966, 122) characterized as an "armchair effect." Ideally, in south China the armchair should face south, with the highest hills, which provide the site with strength, lying to the north. Flanking ridges on the east are referred to as the "azure dragon" or "green dragon" (*qing long*), while those on the western flank are called the "white tiger" (*bai hu*). The two sets of ridges should be complementary in the *yin-yang* fashion. Watercourses are often classified into "trunks" and "branches," the former referring to the main stream, the latter to its tributaries. It is believed that the more meandering branches a trunk possesses, the more potent the *qi*.

Within this topographic and hydrological configuration, the ideal location for human settlement, the *xue*, is situated on slightly raised ground at the back of the chair. In this position, the *xue* is enclosed by the landscape and surrounded in front by many stream branches. A site in front of the chair and near the stream trunk, on the other hand, is believed to be dangerous (Feuchtwang 1974, 130–31). The value attached to the armchair configuration is such that its basic form is often duplicated in the architecture of traditional rural houses and tombs in south China.

Importance also is attached to the pattern of vegetation. Although villages should be open in the front, with their views

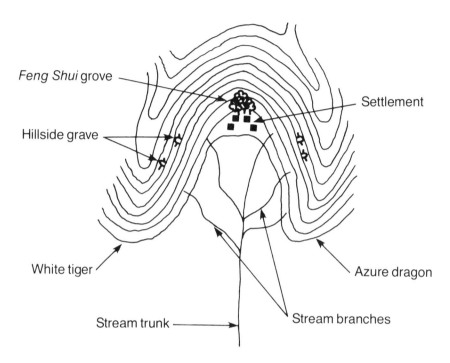

Fig.   14.11   The *feng shui* landscape in south China.

unrestricted, it is recommended that a grove of trees be maintained immediately behind the settlement (Anderson 1973, 131). The location and preservation of these groves were important elements of traditional village life, and today these groves constitute some of the few remaining examples of the natural climax vegetation of south-coastal China (S. L. Thrower 1975).

The siting of ancestral graves, the dwellings of the dead, followed similar principles. Located on the hillsides, the tomb structures resemble small armchairs. The burial urn is placed at the back of the structure in a position that corresponds to the *xue* of the valley (Baker 1965). As is the case with settlements, vegetation behind the graves is considered beneficial. Vegetation in front, on the other hand, is believed to be dangerous and should be removed. Where great concentrations of traditional graves were located, denudation and subsequent erosion of the hillsides usually occurred because vegetation considered beneficial for one grave would often be perceived as dangerous for other graves (de Groot 1897, 945) and therefore would be removed.

The Strategy of Wet-Rice Agriculture and Its Relationship to *Feng Shui*

In many ways the pattern of beliefs associated with *feng shui* is similar to the strategy of wet-rice agricultural settlement in lowland south China and Southeast Asia. Like believers in *feng shui*, lowland wet-rice farmers look for certain kinds of landscapes in which to establish their paddy irrigation systems—those that possess slightly uneven terrain, suitable quantities of nutrients, and substantial water resources (Geertz 1963, 33). The ideal topographic and hydrological configuration in which to establish a wet-rice farming system is " . . . a small but rounded-headed valley containing a small stream that gathers the drainage flow from the uplands above the valley head . . . since the construction of field units can take place below the break in contour which marks the valley head" (Spencer 1974, 60). Spencer's "small but rounded-headed valley" is essentially the same topographic pattern manifested in the "armchair." Both are small, lowland river valleys surrounded on three sides by uplands. Both contain drainage patterns in which the streams begin in the surrounding higher elevations and converge on the valley floor. Although Spencer's description mentions only a single stream, it seems clear that the presence of multiple streams ("branches") meandering across the valley floor would, if controlled, enhance the ease and potential for field expansion and overall production by increasing the number of possible taps, places for discharge, and terraces. The presence of numerous streams also serves to aid in the natural control of the drainage's overall water flow at times of intense runoff, such as during monsoon rainstorms, by dispersing and distributing the overall amount of water.

As in *feng shui*, water in the paddy system should not be allowed to flow too quickly or to stagnate. Instead, a gentle but steady flow is needed. This ensures a constant supply of nutrients that, parallelling the *qi* of *feng shui*, should be allowed to penetrate the soil gently. The gradual movement of water also provides for the conversion of nitrogen into ammonia, creating the neutral or alkaline conditions that make phosphorous available (Hanks 1972, 33).

The southern orientation prescribed for *feng shui* sites in south China appears to be of further importance in arranging effective drainage. The mountainous topography of much of the area is one of the key variables affecting the degree of precipitation (Chin 1971, 11). As the wet monsoon winds blow northward from the sea and cross the uplands surrounding these south facing valleys, the amount of precipitation increases under the orographic effect. Runoff and groundwater percolation feed the hillside streams, which then direct water toward the valley floor and the paddy fields.

Aside from its importance for drainage, the southern orientation recommended by *feng shui* has other advantages as well. Facing south, the armchair receives longer periods of sunlight (Feuchtwang 1974, 117), and the increased photoperiod usually results in higher crop yields (Purseglove 1972, 170–72). The southern orientation further results in the armchair's exposure to southerly breezes that provide relief from the extremely humid conditions of the summer months and protection from the cold north winds that blow from the Asian landmass during the winter months (Feuchtwang 1974, 117).

The locations recommended for settlement and burial by *feng shui* also contribute to the wet-rice pattern. The settlements' locations on slightly raised ground at the back of the armchair reduces the risk of settlement flooding during the wet season as well as the extent to which settlements occupy land usable for cultivation. Similarly, the hillside graves generally are on land that is unsuited to paddy cultivation (Anderson 1973, 130, 140).

Most scholars who have written about traditional systems of cultivation in Southeast and East Asia have commonly drawn a distinction between land-intensive wet-rice agriculture, which relies upon water, and land-extensive swidden cultivation, which employs burning. The employment of burning in rice cultivation has usually been viewed as restricted to either the initial clearing of vegetation so that paddy fields can be constructed or the burning of the adjacent forests to rid them of animals that are perceived as dangerous and predatory (e.g., Tuan 1969, 38). Despite this common characterization, there is evidence to suggest that the two subsistence techniques were not and are not always so distinct, and that it is not uncommon for rice farmers to use burning as a nutrient enriching technique. The earliest historical references to intensive wet-rice cultivators in central and southern China (Han dynasty, 206 B.C.–A.D. 220) suggest that burning was an important part of the early cultivation system (Eberhard 1968, 93; Bray 1980, 60). In the second century B.C., for example, the Grand Historian of China, Si Ma Qian, noted that, "the region of Ch'u and Yueh is broad and sparsely populated and the people live on rice and fish soups. They burn off the fields and flood them to kill the weeds" (*Shi Ji* [*juan* 129], translation by B. Watson [1961, 490]). Similarly, the second-century A.D. Annotator, Ying Shao, described the southern pattern of rice cultivation in the following manner: "They burn the grass, let in water, then sow the rice seed. Grass grows together with the rice. [As the rice] reaches seven or eight *ts'un* in height, cut down all [the grass]. Once again let in water to cover the grass. The grass dies and only the rice grows" (Translation by Hsu [1980, 301]).

References to the burning of preclimax vegetation for increasing the overall supply of nutrients in the paddy fields are also found in more recent times. Hanks (1972, 33), for example, has reported limited burning of this sort in Thailand. Within the past fifty years, the burning of hillside vegetation to enrich lowland fields also has been recorded in Guangdong Province to the north of Hong Kong (Stevenson and Lei 1937, 463). Even in modern Hong Kong, where hillside burning is illegal, it is occasionally practiced in more remote areas of the colony (personal observation). When the burning of hillside vegetation is done before the onset of the wet season, it encourages subsequent erosion and deposition of ash and topsoil on the fields below.

Although these activities would appear to have been conducted consciously, other practices associated with ancestor worship and grave visiting appear to contribute unconsciously to the burning of hillside vegetation. South Chinese villagers traditionally visited the hillside graves of their ancestors twice a year during the *Chong Yang* and *Qing Ming* festivals, which normally occur during the months of October and April. A visit during *Qing Ming*, which is sometimes referred to as "the grave sweeping festival," is particularly important and usually involves repairing and repainting the grave structure as well as the removal of undesirable vegetation in the structure's vicinity in line with *feng shui* considerations. Other important activities that take place during these visits include the burning of incense ("joss sticks") and ritual money and the exploding of large numbers of firecrackers. Both activities are associated with the maintenance of good luck and usually are done with little or no concern for their impact on the surrounding vegetation. Because these festivals occur during drier portions of the year, when the vegetation is more susceptible to burning, hill fires are not an uncommon consequence (Daley 1975, 51). Such fires are especially common during the *Qing Ming* festival in April when the wind pattern is beginning to shift toward the northerly direction of the summer monsoons. With the ensuing summer monsoon rains, the soil, ash, and nutrients of these burned areas are often subject to erosion and downslope deposition.

It might be further noted that such conscious and unconscious practices, which encourage and intensify upland erosion and lowland deposition, also have advantages both for the reclamation of inundated lowlying areas and for the construction of new fields, the former activity requiring the infilling and raising of submerged portions of the landscape, the latter requiring a buildup of clay silt for the purposes of water retention (Claus and Linter 1975, 211–18).

Although lowland deposition can be extremely beneficial for wet-rice agricultural settlement, there also are certain drawbacks if the eroded slope materials are deposited in the settlement rather than upon the fields. This realization provides a possible rationale for the importance attached to the location and preservation of the *feng shui* grove. By recommending the maintenance of a stand of climax vegetation immediately behind and upslope from the settlement, *feng shui* decreases the extent of erosion in areas that would directly affect the settlement, and it provides a barrier against the deposition resulting from erosion in areas further upslope.

It can be seen that a number of rather striking similarities and parallels between the interpretations and practices associated with *feng shui* and those appropriate for wet-rice agricultural settlement in South China are obvious. On the basis of these similarities and parallels, it is possible to conceive of the belief system as a highly codified strategy for managing certain aspects of the settlement-subsistence complex, landscape interpretation, and environmental modification.

That *feng shui* and wet-rice agriculture might be so deeply related is not unreasonable. Both patterns were integral aspects of village life in historic, lowland, South China. Wet-rice agriculture required a sophisticated interpretation of the local environment, a highly organized use of space, and a specialized technology. *Feng shui*, on the other hand, not only provided an appropriate environmental interpretation and instructions for the use of space, but it also encouraged a number of activities that contributed to the continued productivity and viability of lowland wet-rice agriculture. And, just as proper adherence to *feng shui* was believed to foster good luck and well-being for one's self, one's family, and one's descendants, wet-rice agriculture played an essential role in the success or failure of south Chinese villagers and their descendants. If the rice harvests were plentiful, the families, villages, and lineages prospered. Not only was there enough to eat, but surpluses of rice could be sold and wealth accumulated. In contrast, if the rice crop failed and famine was to be averted, supplies would have to be purchased elsewhere, and wealth diminished.

Feng Shui and the Historical Pattern of Settlement

Conceived in this manner, *feng shui*'s role in the settlement and infilling of the inundated lowlands in Hong Kong's northwestern New Territories may now be considered briefly.

The small valleys at the edges of the floodplains occupied during the first stage of settlement generally conform to the armchair landscape recommended by *feng shui*. Although not always facing south, the ideal direction, these small valleys are situated in line with the drainage configurations of the area. The valleys' locations at the edges of the flooded areas not only placed them above the major zones of flooding and malarial infestation, but also in proximity to small tributary streams that were ideal for the construction of irrigation systems by a limited labor force.

As these edge valleys were increasingly occupied through the growth and fission of established migrant families and the continued inmigration of new settlers over time, several developments took place. One was that the increased use of the small drainages that emptied into the inundated floodplain provided a measure of control over the overall water resources. Through control and manipulation of these water resources, both the extent of inundation and the threat of extreme seasonal flooding were reduced. Another development was the increasing depletion of the area's natural vegetation. As the local population grew and small settlements proliferated, substantial deforestation occurred as a result of the clearing of land for cultivation and settlement, the exploitation of vegetation for construction materials and fuel, the burning of vegetation to enrich lower-lying fields, and the removal of vegetation in the vicinity of hillside graves. As deforestation increased, the pattern of upland erosion and lowland alluviation intensified. This, in turn, led to the creation of new lowlands in areas that previously had been inundated. A final development was the growth of extended families, which, combined with increasing population pressure, provided both a larger labor force that could be organized along kinship lines and a need for additional land. The existence of this potential labor force appears to have allowed and encouraged established families to take advantage of this overall situation of increasing water control and lowland alluviation through cooperative land reclamation activities. As a consequence, agnatic groups appear to have been able to extend their fields and settlements onto floodplains.

Although the focus of settlement changed with the move from the small armchair valleys at the edges of the floodplains to the floodplains themselves, *feng shui* was not abandoned. In the case of the settlement of the lower Kam Tin Valley during the late fifteenth century A.D., the Tang lineage took special precautions, and, on the advice of *feng shui* specialists, built a temple to the west to insure the continued good luck and well-being of the lineage (Sung 1974, 170). *Feng shui* also seems to have been involved in the settlement of the lower portions of the inland Ng Tung Ho Basin where the genealogies and oral traditions of the Liu lineage credit the selection of the site and the founding of the new major village at Sheung Shui to two members of the lineage who were well trained in the intricacies of *feng shui* (Baker 1968, 29).

The end product of this pattern of inmigration and wet-rice agricultural settlement was thus a transformation of the early regional environment from a largely natural pattern of tropically forested uplands and inundated lowlands to a highly modified pattern of denuded and eroded hills and expansive lowlands that had been culturally structured for settlement and subsistence.

References

Allen, P. M., and E. A. Stephens
 1971 *Report on the Geological Survey of Hong Kong.* Hong Kong: Government Printer.
Anderson, E. N., Jr.
 1973 Feng Shui: ideology and ecology. In *Mountains and Water: Essays on the Cultural Ecology of South Coastal China,* edited by E. N. Anderson and M. Anderson. Taipei: The Orient Cultural Service. Pp. 127–46.
Baker, H. D. R.
 1965 Burial, geomancy, and ancestor worship. In *Aspects of Social Organization in the New Territories,* edited by M. Topley. Hong Kong: Hong Kong Branch of the Royal Asiatic Society. Pp. 36–39.
 1966 The five great clans of the New Territories. *Journal of the Hong Kong Branch of the Royal Asiatic Society* 6:25–47.
 1968 *Sheung Shui: A Chinese Lineage Village.* Stanford: Stanford University Press.
 1979 *Chinese Family and Kinship.* New York: Columbia University Press.
Balfour, S. F.
 1970 Hong Kong before the British. Reprinted in the *Journal of the Hong Kong Branch of the Royal Asiatic Society* 10:134–79. Originally published in *Tien Hsia Monthly* (1940).
Bard, S. M.
 1976 Chung Hom Wan. *Journal of the Hong Kong Archaeological Society* 6:9–25.
Barnett, K. M. A.
 1967a Hong Kong before the Chinese. *South China Morning Post* (24 April 1967), Hong Kong, p. 12.

1967b   Technological revolution in 900 A.D. *South China Morning Post* (25 April 1967), Hong Kong, p. 12.

Bray, F.
1980   Agricultural technology and agrarian change in Han China. *Early China* 5:3–13.

Chin Ping-cheung
1971   Rainfall in Hong Kong. Ph.D. dissertation, Department of Geography and Geology, University of Hong Kong.

Claus, P. I., and S. Linter
1975   The cultural ecology of a paddy tract. *Tools and Tillage* 11:211–18.

Daley, P.A.
1975   Man's influence on the vegetation of Hong Kong. In *The Vegetation of Hong Kong*, edited by L. B. Thrower. Hong Kong: Hong Kong Branch of the Royal Asiatic Society. Pp. 44–56.

de Groot, J. J. M.
1897   *The Religious System of China*, Volume III. Leiden, the Netherlands: E. J. Brill.

Eberhard, W.
1968   *Local Cultures of East and South China*. Leiden, the Netherlands: E. J. Brill.

Elvin, M.
1973   *The Pattern of the Chinese Past*. London: Eyre Methuen.

Feuchtwang, S. D. R.
1974   *An Anthropological Analysis of Chinese Geomancy*. Vientiane, Laos: Vithagna.

Freedman, M.
1958   *Lineage Organization in Southeastern China*. London School of Economics Monographs in Social Anthropology No. 18. London: Athlone Press.
1966   *Chinese Lineage and Society: Fukien and Kwangtung*. London School of Economics Monographs in Social Anthropology No. 33, London: Athlone Press.

Geertz, C.
1963   *Agricultural Involution: The Process of Ecological Change in Indonesia*. Berkeley: University of California Press.

Grant, C. J.
1961   *The Soils and Agriculture of Hong Kong*. Hong Kong: Government Printer.

1964 The extension of arable land in Hong Kong. In *Land Use Problems in Hong Kong*, edited by S. G. Davis. Hong Kong: Hong Kong University Press. Pp. 55–58.

Hanks, L. M.
1972 *Rice and Man: Agricultural Ecology in Southeast Asia.* Chicago: Aldine.

Hill, D. S., B. Gott, and B. Morton
1975 *Hong Kong Ecological Habitats.* Hong Kong: University of Hong Kong Press.

Hsieh Chiao-min
1973 *Atlas of China.* New York: McGraw-Hill.

Hsu Cho-yun
1980 *Han Agriculture: The Formation of Early Chinese Agrarian Economy.* Seattle: University of Washington Press.

Kendall, F.
1976 High island. *Journal of the Hong Kong Archaeological Society* 6:26–32.

Lance, V. A.
1976 The land vertebrates of Hong Kong. In *The Fauna of Hong Kong*, edited by B. Lofts. Hong Kong: Hong Kong Branch of the Royal Asiatic Society. Pp. 6–22.

Li Hui-lin
1979 *Nan-fang ts'ao-mu chuang: A Fourth Century Flora of Southeast Asia.* Hong Kong: The Chinese University Press.

Lo, C. P.
1966 Some Geographical Aspects of Demographic Change in the New Territories, Hong Kong 1911 to 1961. Master's thesis, Department of Geography and Geology, University of Hong Kong.

1968 Changing population distribution in the Hong Kong New Territories. *Annals of the Association of American Geographers* 58:272–84.

Lovelace, G. W.
1983 Man, Land, and Mind in Early Historic South Coastal China. Ph.D. dissertation, Department of Anthropology, University of Hawaii at Manoa.

March, A. L.
1968 An Appreciation of Chinese Geomancy. *Journal of Asian Studies* 27:253–67.

Marshall, P. M.
  1967   *Wild Mammals of Hong Kong.* Hong Kong: Oxford
          University Press.
Meyer, Jeffrey F.
  1978   *Feng Shui* of the Chinese city. *History of Religions*
          18:138–55.
Needham, J.
  1956   *Science and Civilization in China,* Volume 2. Cambridge:
          Cambridge University Press.
Potter, J. M.
  1968   *Capitalism and the Chinese Peasant: Social and Economic
          Change in a Hong Kong Village.* Berkeley: University of
          California Press.
Purseglove, J. W.
  1972   *Tropical Crops: Monocotyledons.* London: Longman
          Group.
Schafer, E. H.
  1954   *The Empire of Min.* Tokyo: C. E. Tuttle.
  1969   *Shore of Pearls.* Berkeley: University of California Press.
Spencer, J. E.
  1974   Water control in terraced rice field agriculture in
          Southeastern China. In *Irrigation's Impact on Society,*
          edited by T. E. Downing and M. Gibson. Anthropological
          Papers of the University of Arizona, No. 25. Tucson:
          University of Arizona Press. Pp. 59–65.
Stevenson, D. C., and Lei, P. F.
  1937   Some effects of fire on the vegetation of the Loh Fau
          region of Kwantung. *Lingnan Science Journal* 16:463–65.
Sung Hok-p'ang
  1973   Legends and stories of the New Territories: Kam T'in
          (sections 1–3). Reprinted in the *Journal of the Hong Kong
          Branch of the Royal Asiatic Society* 13:111–29. Originally
          published in the *Hong Kong Naturalist* (1935–36).
  1974   Legends and stories of the New Territories: Kam T'in
          (Sections 4–6). Reprinted in the *Journal of the Hong Kong
          Branch of the Royal Asiatic Society* 14:160–85. Originally
          published in the *Hong Kong Naturalist* (1936).
Thrower, L. B.
  1975   The Vegetation of Hong Kong. In *The Vegetation of Hong
          Kong,* edited by L. B. Thrower. Hong Kong: Hong Kong

Branch of the Royal Asiatic Society. Pp. 21–43.

Thrower, S. L.
1975 Floristics of the Fung Shui wood. In *The Vegetation of Hong Kong*, edited by L. B. Thrower. Hong Kong: Hong Kong Branch of the Royal Asiatic Society. Pp. 57–63.

Tuan Yi-fu
1969 *China*. Chicago: Aldine.

Wallace, A. R.
1876 *The Geographical Distribution of Animals*. London: Macmillan.

Wang Chi-wu
1961 *The Forests of China*. Maria Moors Cabot Foundation, Publication No. 5. Cambridge, MA: Harvard University Press.

Watson, B.
1961 *Records of the Grand Historian of China: Translations from the Shi Chi of Ssu-ma Chien*, Volume 2. New York: Columbia University Press.

Watson, J. L.
1975 *Emigration and the Chinese Lineage*. Berkeley: University of California Press.

Wiens, H. J.
1967 *Han Chinese Expansion in South China*. Hamden, CT: The Shoe String Press. Originally published under the title of *China's March Toward the Tropics* (1954), New Haven: Yale University Press.

III.   COMMENTARIES

CHAPTER 15

DEVELOPMENT ISSUES

Gerald C. Hickey

The term "development," like the word "culture," has over time elicited a vast array of definitions, most of which include the desired results of the development process. Putting goals aside, development, one way or the other, involves socioeconomic change that is planned consciously. Any paradigm that purports to reflect the reality of national, regional, or global development must be based on an appreciation of planned change in human history. The chapters in this volume speak to the theme of cultural values and human ecology, and in doing so tell us a great deal about how people plan their adaptation to the physical environment. The authors also remind us that in a setting of endless sociocultural and environmental fluctuations, repeated readaptations are inevitable. This message gets to the heart of developmental change, and with the world in need of a paradigm appropriate for the tasks of efficacious and sustainable development, it is a message worth heeding.

Development and Modernization

In recent years there has been a steadily increasing disillusionment with the vision of development and modernization that has been dominant since the end of World War II. Deeply rooted in the eighteenth-century Enlightenment that gave rise to the idea of progress, this vision saw humankind moving irreversibly in the direction of realizing its full potential: inductive reasoning would increase knowledge of the natural order, enabling modification of customs and institutions so as to eliminate social maladies and create a bright new society. The development/modernization vision was further shaped by the theories of economic progress formulated by

Adam Smith, David Ricardo, Thomas Malthus, John Stuart Mill, and Karl Marx.

Following World War II, the vision fueled efforts to reconstruct the ravaged countries of Europe and Asia. It was the guiding spirit of the movement to bring to the poorer nations of the world, notably those that formerly had been colonies, the breakthrough to modernization that had been attained in the West since the Enlightenment, and thereby close the gap between the rich nations and the poor, declared by Barbara Ward (1962, 36) to be "the most tragic and urgent problem of our day."

Out of this came development/modernization as a distinct field of study, and, since it stimulated lively interest in all the social sciences, the criss-cross of interdisciplinary fertilization resulted in numerous and varied approaches. Understandably, economics played a leading role, and Hainsworth (1982, 3–5) points out that orthodox neoclassical economists and orthodox Marxist economists have "both tended to look with some disdain or disinterest at indigenous agriculture and peasant economies which were expected to have diminishing importance relative to the modern urban-industrial sector, both as a source of output and as a means of employment." Given the economists' lack of experience with the rural economies of what are now called Third World countries, it is not surprising that the "dual economy" concept of Boeke (1942, 1953), based on his Indonesian research, would become influential. It held that there are two economic sectors within one political framework. One sector (originally identified with colonialism) operates according to the principles of modern capitalism. Opposed to this is the traditional peasant economy, conservatively oriented, and concerned with security and continuity rather than change. Cyril Belshaw (1965, 95) notes that for Boeke, "peasant economies were so vastly different in their operations that economic theory could not be applied to them, that they constituted an archaic sector in dual economies, and that they were most unlikely to respond to attempts to transform them on the basis of modern suppositions."

Belshaw (1965, 96–97) criticizes Boeke's model as "too simplistic and easily demolished," but he adds that in 1965 it was still current and "represented, in more subtle forms, in most theories of economic development." Hainsworth (1982, 8) writes that the model was considered "patronizing and gratuitous in the flush of confident expectations accompanying national independence," but that the ideal of duality lived on in economic modelling: "The dichotomy was still between traditional and modern, but was more often formulated in terms of agriculture vs. industry, rural vs. urban, inward- vs. outward-looking strategies, balanced vs. unbalanced growth." Hainsworth traces the emergence of a "conflict of paradigms, between an evolutionary development, conceived of as an attempt to enhance and build upon what the country had inherited as its indigenous socio-economic and cultural structure, and an accelerated modernization, based upon catching up and more closely emulating the MDC (More Developed Country) societies and building increasing interdependencies with the world market"(1982, 8).

During this period (which in many respects was the first phase of postwar development thinking, ending around 1965) anthropological works on socioeconomic change supported the former paradigm with their emphasis on transformations in traditional societies in a setting of modernization. Foster described how his book, *Traditional Cultures*, "attempts to place the phenomenon of planned technological change in the broader perspective of the underlying processes of culture change which occur at all times in all parts of the world" (1962, 7). In *Agricultural Involution*, Geertz (1963) related political, economic, social, and cultural change to ecological transformations within the context of Indonesian history, thereby providing a comprehensive framework in which to project the development of that nation at that particular time. Goodenough's *Cooperation in Change* (1963) emphasized the need for a community's cooperation in bringing about any kind of developmental change. His treatment of customs and values and how they change is still a basic message to development thinkers who regard customs and traditions as static barriers to development.

In retrospect, works such as these marked a high point in anthropological contributions to development thinking because nothing comparable appeared in the phase that began around 1965. Ironically, in his pioneering book, *Transforming Traditional Agriculture*, Schultz (1964) took agricultural economists to task for neglecting traditional agriculture, leaving the subject to anthropologists, some of whose work he found useful. He observed that, "meanwhile, growth economists have been producing an abundant crop of macro-models that are, with few exceptions, neither relevant in theorizing about the growth potentials of agriculture nor useful in examining the empirical behavior of agriculture as a source of growth" (1964, 6). Anthropologists familiar with traditional agriculture, its socioeconomic organization, and changes taking place within it had a contribution to make to this subject, but in the new phase of development, the dominant paradigm emphasized urbanization, industrialization, macro-capital-intensive technology, and central planning. There was little interest in the agricultural innovations that anthropologists were finding in the villages.

Even with the Green Revolution this situation has persisted. Hainsworth (1982, 14) observed that "the urban bias inherent in virtually all experience with economic development has not been repealed by virtue of paying greater attention to agriculture. The Green Revolution has provided a means of accelerating village-level modernization better suited to the international firms and agencies concerned to promote modernization in LDCs (Less Developed Countries), and it has had a mixed bag of effects for local residents."

Misgivings about the notion of progress underlying the predominant paradigm began as early as 1962 when Carson published her *Silent Spring* (1962), which warned against the destructive effects of modern technology. By the early 1970s, the Club of Rome and other organizations and conferences began expressing deep concern about the limits of growth. British economist E. F. Schumacher's work, *Small Is Beautiful* (1973), expressed doubts about modernity and advocated an "intermediate technology" for the "non-modern" sectors of developing countries. World events such as the 1973 war in the Mideast and the subsequent rising power of OPEC caused deadly problems in the developing world and created a generally uneasy atmosphere in which mounting criticism of the dominant paradigm was forthcoming.

At a 1975 conference held at the East-West Center to discuss the use of communication in economic and social development, the participants reviewed the effects of development since 1964 and found mixed results (Schramm 1976). While there had been "spectacular gains" in communication in the developing world, the net economic and social gains had been unevenly distributed between the already privileged and the underprivileged, thereby creating a widening gap between the rich and the poor. Also, the LDCs were becoming poorer in relation to the MDCs. The conference concluded that "the economic development in the last ten years was less than had been hoped for."

Some of the participants addressed the development paradigm more directly. Harry Oshima pointed out that since 1964 there had been "vast changes" in development economics and called for emphasis on a labor-intensive rather than a capital-intensive strategy and greater focus on the agricultural sector than on the industrial. S. N. Eisenstadt declared that the "old paradigm" of development had broken down, and, while he would retain the basic economic assumptions, he felt the new model should emphasize the mobilization of human and social as well as economic resources. He also added that this model must take into greater account the cultural base of change and the change in the cultural base.

Late in 1982, at a Social Science Research Council Area Assembly, Francis Sutton (1982, 53–57), a deputy vice-president of the Ford Foundation, declared that "we are now clearly in a new era when the old development ideology has been eroded." He characterized this ideology as "egalitarian, both in the sense that it proclaimed the right of self-determination of peoples throughout the world and in that it viewed inequalities within and between nations as challenges or problems to be met with deliberate effort towards their removal or mitigation" (Sutton 1982, 50). This egalitarianism required theories that were universally applicable. It also led in practice to a host of general questions about the nature and conditions of economic growth, the possibilities of "take-off," the meaning of modernization, and the concept of human capital. But now, in what he called "the chastened world of the 1980s," the signs of erosion are with us. "The bitterness and intractability of North-South differences are symptoms of a lost consensus," Sutton observed (1982, 52). He added that "the faltering of economic growth in the OECD countries, the rise of unemployment and inflation to levels not known for a generation, the decay of the Bretton Woods framework, debt crises, and a depressed sense that still worse may come in the 1980s preoccupy us" (Sutton 1982, 53). The new ideology, as Sutton envisaged it, would include a great deal more international scholarship in which the humanities and social sciences would converge to give us a "new and comprehensive view of the world."

More recently, in March 1983, a group of development scholars advocated a departure from the "old paradigm" in giving indigenous culture an important role as mediator in introducing new technologies. They observe that:

> The old paradigm suggests a trickle-down, centralized way of planning. The newer concepts suggest bottom-up users initiated changes. The mediation approach respects the needs, knowledge, attitudes and beliefs of the users; it also suggests a much more active role for the people and planners of the Third World than what was suggested by the old paradigm. However, it does not suggest that all the changes should be initiated by users because in many cases, the users may not be aware of, or feel the need for an alternative to their present ways. Changes, therefore, may best be brought about through communication between the planners and the consumers. Only when development is the outcome of interactions, will an appropriate growth of technology take place. (Wang et al. 1983, 22)

All of these views signal an increasing awareness that treatment of differences between traditional and modern, rural and urban, labor-intensive and capital-intensive, and agriculture and industry as dichotomous and mutually exclusive is false and harmful to development thinking. These differences call to mind Redfield's (1960) concepts of folk and urban, or little tradition and great tradition, which he describes as "pairs of conceptions" that provide us with two lenses with which to view the same phenomenon—society in flux. As such, traditional and modern are conceptions that contrast but are intrinsically interrelated. As Redfield (1960, 146) observes, "in every isolated little community there is civilization; in every city there is the folk society."

This dichotomization has had baleful effects on the conceptualization and implementation of development in Asian developing countries. First, it has nurtured the mistaken notion of many decision-makers and planners in these countries and in international donor agencies that tradition is stagnant and a barrier to "progress" (as defined by the Western vision). This in turn has prevented these same people from seeing how traditional values might transform Western-derived concepts of progress and development into concepts better suited to indigenous sociocultural realities, as has been the case in "successes" such as Japan, South Korea, and Taiwan. The urban prejudice also has prevented any real appreciation of the widespread developmental and socioeconomic changes taking place in the rural areas. Hainsworth (1982, 21) notes that "very often the thinking at the two levels—macromodelling and village-level project implementation and problem solving—seem completely disconnected, operating in almost exclusive spheres of activity, involving separate and incongruent paradigms, policies, and visions of the possible."

One result is that, all too often, national- and international-level planning to improve the well-being of human populations in the name of development has occurred predominantly on the "macro" level— employing massive resources and aggregative data with little systemic input from the nonmetropolitan levels of the society. This has led to widespread disjunctures that are costly to the rural populations and are barriers to successful implementation of many projects. These disjunctures also are damaging to the physical environment wherein so many of the natural resources are found.

The Anthropological Perspective

The time has come for development thinking to include more of the anthropological perspective emphasizing cultural values and change. A good place to start is at the primal level of the relationship of cultural values to the physical environment, the subject of the conference upon which this volume is based. Karl Hutterer sets the tone for this topic in his observation that "environment and culture are *intrinsically* and deeply interrelated, and . . . social and environmental processes affect each other in fundamental ways. Neither can be adequately explained without the other, although neither can be explained by reference to the other alone."

Neil Jamieson and George Lovelace approach this topic in chapter 2 with the age-old question, "what is the relationship between ideas and the material conditions of existence?" In examining the largely anthropological literature on this they find a dualistic mind-versus-matter controversy, which they characterize as "ideational" and "materialist." In decrying this dichotomy, they call for a realization that both positions contain some validity in devising a new model for the complex systems with which we are dealing, particularly "when dynamic processes remain poorly understood." Then they touch upon something basic to development as well as to anthropology when they recommend that,

> we need to document and ponder particular instances
> of change in ecological, ideational, and social systems
> with the closest possible attention directed toward the
> specific mechanisms of feedback, decision making,
> and readjustment by which change is either induced
> within any subsystem and then either spreads
> throughout the larger system, inducing further
> change, or is dampened or extinguished by some
> homeostatic process.

Change is the essence of development, and change is very much a primary concern in the chapters of this book that tell us ecology and society interact in a setting of constant flux. Embedded in the society's cultural values are definitions of the ecosystem and strategies for using it. Both are manifest in the society's adaptation to the ecosystem and also in the society's planned change when a readaptation is considered necessary.

Anthropologists should remind development thinkers that a tradition is a record of past changes that members of a society deem important to remember. As such, it is the storehouse for a wide range of very useful social, economic, and ecological information accumulated through countless generations. Aram Yengoyan tells us in chapter 6 how this is done in his discussion of mnemonics, the use of memory for the storage and retrieval of cultural knowledge. He specifies that mnemonic structures are not simply a linear register of spatial and temporal events but must be interpreted as a means by which all cultural information is maintained and transmitted in a society. Legends, myths, history, and ritualized speech are cultural codes of the past that need to be brought into the present as an expression of cultural coherence. Among the Mandaya, mnemonic devices are related to plants (both cultigens and natural growth) or to localities in which significant cultural events have taken place. Changes such as the movement of families from one settlement to another or the identity of a particular swidden planter are linked to cultigens that last for many years (such as durian, mango, and banana trees). Thus, socioeconomic information is preserved and can be retrieved for present reference or used in planning. Historical memory warns against planting a new swidden in the vicinity of a banyan tree associated with evil spirits. Memories of past warfare dictate which clearings and localities are to be avoided as semisacred sites not suitable for farming.

In chapter 14, George Lovelace delves into the early historical settlement of what is now Hong Kong to illustrate how *feng shui* (or Chinese geomancy) provided the settlers with prescriptions for attaining and maintaining harmony with the cosmos through alterations of the natural environment that were at the same time adaptive to wet-rice farming. Here we have an instance of developmental change wherein *feng shui* values guided the planning of land use and settlement patterns. Lovelace also illustrates how developmental readaptations are incessant because of inherent fluctuations in human society and in nature. For the early population, the adaptation was successful and the farm communities thrived. But this in itself led to population increase that strained the ecosystem:

> As the local population grew and small settlements proliferated, substantial deforestation occurred as a result of the clearing of land for cultivation and settlement, the exploitation of vegetation for construction materials and fuel, the burning of vegetation to enrich lower-lying fields, and the removal of vegetation in the vicinity of hillside graves. As deforestation increased, the pattern of upland erosion and lowland alluviation intensified. This, in turn, led to the creation of new lowlands in areas that previously had been inundated.

The planned response was for kin groups to cooperate in land reclamation and for adoption of a variety of paddy suited to brackish-water conditions.

Planned change also was called for when the population expanded to the point where it was not possible to locate settlements in sites adjudged favorable according to *feng shui* precepts, notably small "armchair valleys," forcing selection of less auspicious places such as the floodplains. This involved a risk in that in foregoing the *feng shui* ideal one might lose harmony with the cosmos. Planning therefore included consultation with geomantic experts versed in remedial measures (e.g., construction of a temple on a determined site) to ensure continued good luck.

Pei Sheng-ji's paper (chapter 13) on the Dai (T'ai) people of Yunnan Province also deals with the varying effects of ideological precepts on developmental change, notably on planning and on diffusion of innovation. The ideology affecting the use of the natural environment is a syncretic combination of traditional "polytheistic" beliefs, which appear to be rooted in an ancient animism, and the tenets of Hinayana Buddhism. This ideology prescribes a land-use pattern that places the Dai settlement in the context of a larger agricultural ecosystem comprising paddy fields, home gardens, artificial fuelwood forests, and a naturally forested holy hill. Harmony with the natural environment is vital to attain good fortune and avoid ill fortune, and, while the prescribed land-use pattern provides the basis for this harmony, it is maintained through observation of taboos and sanctions. The holy hill, for example, is sacrosanct, so that flora and fauna must under no circumstances be disturbed. This hill also is the site for rituals honoring the pantheon of dieties whose propitiation will help maintain harmony.

In pursuing his primary interest in botany, Pei investigated the origin of nonendemic plants in the Xishuangbanna (Sip Song Panna) region of Yunnan and concluded that the intrusive ritual and ornamental plants were one result of the spread of Hinayana Buddhism from India and Southeast Asia. From the viewpoint of development, this is a case of innovation diffusing by virtue of religious ideology. Pei points out that Hinayana Buddhist canons related to the founding of a temple call for specified "temple yard plants," and he found that some ornamental plants and all ritual plants introduced from the outside were located exclusively in the temple yards. "The presence of nonendemic plant species from Southeast Asia and India, as well as from the more eastern portions of China, suggests the possibility that the plants were introduced to Xishuangbanna from both the south-southwest and the east." Since recent scholarly opinion (expressed in the October 1982 Academic Symposium held in China) held that Hinayana Buddhism was introduced to the Dai from India via Thailand or from central or northeastern Burma, Pei speculates that the agents of diffusion might well have been monks travelling along a little-known early land route, the nature of which should be revealed with continued historical research.

June Prill Brett in chapter 5 takes us to the heart of a situation where the Tukukan Bontok have successfully sustained an adaptation to the ecological constraints of mountainous northern Luzon for a relatively long period without major readjustments. This adaptation is based on an elaborate system of effectively irrigated rice terraces, in which, despite rigid social stratification (with an identifiable elite), all members participate beneficially. Prill Brett focuses on the equitable distribution of an essential resource—water—attained through deeply rooted cultural values manifest in interlocking social, religious, and economic institutions centered on agriculture.

Equity is linked to a "subsistence ethic," which "disallows any member of the village from depriving another member of resources that ensure—and indeed are required for—survival." This ethic is validated by Bontok mythology (functioning as Yengoyan's mnemonic device) which recalls the origin of the rice terraces and irrigation channels built by labor expended by their ancestors. It provides the basis for the equal right of all members of the community, regardless of status, to subsist. While this does not translate into an equitable distribution of land, it does mean that everyone has an equal right to water, a principle that is woven through religious ideology and socioeconomic practices and is supported by sets of taboos and sanctions.

Prill Brett notes only one innovation: a new law, approved by members of the community, that imposes a fine on anyone proved to have attempted to deprive another of water rights. She recognizes that change in the adaptation is inherent. "As in the example presented in the text, new rules that govern ecological behavior between or within social groups will be devised in accordance with changes in resource availability—changes that in turn prompt alterations in human behavior and belief over time."

Chavivun Prachuabmoh (chapter 11) presents us with a situation where socioeconomic changes in the existing agricultural adaptation have arisen, but constraints, notably those preventing effective improvement in water management, have forced members of the community to leave farming and take up trading. In her Thai-Muslim community, population increase rendered incomes from traditional paddy farming inadequate, but increasing agricultural yields was inhibited by several factors. Family landholdings were small and usually dispersed, sometimes located within the boundaries of neighboring villages. This land-tenure situation contributed to the failure of attempts to improve water management, as did the collapse of a water users' association (introduced by outside government agents). Prachuabmoh sums up the situation in her observation that "the population under discussion here cannot satisfy its needs by exploiting resources provided by the local natural environment alone . . ."

For some members of the community, the adaptation strategy was to engage in full-time trading. Prachuabmoh points out that these villagers retained their value for land ownership along with other values related to reciprocity and the maintenance of close social bonds with kin and neighbors. They also kept their Islamic practices and beliefs, which they felt sanctioned trade as an economic endeavor (although Prachuabmoh does not regard this sanction as the primary reason for deciding to go into trade). Other basic values, however, underwent change. Skills needed for using the physical environment gave way to new skills regarding money, profit, use of time, and personal relationships geared to the market world of trade. Prachuabmoh notes that with market values at variance with the traditional village values concerning socioeconomic relationships, those who elected to trade did so outside the community.

These chapters have taken us from the highly rural world of early Chinese settlements in Hong Kong to a community in southern Thailand where rural and urban influences blend. This continuum succinctly traces the evolution of socioeconomic development from the past when adaptation involved subsistence farming geared to the natural surroundings, to the present, when adaptation invariably involves the market. The market, however, has not been a primary focus for most anthropologists, who have left it to the skills of the economists. Alice Dewey (chapter 7) is an exception, and after many years of studying peasant markets in Java, she finds herself questioning "the usual Western economic view that they consist of small individualistic competitive units perennially short of capital and unable to mount cooperative efforts that would enlarge the scale of operations."

Dewey found competition among the small, self-sufficient firms in the market, but she also noted that "the barrier of competition becomes a tie of cooperation, shifting the competitive boundary outward so that opposed units fall within a larger grouping." When a group of small traders saw the need, they lowered their boundaries, pooled their resources, and functioned for a period as one unit, after which, up went the boundaries and competition was restored. Here we have planned, socioeconomic, structural change adapted to the changing needs of the market (which, as Dewey points out, may be related to such things as the supply-and-demand fluctuations of the agricultural cycle or to specialization). Clearly the adaptation is rooted in Javanese cultural values having to do with altering "focus/boundary relationships," which are intriguingly manifest in batik patterns, classical gamelan music, dance dramas, and shadow plays. It is possible, nonetheless, that values for sharing and pooling, so common in Asian villages (particularly among kin and neighbors), may be more a part of market-town culture than has hitherto been thought. The market town is, after all, in the middle ground, and we have yet to sort the village values from the urban values in the blend.

In her discussion of the focus/boundary relationship, Dewey points out that the Western frame of reference would lead one to see ambiguity in the batik patterns, whereas in the Javanese approach a greater complexity of boundary relationships is possible, so that shifting identities are not seen as ambiguity. "Theirs is a view that allows the world to keep more of its holistic integrity. Things are seen in context with first one element or pattern highlighted, then another. Foreground and background interchange and boundaries shift so that elements bond or break up and interdefine each other."

Otto Soemarwoto (chapter 9), drawing on his rich experience with development projects in Indonesia, presents a holistic version of the "sociobiophysical ecosystem," with its background of change and its foreground of constancy, moving the focus so that relationships within boundaries say varying things about cultural values and the ecology. One result is some unusual insights on the effect of values on planned change, or lack of planned change on ecological adaptation.

Soemarwoto focuses first on wet rice and the environment. The traditional pattern—wet rice farmed during the rainy season followed by fallowing or secondary crops in the dry season—was a suitable adaptation (constancy). Rising population (change) required additional food production, and the response by both central planners and farmers was guided by the strong and persisting social value placed upon rice (constancy). The government's planned adaptation was influenced not only by this traditional value but also by the modernist value for macro, capital-intensive technology, which in this case took the form of vast water-management projects and the introduction of Green Revolution "miracle" rice, using not only persuasion but also coercion (change). The farmers' planned adaptation was to replace the traditional dry-season crops with rice (change). Within the boundaries of the focus, the various elements have interrelated so that the social value of rice is strengthened (constancy), while the nation has become more dependent on one staple food, rice, so that "ecologically it has become more vulnerable" (change). Rising ecological problems such as soil erosion and increase in pests raise questions concerning the sustainability of the increased yields. Within the boundaries of the focus, the social value of rice, which previously had been an asset, has become a liability (change).

Soemarwoto then turns to home gardens and another example of interplay between constancy and change. The traditional Javanese home garden reflected accumulated knowledge (change) that led to an ecologically effective adaptation with layering that provided the right light for various plants, a rich genetic diversity, a suitable microclimate, and protection against soil erosion (constancy). The garden served social needs (playground, gathering place, ornamental site) as well as economic needs (source of cash and noncash income). Focusing on these aspects of the home garden, Soemarwoto sees an element of change in increasing urban influences in Java, and this in turn produces some changes in traditional values, with many gardens now being planned for commercial production (fruit groves) and some gardens near the urban centers being planned primarily for aesthetic functions. These changes have altered subsistence and social functions, and they have wrought such negative ecological effects as the lessening of layered structuring, of genetic diversity, and of soil conservation.

Soemarwoto then turns his focus to another aspect of home gardens—fishponds—and to constancy in cultural values against a background of ecological change. With latrines built over them, the fishponds were ecologically efficacious in the local plant-animal-human food chain (constancy). As population density increased (change), the more rapid use of fishpond water prevented natural purification from occurring, so the latrines in effect became a source of pollution (change), and one result has been increased outbreaks of cholera (change).

With his holistic view of constancy and change, Soemarwoto sees these focused sets of relationships in the broad spectrum of modernization with its burgeoning of urban centers and its population increase. Within this, everything is related—urban and rural, modern and traditional—in a setting of change.

> Sociobiophysical ecosystems, even so-called tradi- tional ones, should therefore not be considered as homeostatic, but rather dynamic. They are in constant flux, sometimes slow and sometimes rapid, sometimes leading to successful new adaptations, but not seldom also to maladaptations. In a world of changing environmental conditions, changes are necessary for the sake of survival, although there is never an assurance of survival in the long term. To remain constant, however, would assure extinction.

Working with hunting-and-gathering and swidden-farming peo- ples, Geoffrey Benjamin (chapter 10) takes a longer-term view of ad- aptational change. "It should be noted, however, that people every- where can at best only guess at the environmental consequences of their actions, for those consequences usually take far too long to manifest themselves." This to him suggests that "throughout most of history, people's actions with regard to the environment have usually been aimed at holding on to what they already think of as the advantages of their current way of life, rather than at changing their way of life so as to produce some hoped for, but as yet unexperienced, advantage."

In his oral presentation Benjamin touched upon the varying notions of change that are basic to the modern-traditional dualism: "Indeed, one of the ways in which modernism has come to differ from other, older, orientations is in its reversal of this evaluation: change is now desired for what it will supposedly produce in the long run; but holding to what one already does is regarded as bad simply because it is old-fashioned."

It is clear that, for development to become more effective and more beneficial to a maximum number of people in any given country, the modern-traditional dualism must be resolved with the creation of a new synthesis wherein the notion of planned socioeconomic structural change draws contributions of knowledge from the modern and traditional sectors alike. There would be in such a synthesis an explicit recognition that there is essential knowledge about ecology contained in both the modern techno-science-environmentalist tradition and the village tradition, knowledge that should be shared in planning develomental changes. The farmers, for example, have a wealth of ecological information accumulated through many generations, but this remains for the most part untapped by the technicians and scientists whose fund of knowledge is not available to the farmers. Both bodies of knowledge share the characteristic of having been derived from a process of trial and error.

These chapters give us valuable insights about the kind of knowledge that could be shared in order to improve planning and implementation of development efforts. Societies with ecological adaptations that have survived for a relatively long period have much to tell technicians and scientists about socioeconomic practices well suited to the physical environment. A scientific focus such as Soemarwoto brings to bear on traditional wet-rice farming and home gardening in Indonesia would undoubtedly reveal extremely useful information about the adaptations of the Dai people of China or the Bontok of the Philippines to their respective ecosystems. Surely knowledge obtained on such subjects as pest control, soil conservation, and genetic diversity would enable the central planners to gauge some of the environmental effects of their development plans.

Central planners also could anticipate some of the socioeconomic effects if they took into consideration the patterns of developmental planning that occur in rural societies. The chapters tell us that members of traditional societies by and large will seek adaptational change when ecological imbalance gives rise to perceivable problems. Changes among the early Hong Kong settlers were made in reference to *feng shui* precepts and needs related to wet-rice farming. The Dai people appear to have adopted alien plants suited to their ecology and their needs. The Bontok instituted a legal practice (cash fines) to maintain their successful water-management system. But not all planned changes at this level are suited to all members of the society; villagers in the Thai-Muslim community with a penchant for trade could turn to it, but those not so inclined were left in the constrained agricultural sector. Then there are the locally planned changes that may be ecologically adaptive to the environment in some respects and maladaptive in others. The increase in rice production and changes in home gardening were adaptive to growing populations and urbanization, but both had ill effects on the ecosystem.

Information accumulated in the technoscience tradition could be readily of use to persons in the traditional community when they plan their changes. New technologies and scientific information about ecological conditions could be drawn upon to plan more efficacious readaptations from which a maximum number of people would benefit and which also would lead to restoration of balance with the ecosystem. Could modern technology, for example, be used to make water management more effective among the Thai-Muslims? Conceivably, a number of the Indonesian villagers, made aware of the negative ecological effects of their new rice-farming and home garden innovations, might seek scientific information on how to restore balance. Also, if the West Javanese were informed that the cause of cholera outbreaks was latrine pollution, they probably would plan new arrangements for their fishponds.

Soemarwoto points out that, in the context of world development, environmental benefits are always accompanied by environmental risks, and both should be considered. Changes affecting human society and ecology are occurring at an ever-increasing rate, and he warns that "time is running short," so that "we must work fast and be willing to face the realities of the world."

Ho Ton Trinh (chapter 12) sees the same problem with the proliferation of what he calls "artificial" ecosystems associated with the technoscience, urban, industrialized world. "How is it possible," he asks, "to organize life under these conditions in such a way as to maintain an equilibrium, a harmony, in the relations between humans and the environment?" For him the solution lies in the creation of a "cultural medium"—a new perspective—in which a "new field" dealing with the complexity and dynamic character of human and natural environments can be better understood. "By cultural medium I mean a social atmosphere involving a wholesome and vibrant way of life, based on disciplined and conscious labor as well as good relations between human beings." It would embody "a political, cultural, and moral life that is rich and varied and helps people express their values and satisfy their needs as masters of the environment, pursued in the interest of all and not to the prejudice of anybody." For Trinh, culture in the anthropological sense holds the key. "Thus," he says, "in concrete historicosocial terms, we have to know how to use cultural values as a dynamic, generative force to raise the quality and way of life. These in turn will help us transform the physical and material base of the human environment according to our goal."

References

Belshaw, Cyril S.
   1965   *Traditional Exchange and Modern Markets.* Englewood Cliffs, NJ: Prentice-Hall.
Boeke, J.H.
   1942   *The Structure of the Netherlands Indian Economy.* New York: Institute of Pacific Relations.
   1953   *Economics and Economic Policy of Dual Societies.* New York: Institute of Pacific Relations.
Carson, R.
   1962   *Silent Spring.* New York: Houghton Mifflin.
Foster, G. M.
   1962   *Traditional Cultures and the Impact of Technological Change.* New York: Harper & Brothers.
Geertz, C.
   1963   *Agricultural Involution: The Process of Ecological Change in Indonesia.* Berkeley: University of California Press.
Goodenough, W. H.
   1963   *Cooperation in Change.* New York: Russell Sage Foundation.
Hainsworth, G. B.
   1982   Beyond dualism? Village-level modernization and the process of integration into national economies in Southeast Asia. In *Village-Level Modernization in Southeast Asia: The Political Economy of Rice and Water*, edited by G. B. Hainsworth. Vancouver: University of British Columbia Press.
Redfield, R.
   1960   *The Little Community and Peasant Society and Culture.* Chicago: University of Chicago Press.
Schramm, W.
   1976   An overview of the past decade. In *Communication and*

Change: The Last Ten Years—and the Next, edited by W. Schramm and D. Lerner. Honolulu: The University Press of Hawaii.

Schultz, T. W.
1964 *Transforming Traditional Agriculture.* New Haven: Yale University Press.

Schumacher, E. G.
1973 *Small Is Beautiful.* New York: Perennial.

Sutton, F. X.
1982 Rationality, development, and scholarship. *Items* (Social Science Research Council) 36(4):49–57.

Wang, G., Wimal Dissanayake, and B. Newton
1983 *A Cultural Mediation Approach to Development and Change.* Nagoya, Japan: United Nations Centre for Regional Development, Working Paper No. 83-1.

Ward, B.
1962 *The Rich Nations and the Poor Nations.* New York: W. W. Norton.

CHAPTER 16

PARADIGMS, PERCEPTIONS, AND CHANGING REALITY

Neil L. Jamieson

Both the scope and the intensity of change in Southeast Asia are vividly illustrated by the chapters in this volume. Change is clearly shown to be a major theme in this region now, as it has been in the recent past. We also are reminded that the environments and human inhabitants of Southeast Asia have always been changing. Change is inherent in existence. But these chapters also point out our common failure to realize the extent to which our perception of change and our attitudes toward it are shaped by the words we use to talk about it.

Our normal use of words like "evolution," "adaptation," "history," or "development" in talking about change and organizing our analysis of it involves a number of implicit assumptions about both the scale of time to be considered and the level at which analysis is conducted. These terms involve implicit cultural values and beliefs, epistemological assumptions, and emotionally charged attitudes toward particular classes of events. They are "paradigmatic," combining a cluster of interrelated cognitive and emotional components, and, like most paradigms, they are transmitted and reinforced by powerful exemplars.

Thus, the words we use to describe change can exert a subtle but powerful influence upon the kinds and levels of change that we perceive and the attitudes we assume toward these changes. And, since our attitudes can in turn influence behavior in ways that may promote, inhibit, or alter the kinds and extent of changes that take place in our society, our environment, and even in our private lives, these labels for the process of change deserve the closest possible scrutiny. To the extent that we can become more fully aware of the

values, beliefs, and assumptions that underlie our thinking and our
work, we can partially free ourselves of the sometimes unnecessary
and often harmful constraints they impose upon our thinking.

Adaptive Change

Environmental changes pose new sets of constraints for various
biological components of affected ecosystems. For any given population, traits that were previously neutral or even disadvantageous
may become useful under new conditions. Through the process of
natural selection, these newly advantageous traits are assumed to
become more common over time; other traits, once useful, may
become a liability, and the frequency of their distribution in the
population will be reduced over time. This is called "adaptive
change." It occurs in response to shifting needs, problems, stresses,
and opportunities—change in response to change, if you will. The
critical unit of time in such adaptive biological change is a
generation. The chronological span involved varies, of course,
depending upon the species under consideration. Ladybugs, human
beings, walruses, and elephants have generations of quite different
durations.

Where human beings are concerned, the situation is more
complicated, in part because adaptation takes place simultaneously
at multiple levels and at very different rates of speed. Human
generation times are relatively long, and genetic change is therefore
relatively slow. Although genetic and physiological adaptations
remain important for human populations, the most significant
adaptive mechanisms are cultural. Instead of relying upon such
physiological mechanisms of response to environmental stimuli as
shivering in the cold, we typically interpose cultural devices between
ourselves and our environment. We build a fire, put on warmer
clothing, close a window, or simply turn up the thermostat on central
heating units in our homes.

Adaptation provides a conceptual framework for much research in
the natural and the social sciences and furnishes guiding principles
for policymaking in the world of today. It also entails an implicit
time frame for use in decision making by individuals, corporations,
and nations. Change generates choices between new sets of problems
and opportunities and different forms and degrees of costs and
benefits. These choices are complicated by the fact that ecosystems

entail interactions among many different factors and biological components. Through these interactions, changes in one element may reverberate throughout the system. The complexities of the interactions often make the ultimate results unpredictable.

Human responses to malaria furnish a good example of this situation. Malaria is a tropical parasitic disease introduced into the human bloodstream through bites by anopheles mosquitos. The malaria parasite itself evolved in interaction with anopheles mosquitos and homeothermic mammals in the tropics. It probably became a significant problem for human communities with human expansion into rain-forest habitats and the clearing of tropical forests for agriculture, which inadvertently brought about an expansion of breeding grounds for mosquitos. Human groups with long residence in the tropics have evolved genetic (sickle-cell anemia) and immunological adaptations to endemic malaria. It is well known, of course, that these adaptations, particularly sickle-cell anemia, not only provide some protection from malaria but also exert heavy costs in other areas of human health.

In recent decades, massive efforts have been made to control malaria through culturally induced means. Vast, worldwide programs were instituted to eradicate mosquitos through pesticides, particularly DDT. In spite of initially encouraging results, however, these programs have not been successful in the long term and have created new problems: pesticide-resistant strains of mosquitos have emerged and have led to a resurgence of mosquito populations in some areas; through the food chain, high levels of pesticide toxins began to accumulate in organic tissues of humans and some animals; and other important elements of ecosystems were often damaged or destroyed by the pesticides although they had not been targeted. In some cases, attempts to destroy one pest led to the emergence of new ones that were even worse, examples of which were described by Rachel Carson in *Silent Spring* (1962).

Even after *Silent Spring* had become an international bestseller, health workers operating under a World Health Organization program in Borneo sprayed DDT in Dayak longhouses as part of the continuing effort to eradicate mosquitos that spread malaria. The DDT killed or heavily toxified large numbers of mosquitos, cockroaches, and other insects and small pests that were then eaten by lizards, who built up high levels of DDT in their fatty tissues. The DDT-bearing lizards were in turn eaten by cats until the cats consumed lethal doses and died. As the cat population decreased, the rat population increased. Fleas, lice, and other parasitic populations increased along with the rats, spreading, among other things, sylvatic plague. Meanwhile, the caterpillar population also increased, because its natural predators had been destroyed, and hungry caterpillars began consuming the thatched roofs of the longhouses. The collapsing roofs provided an apt metaphor for the entire ecosystem.

What is the situation now, after nearly forty years of massive campaigns to eliminate anopheles mosquitos and other disease vectors and crop pests through chemical spraying? In 1984 nearly one billion people had malaria, about one in every five people in the world. And the number of victims is rising rather than decreasing! Not only have the mosquitos become resistant to the pesticides, but the parasites themselves have developed immunity to the wonder drugs developed to treat the disease. The case of malaria illustrates the fundamental proposition that there is no such thing as an absolute, permanent, or totally successful adaptation. It seems to be inherent in the nature of things that we alter that which we adapt to and plant the seeds of future failure in the very process of adapting to that which now exists. This makes permanent victories exceedingly difficult to come by.

When we talk or think about adaptive change we generally mean a change in one variable (such as incidence of malaria) in response to a change in another variable (mosquito populations). As a species that employs culture as a primary means of adaptation, we tend to engage in specific practices (spraying DDT) to achieve specific goals (eradicate malaria, increase food production) by manipulating a particular variable (anopheles mosquitos) that we perceive to be causing a problem. But in a world composed of interlocking and interacting systems and subsystems of great complexity, a change in one variable inevitably leads to changes in others. Not only do mosquitos adapt to pesticides, but a host of other organisms adapt to the changes in their environment induced by both the intended and unintended, the direct and indirect, effects of spraying the pesticides.

Most of our problems in the future will be, like those of today, anthropogenic. Our solutions to the problems that now confront us will generate a new set of problems for our children and grandchildren, just as many of our problems today are products of the actions of our ancestors, including those actions and innovations that are considered to be their greatest achievements. It is clear that every change has costs and benefits. But purposeful change (like adaptive cultural responses) is often difficult to evaluate in advance because some benefits and costs are indirect and delayed. Thus, every major intervention in a complex system constitutes a gamble. Determinations of what is or is not adaptive are meaningful only in terms of specific values, for a particular population, and within a specific time frame. Reliable judgments can be made only retrospectively, in the much longer time frame of evolution. Evolution may be thought of as the long-range, aggregate outcome of adaptive changes.

Adaptation and Evolution

The relationship between evolution and adaptation is exceedingly complex and often misunderstood. Even if we restrict our discussion to biological evolution, which we certainly cannot do, it can be misleading to use the term "aggregate outcome" to describe the relationship of evolution to adaptation. The chief difficulty lies in the tendency to see evolution as consisting of one "improvement" being added to another to make whatever is said to have evolved get better and better. This is thought to constitute progress. But evolution and progress are not at all the same thing. When one asks "better for what?" the answers are usually tautological, on the order of "better at being what it is." Human beings become more like what human beings already are. This is evolution and it is good, and so it is progress.

By this line of thought, we are seen as having acquired opposable thumbs, stereoscopic vision, bipedalism, a large and complex cerebral cortex, skill in toolmaking, language, and finally religion, art, and science one by one to make us better and better. We progressed step by step to become what we are now, as if this had been our goal from the beginning. This accretive model of evolution is intimately related to the equally mistaken notion of adaptation as something that is or can be an absolute good. Even those of us who should know better by virtue of our education and experience sometimes lapse into confusing "fittest" with "best." But "survival of the fittest," as we all know, means survival of the most appropriate. Appropriate to what? To the context of which the evolving organism is a part. And since contexts change in ways that are unpredictable and often uncontrollable, the definition of what is best, in the sense of best suited, is always in a state of flux. Fitness is always relative to context.

This, of course, is why that which is "adaptive" is so relative and transitory. It is why that which is good for one group may be bad for another and why what is good today may be bad tomorrow. I do not wish to advocate any ultimate relativism, nor do I wish to be an apostle of change, but since change is inherent in existence, a sense of relativism is an implicit and necessary part of what we know about the world and about ourselves as a species. To see this is essential if we are to make the complexity and diversity of the world comprehensible to ourselves and to others or to achieve mutual understanding. But, as Aram Yengoyen (chapter 6) reminds us, there is no sound philosophical basis for relativism in Western thought, nor in the thought of most other peoples in the world. This is a serious and urgent problem that deserves much more attention than it has received.

The related notions of relativity and context serve to pull together many of the papers in this volume and much of the discussion that took place following their presentation in Honolulu. We have all been talking about various aspects of diverse, partially overlapping, and variously organized contexts. Lovelace (chapter 14) provides us with an example of the dialectical relationship between culture and the environment taken from the early history of southern China. Benjamin (chapter 10) has analyzed an extensive network of relationships in the highlands of Malaysia over the past few centuries. Prill Brett (chapter 5) presents an unusually holistic overview of the complex web of well-established and mutually adjusted relationships between ideology, society, and an environment already substantially transformed by culture in the Bontok region of northern Luzon, a traditional system that now must adapt to external changes arising from bureaucratic decision making in Manila. Prachuabmoh (chapter 11), on the other hand, describes the interplay between religious values and economic strategies in a Muslim community in southern Thailand that has chosen to reach out to exploit the resources of the larger national system to compensate for the effects of increasing population pressure upon limited local resources. At another extreme, Trinh (chapter 12) offers us an exposition of a contemporary nationalistic perspective on the relationship between planned, state-level intervention in cultural and social matters and the attainment of more viable relationships in the future between people and between humans and their environment.

These chapters provide us with a vivid illustration of the many ways in which human choice based upon cultural values can be interposed between the discrete events of cultural adaptation and the larger process of social evolution. Taken together, they reveal that different kinds of choices are made on different levels of the social system. It is a difference between levels of choices made by individuals or by groups (to be vegetarian or carnivorous, to hunt or to domesticate animals as a source of meat, to make pork taboo, or to eat lamb instead of beef for supper tomorrow), and there is a difference between levels of the social hierarchy at which choices are made (by a company commander or the minister of defense; by the king, the local leader, or the woman next door). Both the level of choice being made and the level at which choice is made significantly influence the degree of constraint exerted upon other potential options at lower, more specific levels. As Benjamin (chapter 10) has demonstrated, some choices lock both individuals and groups into certain larger strategies by greatly increasing the constraints upon other options, both for oneself and for others. Such decisions—very general or abstract decisions made at the upper levels of hierarchies, for example—become an important part of the new context within which choices will have to be made in the future. Such changing patterns of constraint upon choice will, over time, ramify through the system, always generating new sets of constraints in various hierarchies of transformed contexts. The original decision maker will thus sooner or later be part of a context that may be very different from that in which the original, locking-in, "adaptive" decision was made.

The inescapable implication of this is that the deepest meaning of any change in environment, society, or culture is to be determined by examining its effects upon relationships. Neither people nor plants nor animals exist "in" contexts (seen to be something external to them), but always as an integral part of a context that is composed of, and in turn a part of, still other contexts. One of our main analytical concerns, therefore, must always be to sort out the contextual background of what we are talking about.

Gregory Bateson remarked in 1972 that what he had failed to see in his own earlier work was that

the evolution of the horse from Eohippus was not a
one-sided adjustment to life on the grassy plains.
Surely the grassy plains themselves evolved *pari
passu* with the evolution of the teeth and hooves of
the horses and other ungulates. Turf was the
evolving response of the vegetation to the evolution of
the horse. It is the *context* which evolves. (1972, 155)

Bateson's statement sums up much of the earlier discussion about
adaptation and raises a crucial issue about the relationship between
adaptation and evolution. It is the total ecosystem that survives over
time and slowly evolves in ways and in directions not visible during
the span of a human lifetime. In this process, the relata (teeth,
hooves, grass, and soil in one case, for example, or parasites,
mosquitos, people, and forests in another) each undergo changes that
may be adaptive in the relatively short period of adaptive change and
natural selection. But, whether any specific change is adaptive or
not, it still may be expected to induce change in other parts of the
ecosystem of which it is a part. Without some compensatory change
in the other part or parts of a relationship, any change can be
disruptive to a system.

The process of natural selection does not necessarily lead to
greater adaptedness for individual species. On the contrary, it can
lead to extinction. It was the notion that evolution is not a
goal-directed process that constituted one of Darwin's most notable
contributions to modern thought. Yet the intellectual framework that
was constructed upon this crucial insight has often been distorted in
the process of applying it to the development of human society and
incorporating it into a generalized paradigm for thinking about
change in the world and our relationship to it. We employ concepts
like evolution, progress, and development as if they implied
movement toward predetermined, universal goals. Darwinian
evolution, however, offers no valid support to this widespread notion
of incremental progress toward a specific goal.

The horse, in other words, is not to be viewed as an improved
version of the Eohippus. It is simply a product of changing patterns
of constraints in the context of which the Eohippus was once a part.
One cannot meaningfully compare the Eohippus and the horse in
absolute terms because they have no common context. One was
adapted to browsing in an environment in which this was an

appropriate thing to do; the other is adapted to grazing in an environment where this has been an appropriate thing to do. If the modern-day horse and the prehistoric Eohippus could be brought together by a time machine, or failing that, in our imagination, to share a common habitat, they would not even be competing for the same niche. In any case, we could—as part of our cultural adaptation now—raise either animal on a ranch by transforming the environment to meet their needs or by bringing in appropriate food from another ecosystem altogether, perhaps from one many miles away.

We know that the typical horse differs from the typical Eohippus in various ways, the horse being bigger, among other things. But it would be an arbitrary value judgment to say that being bigger is better or worse than being smaller. We can say that one is better, however, if we specify "better for what?" The horse, being bigger, is better for carrying heavy loads. I would not care to ride an Eohippus, even in my imagination. But since I do not especially enjoy riding horses either, and have no compelling reason to do so, the Eohippus and the horse have equal worth to me. I have the same attitude toward both. But when we speak of goals, values, and attitudes, we have left the proper realm of evolution to enter the value-oriented realm of history.

History and Adaptation

Clearly there is more to history than the sort of adaptive change that is involved in altered distributional frequencies of certain types of red blood cells in tropical populations. Most of us do not perceive our past or have visions of our future as merely a series of essentially passive and mechanical responses to shifting environmental pressures. We attribute our historical trajectories to deliberate, conscious, purposeful human action. We have a teleological view of history. With some justification, we see conscious purpose and human choice in history. We believe ourselves to be possessed of a distinctive ethnic or national spirit, a *geist*, which implicates cultural values and beliefs in our self-definition both as individuals and as groups. We see our evolutionary path as guided by historical and mythical exemplars that retain a timeless, non-negotiable value as indicators of both the ends and the means of action.

Events in the history of a people or a region tend to be perceived both by the people involved and by others as either specific instances of general progress or as pathologies. Within this framework, change is seen to be directed and constrained by cultural values. It is further believed that change should be, and often is, either an extended expression of *geist* or a disruptive force to 'be resisted and suppressed. Evolution is progress, an incremental growth of approved adaptations characterized by this unique *geist*. The *geist* itself generally is seen as unchanging, but its expression is continuously improving through the accretive growth in adaptive exemplars, techniques, and superior social, cultural, and technological innovations.

This view of history, while plausible, is not amenable to empirical verification in the strict sense. From a more empiricist perspective, one may usefully conceive of history as the aggregate outcome of chains of adaptive changes that represent stochastic processes—interacting chains of probabilities that are patterned, comprehensible, and explainable on the whole, at least within the context of natural selection and systemic interaction. Processes need not be "contained" to be "constrained." And they need not be highly predictable for a systems approach to be useful. The cultural values and beliefs that constitute the core of a people's *geist* constitute one set of constraints upon choice, a set that interacts with other sets of constraints that arise from the environment, the behavior of surrounding groups, and so forth.

When all these various kinds of constraints, both material and ideational, are taken into account, all conceivable outcomes are not equally probable. It is probably true, as Benjamin has argued, that history is primarily about things that could just as well have turned out differently and that there is nothing in the circumstances we normally study that could, by itself, have decided the outcome of historical developments. But this is not to grant complete randomness to history. There are some discernible patterns. Historical change, like biological evolution or the trajectory of a falling raindrop, is shaped by various kinds of contextual constraints. Many of the more important of these constraints can be identified and investigated, however imperfectly. We can, to a greater or lesser extent, gain some understanding of the mechanisms and processes of historical change by not viewing history, in the narrow sense, as a phenomenon of an entirely different order from other diachronic processes involving changes in system states. We can talk meaningfully about ecosystems, social systems, or cultural systems while granting their indeterminancy. Such an approach does not prejudge the fiercely debated questions concerning the extent to which history is determined (as opposed to being an expression of the free will of the actors) and what sorts of things it is determined by.

Of course, both historical determinism and the "free will" or *geist*-directed schools of thought are cultural models that may themselves usefully be viewed as products of adaptive change. They arose and gained currency in response to shifting needs and opportunities in the emerging contexts of intellectual life in civilized society. The question is, to what extent do such theories of historical change exert influence upon historical change itself? The answer depends upon the relationship between a given theory and other components of the context within which it is put into practice.

It seems to me that the real problem with theories of political action, or *geist*-directed approaches, or any other approach to change, be it materialist or ideational, is that such constructs are generally assumed to retain their validity across time and space. In other words, there is a tendency to ignore or seriously underestimate the importance of the context in which a given construct arose and the relativism such a consideration necessarily entails. Whether we are talking about *feng shui* or dialectical materialism or Keynesian economics, we are discussing intellectual constructs that arose in the context of adaptive change (like the use of DDT) at some earlier point in time. But because we seldom view our theories or values in that light, we tend to cling to them after a changing context (e.g., resistant mosquitos) may have rendered them ineffective, no matter how "adaptive" they may have been at some earlier time.

I think that cultural values and beliefs (and the theories and models based upon them) do influence change in history and in the physical environment, and I gladly grant that this can be a very good thing indeed. But, committed as I may be to particular values or theories or ways of life, I am forced to accept the fundamental premise that virtues I attribute to them are conditional, relative, and contextual in some ultimate sense. They are good to me within certain contextual parameters, and in the very act of having been put into practice they may themselves have contributed to unanticipated changes in that context that might render them less useful now or in the future.

History may be seen as the study of processes that are value oriented, very loosely determined, and not very predictable. But these processes also may be viewed as changes that take place within an evolving context and under shifting sets of constraints within that context. There are multiple processes at work within history that are conceptually distinct but nevertheless interrelated. The relationship between the teleological dimension of history and the stochastic processes of evolution is a dialectical one. Development, which is a major cause and shaper of change in the world today, represents a special case in which issues of adaptive change, evolution, and voluntaristic theories of history confront each other. In most tropical countries in particular, development programs constitute important frameworks within which cultural values and ecological issues are articulated with each other. We must now turn our attention to this.

Adaptation, Context, and Development

"Development," as the term is widely used today, involves a distinctive way of looking at and acting upon the world that is less than forty years old, but its intellectual roots extend back to the Renaissance and rest heavily upon the Enlightenment in Western European thought. The Renaissance was characterized by trends toward greater secularity, closer observation and reporting of natural phenomena, and a general skepticism toward all manner of ancient truths. Fragmented and diffuse tendencies in fields as disparate as painting and literature, philosophy, medicine, and astronomy were systemized and popularized by Francis Bacon (1561–1626) and René Descartes (1596–1650) to form the popular intellectual bases of an emerging new paradigm that was further developed and legitimated by Newton, Locke, Bentham, and others to shape the Industrial Revolution and the modernization of Western Europe and North America.

This paradigm was based upon a belief in an unchanging natural order in the world and an objective, material reality that could be observed, described, and analyzed. During the eighteenth century, scholars believed that they were progressively increasing our understanding of the universal natural order through careful observation and inductive reasoning. They thought themselves to be discovering Natural Law. From this followed the conviction that human customs and institutions could and should be modified piecemeal to conform ever more closely to this natural order as human knowledge of it was being incrementally enlarged and confirmed by empirical, positivistic methodology. Natural Law, it was believed, could empower us to master both society and nature. Prosperity, peace, and even happiness were thought to be obtainable through proper modification of social institutions (such as government and law) accompanied by the steady expansion of technological control over the physical world.

At the same time, exploration and discovery confronted thoughtful Europeans with a new view of the earth and their place in it. Awareness of previously unsuspected diversity among peoples raised the question of how such vast differences among human groups could be reconciled with the operation of Natural Law and universal order. Mere extension of the concept of Natural Law was required to supply the answer: the theory of evolution based upon natural selection. Following the publication of Darwin's *Origin of Species* in 1859, many writers applied the concept of evolution to the study of human societies.

In general, the early and most influential writers on social evolution postulated that all humanity shares a common developmental sequence and that major differences among human groups in the cultural sphere reflected their relative rates of progress up this universal evolutionary ladder. It seemed self-evident to Western Europeans that their own rapidly industrializing societies represented the highest level of human achievement, and all other peoples were ethnocentrically ranked below them as representatives of earlier stages of social evolution. All human groups were assumed to be moving inexorably, albeit at different rates, toward the higher existence typified by industrialized Europe.

The views of Herbert Spencer (1820–1903) and other social-evolutionary theorists spread through educated elites around the world, to Hanoi and Bangkok, to Manila, Jakarta, and Kuala Lumpur, to Tokyo and Peking. At the same time, these ideas were becoming legitimated as plain common sense in the general intellectual milieu of the Western world. At a time when colonialism, increasing ease of travel, and the rapid growth of the print media were combining to provide all peoples of the earth with many vivid and detailed illustrations of the large and ever-increasing inequality of wealth and power in the world (and a notion of how often the two went together), these theories of social evolution simultaneously offered legitimation to the colonizers, a remedy and blueprint to the colonized, and a powerful object lesson to those nations that were neither.

While colonial school-systems inculcated the values and assumptions of this paradigm of progress into the minds of future leaders and opinion makers throughout much of Asia, Africa, and South America, countries such as Thailand and Japan initiated independent efforts to imitate and catch up with the technological and scientific marvels of the West in order to maintain independence and national pride. The presumed merits of this strategy were dramatically reinforced by Japan's victory over Russia in 1905. Patriots from many countries in Asia flocked to Japan to learn from the miracle of the rising sun.

Meanwhile, Karl Marx (1818–83) had been appalled by the conditions he observed in England, the most industrialized nation in the world at that time. He came to view free-enterprise capitalism as a transitional phase in social evolution that contained internal contradictions causing various social ills that would eventually bring about its replacement by communism. In general, however, his scheme was very much within the dominant paradigm. Like all the other variants of nineteenth-century social-evolutionary thought, it was universalist, unilineal, deterministic, utilitarian, and based upon a world view that posited a mechanical, Newtonian universe. With the advent of the Russian Revolution some thirty-five years after Marx's death, his ideas formed the basis of a radical experiment in planned change on a grand scale. The Soviet Union's first Five-Year Plan, adopted in 1929, was the first formal, national, development plan.

Development on a global scale, as now commonly conceived, began only in the aftermath of World War II. In many ways, the past four decades constitute a unique era in our existence as a species. In 1949, the Universal Declaration of Human Rights proclaimed a formal commitment by all nations to eliminate poverty and injustice from the face of the earth. True to the intellectual heritage of Bacon, Descartes, Newton, Darwin, and earlier social-evolutionary thought, the development planning of the 1940s and 1950s assumed that all peoples were moving along a progressive evolutionary path. It was widely believed that by transferring capital and technological know-how from the richer industrialized nations to the poorer ones, in a massive program of planned and controlled change, newly independent governments would be able to lead their citizens into genuine material and spiritual progress.

Despite the ideological and political schisms of the time, all factions shared those "evolutionary," positivistic, and utilitarian assumptions to a remarkable degree. They shared a common faith in the capacity of technoscientific bureaucracies to shape the world in desirable ways. Capitalists, communists, and the ideologically uncommitted all held the same fervent and largely unexamined beliefs in universalism, industrialism, and progress!

Within this context, traditional knowledge systems throughout the world were viewed not as valuable resources but as impediments to progress. Following the fundamental misunderstanding of the concepts of "adaptation" and "evolution" that was inherent in nineteenth-century social-evolutionary thought (and became the "common sense" of the twentieth century), post-World War II development schemes sought to modify on a piecemeal basis not just systems of production but even systems of meaning to achieve greater conformity with the putatively higher forms associated with rapid industrialization and economic growth in Western Europe and North America.

Deliberately induced technological and social change, promoted in the name of development, has indeed led to results that are commonly viewed as measures of "progress." In Southeast Asia and around the world, mortality rates have been significantly reduced over the past four decades. The population of Southeast Asia has doubled in less than forty years. Educational opportunities have improved considerably for the majority of the people. Literacy rates have soared. Transportation and communication facilities have dramatically increased both in quantity and quality. Both agricultural and industrial productivity have increased substantially, and there has been a real, and in some areas quite considerable, increase in per-capita income despite population growth. In other words, very real benefits have accrued to a large number of people, so the belief in progress is still strong.

On the other hand, the number of very poor, undernourished people in the world today is roughly equal to the total world population of 150 years ago: approximately one billion human beings, many of whom live in Southeast Asia. In addition, the gap between rich and poor nations has increased rather than diminished, along with the gap between rich and poor people within nations. At the same time, deforestation, soil erosion, and environmental pollution are proceeding at a terrifying rate that cannot long be sustained. The quest for rapid modernization, with emphasis upon rates of growth in the gross domestic product (GDP), in order to "catch up" with "more developed" nations—i.e., to "progress"—has exacted high social and ecological costs from the land and people in the Third World, including Southeast Asia.

The cultural and psychic costs of "progress" are more difficult to measure but certainly no less real. Nor are they, in terms of adaptation and evolution for a species whose mode of adaption relies so extensively on cultural facilities, a whit less important. Rapid transformation of familiar ways of life in the name of development has provided the world with numerous examples of the pain and indignity that human beings experience when accustomed patterns of being are altered in ways that deprive life of meaning, that make individuals the "objects" of external processes in which they do not participate in any significant way. Feelings of anxiety and alienation become part of the emerging context. They can result in behavior, as in Iran or Cambodia, that reduces the technoscientific plans for development to a shambles.

The one certain effect of development is a continuous acceleration in the rate of change. Since the modern technological coefficient was added to human life in the seventeenth century, we have been increasingly simplifying the world in which we live—biologically, socially, and culturally. One might go so far as to say that, since the onset of deliberate development on a vast scale in Southeast Asia and elsewhere, we have been experiencing a fundamental change in the nature of social and technological change itself, one that has both quantitative and qualitative dimensions as well as profound evolutionary implications for life on this planet in general.

Biological diversity in Southeast Asia is being inexorably reduced by numerous dimensions of the development process. Urbanization, extensive logging and resultant deforestation, a monocultural trend in agriculture, the use of mass-produced commercial seeds, use of chemical pesticides, and other changes intended to promote economic growth simultaneously promote the process of simplification and a concomitant increase in biological vulnerability. The trend throughout the region is unmistakably toward a reduction in the number of species of plant and animal life and a reduction in genetic diversity within species. This may be of particular significance in the tropical environments of Southeast Asia since their natural diversity is thought to be directly related to their resilience and productivity.

This process of simplification is not limited to biological life. If we look at political units on Earth, we see a similar trend. It has been estimated that just 3,000 years ago there were several hundred thousand separate political units in the world. Just 1,500 years ago there were probably still more than 100,000. Now there are fewer than 200 separate political units on Earth for a world population approaching 4.5 billion. And within each unit the trend is toward increasing assimilation (that is to say, homogenization) of all segments of the population, at least with regard to certain aspects of behavior. Schools, mass media, law enforcement, development programs—these and many other factors promote political integration and control and reduce diversity over time.

There is, it must be noted, evidence of opposing tendencies, such as an apparent increase in separatist movements and in demands for greater local, cultural, and political autonomy. Among groups advocating greater political and cultural pluralism, one may, for example, point to the Quebecois in Canada, the Basques in Spain, the Ibo in Nigeria, the Druse in Lebanon, the Sikhs and the Assamese in India, or the Tamils in Sri Lanka. Or, more to the point, there are Moslem groups in the Philippines, the Shan and Karen in Burma, the tribespeople of Irian Jaya and Timor in Indonesia, the FULRO coalition of highland peoples in Indochina, and organizations like the Asian Cultural Forum on Development in Bangkok or the Consumers Organization of Penang.

The advocacy by such groups of greater local autonomy in decision making must be understood as specific local manifestations of a more general reaction to the process of homogenization noted previously. The proliferation of specialization in productive activities combines with the simplification of our biological and social environments to generate an exponential increase in the genuine need for coordination and control at higher organizational levels. It seems that increasingly we all are being drawn (or pushed) into a higher shared context for which our experience (as summarized by our culture) has not prepared us. We are all tied into a very complex and intricately interrelated set of economic, ecological, and social systems that are global in nature. We live, unquestionably, in a world of mutual dependence. Development programs themselves, by their very nature, foster this situation.

No single ethnic, religious, or political group, no nation or group of nations, can now exert any meaningful degree of fully autonomous control over the complex changes that are occurring in these massive and poorly understood systems of interdependence in which we all are enmeshed. We do not possess the theoretical understanding, the political will, or the operational abilities for effective collective action to counter these developments. We now need the massive technoscientific bureaucracies of nation-states and international organizations to keep goods, money, information, and people flowing across ethnic and political boundaries and geographical barriers. Thus, development is creating dependency, not just of "undeveloped" upon "developed" nations, as "dependencia" theorists suggest, but of *all* parts upon the effective functioning of the total megasystem. This applies to the relationship between local communities and central governments as well as to those between nations and international organizations. Yet these same bureaucracies, indispensable as they have become, often restrict options at lower levels in unnecessary and inappropriate ways. They make decisions that are rational from their macro-level perspectives, based upon highly aggregated data or general principles, but which sometimes make no sense at all in concrete contexts of particular cultural groups living in specific environmental settings.

Development, then, generates more mutual dependence and greater need for centralized coordination and control. Yet such macro-level planning and decision making can never, by its very nature, take adequate account of local conditions, local needs, or local attitudes and values. All over the world, but especially amidst the rapid change affecting the traditional ecological and cultural diversity of Southeast Asia, development is generating an imperative for simultaneously achieving both more effective planning and coordination at national, regional, and international levels *and* a greater capacity for planning and decision making that is appropriate and meaningful in the context of local conditions (i.e., more local participation and autonomy). Those who urge more centralization and those who urge or demand more decentralization are therefore both correct. Each side has incontrovertibly valid arguments. The relationship between centralized and decentralized processes must be seen as a dialectical one, as both competitive and complementary. Too little recognition is now given to the complementarity of the two processes, and that is what must now be emphasized.

Paradigms, Perception, and Scale

As Bertrand Russell once observed, the sense of stability in the physical surroundings we take for granted is due largely to our relative size. If we were the size of an electron, for example, what we now see as a carefully designed building would be far too large for us to even perceive it as a structured whole. We would see instead only tiny, constantly moving fragments of matter.

On the other hand, as Russell noted, if we were as large as the sun and lived as long, with a corresponding slowness of perception, we also would find an unstable universe, a world of change and impermanence in which "stars and planets would come and go like morning mists, and nothing would remain in a fixed position relative to anything else" (1958, 12). Our perception, indeed, in some genuine sense the very nature of our reality, is relative to the scale of magnitude with which we are dealing and the time scale in which we view it.

In the fifteenth century people got along very well believing that the earth was flat. Few people traveled more than a few dozen miles from their place of birth, going usually by foot or ox cart. Both transportation and communication were extremely limited, by today's standards, but there was much more local autonomy and self-sufficiency. In the time-space scale of fifteenth-century life, the earth was indeed close enough to being flat as to make no practical difference. Although the achievement of Christopher Columbus, seemingly confirming that the world is not flat, startled many people at that time, it had little if any impact upon the lives of most Europeans or anyone else. Similarly, the Copernican revolution that preceded Columbus's voyage, though it upset the world view of Christendom and stirred up a great furor, met no practical needs and provided little immediately useful information.

In contrast, the Newtonian physics of the sixteenth century led rather quickly to changes in people's lives and their views of the world and their place in it. It gave rise to a powerful new paradigm. What is significant here is that the many applications of Newtonian physics have continued to be immensely useful—even though Einstein convincingly refuted some of the fundamental assumptions upon which Newton's model of the universe was constructed—and the scale of time and space in which we live, the tremendous conceptual distinctions between Newton's universe of straight lines and absolutes and Einstein's curved and relativistic world are for all practical purposes indistinguishable.

It should be noted further that Einstein also has presented us with a totally new perspective in which to interpret Copernicus's revolutionary reversal of the Ptolemaic view of the relationship between the Earth, Sun, Moon, and other planets. From the perspective of Einstein, it makes absolutely no difference whether we say the Sun moves around the Earth or the Earth moves around the Sun. The same repeating cycle of spatial relationship is described equally well by either statement. We now realize that it is this kind of constancy in relationships that is crucial to understanding the functioning of the larger system of which both the Earth and the Sun are a part, whether we choose to talk about it as a solar system or a planetary system.

Yet we still think egocentrically, ethnocentrically, and anthropocentrically, rather than in terms of sets of relationships, in the conduct of our affairs. We still think of ourselves as purposeful and highly autonomous actors within a context that is external to us, rather than as ever-adapting, integral parts of larger evolving contexts. Despite protestations to the contrary, and the existence of clichés like "environmental impact assessment" or "integrated development," almost all development projects—whether by local entrepeneurs or international agencies—are still basically conceived within a lineal, Newtonian framework, confusing adaptation and evolution with progress, and essentially constituting single cause-single effect interventions in complex systems (like spraying with DDT to eradicate malaria or building big dams to produce cheap electricity). We developed the habit of thinking and acting this way when the temporal and spatial effects of most human actions were far more restricted, in a time-space matrix that was very different from the one we face today.

The world population is today much larger, and our technological capability much greater, than when James Watt produced a steam engine that captured the imagination of a generation intent upon making its dream of progress into a reality. The technological extension of human thought and action has so quickly and so powerfully been magnified and multiplied by modern science and technology that our traditional cultural constructions of time and space are no longer fully adequate.

As Karl Hutterer explains in chapter 3, tropical environments such as those found in Southeast Asia are extraordinarily diverse and complex systems. They are highly productive and relatively resilient. Consequently, as Benjamin has observed in chapter 10, the inhabitants of the Asian tropics could until quite recently interfere with their environments without major negative effects. Relative to the cultural time frames through which both individuals and groups perceived the world, environmental degradation was so slow that it was virtually imperceptible.

Change has always been an important part of Southeast Asian realities, but the worrisome, problematic dimension of change for Southeast Asian peoples almost always has been related to ethnic, social, political, or cultural concerns. Environmental problems have generally been subsumed under a larger category of events considered to be caprices of nature. It was believed that by submitting to cultural rules that provided guidance for dealing with one's social, natural, and supernatural environment, an individual could induce a benificent response from all external forces. The traditional peoples of Southeast Asia can be said to have lived in harmony with nature in the sense that they transformed their natural environment according to cultural precepts and submitted to it by the same token.

The perspective from which we as a species question the meaning and significance of change has shifted radically over recent centuries. Our distant ancestors in prehistoric times died in a world much like that into which they were born. With the coming of agriculture and the domestication of animals, as human groups organized themselves into tribes and then into chiefdoms, between 5,000 and 10,000 years ago, and as human population began to grow at a perceptibly increasing rate, the tempo of change increased, but not to an extent that would have appeared to be dramatic or fundamental from within the existence of an individual human observer.

Only with the coming of the state, with a pervasive division of labor, do we encounter sustained inquiry into the problematic nature of change in human existence. But civilization was, on the whole, seen to be a mechanism for maintaining stability rather than a source of continuous change. Democritus and Lao-Tse, on opposite sides of the Earth, recognized change as inherent in existence, but to both the Greeks and the Chinese change was, although constant, essentially epiphenomenal. Both posited a rather static equilibrium system at work in the world, whereby the more things change, the more they remain the same, in the social realm as well as in the physical environment.

Not until the middle of the seventeenth century in Western Europe did a radical and sustained shift away from relative stasis toward accelerating lineal change become a basic characteristic of the human condition. Simultaneously, both human population-size and the technological capability to manipulate the physical environment began to increase at an exponential rate. This occurred first in England, then in the rest of Western Europe, in Asia, Eastern Europe, and the Americas, in the Pacific, and in Africa. By the end of the nineteenth century the twin thrust of rapid population growth and fast-paced technological innovation had covered the globe. The world population reached the one billion mark only about 200 years ago. It is now racing toward five billion and still soaring upward at a rapid rate.

Population growth and technological expansion are highly related to several other phenomena, chief among them the depletion of natural resources, the simplification of biotic communities, pollution, and the reduction of cultural diversity. Together these interactive processes produce a totally new and ever-more-rapidly changing ecological context for human existence. Both sides of the equation have changed radically since Bacon and Descartes enunciated the doctrine that humans could change and control their physical environment at will and that this was good. At this juncture nothing less may be demanded of us than a fundamental change in our symbolic view of the world, in our myths of reality, in—to borrow a phrase from Yengoyan (chapter 6)—the moral and ontological principles that provide structure to our behavior. The most basic problems in the modern world, I suggest, may be metaphysical rather than technical or organizational. Adaptation in the emerging context may urgently require of us change in the metaphysical underpinnings of our existence.

If, in Yengoyan's terms, our behavior is structured and "the existence and source of the structure are to be determined, not in terms of behavior, but in the moral and ontological principles that provide content to structure," then it is precisely in those moral and ontological principles, particularly as expressed and transmitted by our myths, that this reorientation must occur. Our myths must adapt to conform to the radically new scales of time and space associated with our new and expanding capacity and determination to transform our physical environment. The rate of change exceeds the capacity of the biological and social systems to respond to a sufficient degree through self-regulation.

This appeal to mobilize our creative energy to achieve a new paradigm that will be more appropriate in the context now discernible does not constitute an unwarranted attribution of free will to the intellect, nor is it incompatible with the perspective of systemic organization espoused here. We cannot plan such change. But we can contribute to the process whereby constraints upon change get transformed. Ultimately, we have to change or we will perish in a changing world. I think we are capable of efforts that can hasten, and deliberately contribute to, this process of adapting to a changing context. I think this volume, and the Conference on Cultural Values and Tropical Ecology that gave rise to it, constitute a small but useful step toward becoming a more fit species.

References

Bateson, G.
   1972   *Steps to an Ecology of Mind.* San Francisco: Chandler.
Carson, R.
   1962   *Silent Spring.* New York: Houghton Mifflin.
Russell, B.
   1958   *The ABC of Relativity.* Revised edition. London: George Allen and Unwin.